SUSTAINABILITY IN POLYMER TECHNOLOGY AND PLASTICS ENGINEERING

Concepts, Strategies, and Opportunities

SUSTAINABILITY IN POLYMER TECHNOLOGY AND PLASTICS ENGINEERING

Concepts, Strategies, and Opportunities

Tamara Tatrishvili, PhD
Neha Kanwar Rawat, PhD
Swati Gokul Talele, PhD
A. K. Haghi, PhD
Editors

AAP | APPLE ACADEMIC PRESS

First edition published 2025

Apple Academic Press Inc.
1265 Goldenrod Circle, NE,
Palm Bay, FL 32905 USA

760 Laurentian Drive, Unit 19,
Burlington, ON L7N 0A4, CANADA

CRC Press
2385 NW Executive Center Drive,
Suite 320, Boca Raton FL 33431

4 Park Square, Milton Park,
Abingdon, Oxon, OX14 4RN UK

Library and Archives Canada Cataloguing in Publication

CIP data on file with Canada Library and Archives

Library of Congress Cataloging-in-Publication Data

CIP data on file with US Library of Congress

ISBN: 978-1-77491-762-6 (hbk)
ISBN: 978-1-77491-763-3 (pbk)
ISBN: 978-1-00356-376-1 (ebk)

About the Editors

Tamara Tatrishvili, PhD
Assistant Professor, Main Specialist, Office of the Academic Process Management, Ivane Javakhishvili Tbilisi State University (TSU); Director of the Institute of Macromolecular Chemistry and Polymeric Materials at TSU, Tbilisi, Georgia

Tamara Tatrishvili, PhD, is an Assistant Professor and the main specialist at the Office of Academic Process Management (Faculty of Exact and Natural Sciences) at Ivane Javakhishvili Tbilisi State University, as well as the Director of the Institute of Macromolecular Chemistry and Polymeric Materials at TSU, Tbilisi, Georgia; DAAD alumni; and a member of Georgian Chemical Society. Her research interests include polymer chemistry, polymeric materials, and chemistry of silicon-organic compounds. She is the executive editor of *The Journal of the Georgian Chemical Society.* Dr. Tatrishvili is the author of more than 190 scientific publications, 12 books, and monographs.

Neha Kanwar Rawat, PhD
Researcher, Department of Materials Science and Engineering, Indian Institute of Science, Bangalore, Karnataka, India

Neha Kanwar Rawat, PhD, is a recipient of a prestigious DS-Kothari Postdoctoral Fellowship and DST Young Scientist Postdoctoral Fellowship and is presently a researcher in the Department of Materials Science and Engineering, Indian Institute of Science, Bengaluru, India. She received her PhD in Chemistry from Jamia Millia Islamia (a Central University), India. Her main interests include nanotechnology and nanostructured materials synthesis and characterization, with her main focus on green chemistry along with novel sustainable chemical processing of nano-conducting polymers/nanocomposites; conducting films, ceramics, silicones, and matrices: epoxies, alkyds, polyurethanes, and others. She is also pursuing her interest in fusing new technology in areas that include electrochemistry and organic-inorganic hybrid nanocomposites for biomedical applications, which also includes protective surface coatings for corrosion inhibition and MW shielding materials. She has published numerous peer-reviewed research articles in journals of high repute. Her contributions have led to many books and chapters in international books published with the Royal Society of Chemistry, Wiley, Elsevier, Apple Academic Press, Nova US, and many others in progress. She is a peer reviewer for many international books and is a member of many groups, including the Royal Society of Chemistry and the American Chemical Society (USA), and a life member of the Asian Polymer Association.

Swati Gokul Talele, PhD
Associate Professor, Department of Pharmaceutics, Sandip Institute of Pharmaceutical Sciences, Savitribai Phule Pune University, Pune, Maharashtra, India

Swati Gokul Talele, PhD, is currently serving as an Associate Professor, in the Department of Pharmaceutics at Sandip Institute of Pharmaceutical Sciences. She has 22 years of experience in research along with teaching. She has published more than 35 research papers in various reputed international and national journals as well as more than 30 review papers. She has also presented research work at several conferences and has received several awards. She has authored the books *Natural Excipients and Laboratory Manual of Community Pharmacy and Management* and is co-editor of the books *Microbial Technologies and Phyto-Pharmaceuticals in Drug Development; Applied Pharmaceutical Science and Microbiology: Novel Green Chemistry Methods and Natural Products,* among others. She has also published many book chapters. She has supervised many MPharm students and is currently associated with many research projects and guiding PhD students. She also worked as a college examination officer (CEO) for more than three years and is a member of the university examination committee. Dr. Talele has delivered interactive talks at continuous education programs for registered pharmacists and is a life member of the Association of Pharmacy Teachers of India (APTI) and a member of the Indian Pharmaceutical Congress Association (IPCA).

A. K. Haghi, PhD
Research Associate, Department of Chemistry,
University of Coimbra, Portugal

A. K. Haghi, PhD, is a Retired Professor and has written, co-written, edited, or co-edited more than 1000 publications, including books, book chapters, and papers in refereed journals with over 3800 citations and an h-index of 32, according to the Google Scholar database. He is currently a research associate at the University of Coimbra, Portugal. Professor Haghi has received several grants, consulted for several major corporations, and is a frequent speaker to national and international audiences. He is the Founder and former Editor-in-Chief of the *International Journal of Chemoinformatics and Chemical Engineering* and *Polymers Research Journal.* Professor Haghi has acted as an editorial board member of many international journals. He has served as a member of the Canadian Research and Development Center of Sciences & Cultures. He has supervised several PhD and MSc theses at the University of Guilan (UG) and co-supervised international doctoral projects. Professor Haghi holds a BSc in urban and environmental engineering from the University of North Carolina (USA) and holds two MSc degrees, one in mechanical engineering from North Carolina State University (USA) and another one in applied mechanics, acoustics, and materials from the Université de Technologie de Compiègne (France). He was awarded a PhD in engineering sciences at Université de Franche-Comté (France). He is a regular reviewer for leading international journals.

Contents

Contributors

Ketevan Archvadze
TSU Petre Melikishvili Institute of Physical and Organic Chemistry, Tbilisi, Georgia

M. R. Badwar
Department of Quality Assurance, Sandip Institute of Pharmaceutical Sciences, Mahiravani, Nashik, Maharashtra, India

Smriti Bansal
Department of Chemistry, Netaji Subhas University of Technology (N.S.U.T), Azad Hind Fauj Marg, Dwarka, New Delhi, India

Laxmikant B. Borse
Department of Pharmaceutics, Sandip Institute of Pharmaceutical Sciences, Mahiravani, Nashik, Maharashtra, India

S. R. Chaudhari
Department of Quality Assurance, Sandip Institute of Pharmaceutical Sciences, Mahiravani, Nashik, Maharashtra, India

Ia Chitrekashvili
TSU Petre Melikishvili Institute of Physical and Organic Chemistry, Tbilisi, Georgia

E. Chkhaidze
Assistant Professor, Faculty of Chemical Technology and Metallurgy, Georgian Technical University, Tbilisi, Georgia

Iryna Danylo
Department of Organic Technology, University of Chemistry and Technology, Prague, Czech Republic

Nora Dokhturishvili
TSU Petre Melikishvili Institute of Physical and Organic Chemistry, Tbilisi, Georgia

Eter Gavashelidze
TSU Petre Melikishvili Institute of Physical and Organic Chemistry, Tbilisi, Georgia

Nazi Gelashvili
TSU Petre Melikishvili Institute of Physical and Organic Chemistry, Tbilisi, Georgia

Rinky Ghosh
Department of Materials Science and Engineering, Indian Institute of Technology Kanpur, Kanpur, Uttar Pradesh, India

Priyanka R. Gupta
Department of Pharmaceutics, Sandip Institute of Pharmaceutical Sciences, Mahiravani, Nashik, Maharashtra, India

Marina Gurgenishvili
TSU Petre Melikishvili Institute of Physical and Organic Chemistry, Tbilisi, Georgia

Purnima Jain
Department of Chemistry, Netaji Subhas University of Technology (N.S.U.T), Azad Hind Fauj Marg, Dwarka, New Delhi, India

Shilpa M. Jumde
Department of Pharmaceutics, Sandip Institute of Pharmaceutical Sciences, Mahiravani, Nashik, Maharashtra, India

Viacheslav Kaulin
Department of Chemical Technologies and Chemical Engineering, Donetsk National Technical University, Pokrovsk, Ukraine

R. Kavya
Department of Chemistry, Sri Jayachamarajendra College of Engineering, JSS Science and Technology University, Mysore, Karnataka, India

Nanuli Khotenashvili
TSU Petre Melikishvili Institute of Physical and Organic Chemistry, Tbilisi, Georgia

Iryna Krutko
Department of Chemical Technologies and Chemical Engineering, Donetsk National Technical University, Pokrovsk, Ukraine

K. M. Anil Kumar
Department of Environmental Science, JSS Academy of Higher Education and Research, Mysore, Karnataka, India

O. Kychkyruk
Ivan Franko Zhytomyr State University, Zhytomyr, Ukraine

Riva Liparteliani
TSU Petre Melikishvili Institute of Physical and Organic Chemistry, Tbilisi, Georgia

Zaur Lomtatidze
The Botanical Institute of Ilia State University, TSU Petre Melikishvili Institute of Physical and Organic Chemistry, Tbilisi, Georgia

B. S. Madhukar
Department of Chemistry, Sri Jayachamarajendra College of Engineering, JSS Science and Technology University, Mysore, Karnataka, India

K. N. Madhusoodanan
Rubber Technology Division, Rubber Research Institute of India, Kottayam, Kerala, India

Sajid S. Mansuri
Department of Pharmaceutics, Savitribai Phule Pune University, Sandip Institute of Pharmaceutical Sciences, Mahiravani, Nashik, Maharashtra, India

Lela Mirtskhulava
Department of Computer Science, Ivane Javakhishvili Tbilisi State University, Tbilisi, Georgia

Aniket A. Nerkar
Sandip Institute of Pharmaceutical Sciences, Nashik, Maharashtra, India

Diksha K. Nunse
Department of Pharmaceutics, Savitribai Phule Pune University, Sandip Institute of Pharmaceutical Sciences, Mahiravani, Nashik, Maharashtra, India

Givi Papava
TSU Petre Melikishvili Institute of Physical and Organic Chemistry, Tbilisi, Georgia

Ketevan Papava
TSU Petre Melikishvili Institute of Physical and Organic Chemistry, Tbilisi, Georgia

V. T. Pathan
Department of Pharmaceutics, Sandip Institute of Pharmaceutical Sciences, Mahiravani, Nashik, Maharashtra, India

Harshal S. Patil
Department of Pharmaceutics, Savitribai Phule Pune University, Sandip Institute of Pharmaceutical Sciences, Mahiravani, Nashik, Maharashtra, India

H. S. Nagendra Prasad
Department of Chemistry, Sri Jayachamarajendra College of Engineering, JSS Science and Technology University, Mysore, Karnataka, India

Praveen C. Ramamurthy
Department of Materials Engineering, Indian Institute of Science, Bangalore, Karnataka, India

Kiran Rana
Department of Textile Fiber and Engineering, Indian Institute of Technology Delhi, New Delhi, India

Neha Kanwar Rawat
Materials Research Laboratory, Department of Chemistry, Jamia Millia Islamia, New Delhi, India; Department of Materials Engineering, Indian Institute of Science, Bangalore, Karnataka, India

I. Savchenko
National Taras Shevchenko University of Kyiv, Kyiv, Ukraine

Hitesh V. Shahare
SNJBs Shriman Sureshdada Jain College of Pharmacy, Chandwad, Nashik, Maharashtra, India

Shweta H. Shahare
Sandip Institute of Pharmaceutical Sciences, Nashik, Maharashtra, India

L. Shamanauri
R. Dvali Institute of Machine Mechanics, Tbilisi, Georgia

Deepa Sharma
Department of Chemistry, Netaji Subhas University of Technology (N.S.U.T), Azad Hind Fauj Marg, Dwarka, New Delhi, India

Ashwini Shelke
Sir Dr. M.S. Gosavi College of Pharmaceutical Education and Research, Nashik, Maharashtra, India

Natia Shengelia
Sokhumi State University, Tbilisi, Georgia

C. K. Shinde
Department of Pharmaceutics, Sandip Institute of Pharmaceutical Sciences, Mahiravani, Nashik, Maharashtra, India

M. B. Shivaswamy
Department of Chemistry, Sri Jayachamarajendra College of Engineering, JSS Science and Technology University, Mysore, Karnataka, India

D. Sternik
Maria Curie-Skłodowska University, Lublin, Poland

E. Sumitha
Department of Microbiology, JSS Academy of Higher Education and Research, Mysore, Karnataka, India

Swati G. Talele
Department of Pharmaceutics, Sandip Institute of Pharmaceutical Sciences, Mahiravani, Nashik, Maharashtra, India

Swati T. Talele
Department of Pharmaceutics, Sandip Institute of Pharmaceutical Sciences, Mahiravani, Nashik, Maharashtra, India

G. Vidya
Department of Chemistry, Dayananda Sagar College of Engineering, Bangalore, Karnataka, India

L. Vretik
National Taras Shevchenko University of Kyiv, Kyiv, Ukraine

E. Yanovska
National Taras Shevchenko University of Kyiv, Kyiv, Ukraine

Preface

Polymer and plastic materials have grown widely in recent years due to their wide-ranging applications in both science and engineering. This new volume covers the characterization of modern polymer and plastic materials with functional and sustainable applications in various sectors.

In this era of rapidly advancing technology and global challenges, it has become crucial to adopt an integrated approach that bridges the gap between scientific principles and their practical applications. The chapters provided in this book reflect this need for sustainability in polymer technology and plastic engineering.

In this valuable research-oriented book, diverse perspectives and shared experiences have created a comprehensive resource that will undoubtedly inspire researchers, engineers, policymakers, and students to further explore the field of sustainable development.

This new volume covers a comprehensive overview of the engineering properties of polymer composites and plastic materials. It also brings together subjects recently dispersed throughout the most current literature and offers viable and environmentally friendly alternatives to conventional polymeric composite materials and plastics. Chapters also review the engineering properties and applications of modern polymeric composites and plastics with new techniques.

This book:

- Provides a thorough review of the engineering properties of polymeric composites and plastics with their applications.
- Reports the practical use of sustainable polymer composites and plastics in different sectors.
- Presents case studies on environmentally friendly polymers and plastics for sustainable development.
- Reviews the influence of additives on different properties of the polymer-based composites and plastics.

CHAPTER 1

Solid Carbon Foams Based on Pitch-Thermoplastic

IRYNA KRUTKO,[1] VIACHESLAV KAULIN,[1] and IRYNA DANYLO[2]

[1]*Department of Chemical Technologies and Chemical Engineering, Donetsk National Technical University, Pokrovsk, Ukraine*

[2]*Department of Organic Technology, University of Chemistry and Technology, Prague, Czech Republic*

ABSTRACT

The development of solid foam synthesis based on a cheaper raw material – pitch-thermoplastic can provide an economical way for the production of light porous carbon materials. Pitch-thermoplastic is a product of the coal tar pitch interaction with active polar polymers at low temperatures (150–170°C). It is a thermoplastic material with polymeric properties. Solid foams based on pitch-thermoplastic (pitch-thermoplastic foams) are gas-filled composite materials in which pitch-thermoplastic is used as a matrix.

The kinetics of pitch-thermoplastic foaming at T = 150°C under pressure have been studied. The influence of pressure, the amount of a complex blowing agent (a mixture of azodicarbonamide and zinc stearate), and technological additive $ZnSt_2/CaSt_2$, as well as the viscosity of the pitch-thermoplastic on the maximum foaming ratio of hard foam samples and their properties, were determined.

The porous structure of solid foams, depending on viscosity and pressure, was characterized by ellipsoidal cells ranging in size from 20 μm to 250 μm, elongated in the direction of material movement from the extrusion chamber opening. With an increase in viscosity and pressure, the cells acquired an irregular (torn) shape.

Sustainability in Polymer Technology and Plastics Engineering: Concepts, Strategies, and Opportunities. Tamara Tatrishvili, Neha Kanwar Rawat, Swati Gokul Talele, and A. K. Haghi (Eds.)

The authors proved the possibility of using pitch-thermoplastic as a precursor for obtaining solid foams using the chemical blowing agent at a relatively low foaming temperature (150°C). Solid foam obtained by pitch-thermoplastic foaming (MFI=4.76 g/10 min) under a pressure of 0.265 MPa was characterized by a porosity of 36–42%, the apparent density of 760–850 kg/m^3, thermal conductivity coefficient of 0.068–0.073 W/(m·K), and compressive strength in the range of 3.4–3.6 kgf/cm^2.

The received solid foams based on pitch-thermoplastic belong to non-carbonized materials, have heat-insulating properties, and are resistant to the action of aggressive environments.

The prospect of pitch-thermoplastic use for the production of new material – pitch-thermoplastic foams is due to its low cost and excellent ability to be modified with various polymers and additives.

1.1 INTRODUCTION

Given the high demand for porous carbon materials in various industries, the relevance of the work is enhanced by the growing demand for modern production of cheap raw materials. The development of the synthesis of solid foams based on a cheaper raw material – pitch-thermoplastic, will provide an economical way for the production of light carbon materials. Pitch-thermoplastic is a thermoplastic material based on coal tar pitch, which is characterized by polymer properties [1].

Solid foams based on pitch-thermoplastic (pitch-thermoplastic foams) are gas-filled composite materials in which pitch-thermoplastic is used as a matrix. As known [2–5], for obtaining gas-filled carbon materials from coal tar pitch, it is necessary to improve its rheological characteristics. The developed method for producing pitch-thermoplastic allows, by changing the conditions for modifying coal tar pitch, as well as the component composition of active polymer additives, to control its viscosity and thus provide the required characteristics for the foaming process [6].

Pitch-thermoplastic is a product of the interaction of coal tar pitch with active polymers PVC, and PMMA in various ratios at a temperature up to 170°C. The properties of pitch-thermoplastic depend on the time of heat treatment, amount, and the ratio of polymer additives [7, 8].

Gas-filled pitch-thermoplastics were obtained by foaming and subsequent curing of a plastic-viscous composition. The exothermic chemical blowing agent azodicarbonamide (ADCA) was used to foam the pitch-thermoplastic (PTP). Exothermic chemical blowing agents mainly release nitrogen, which is a more efficient blowing gas than CO_2 due to its slower diffusion rate through polymers.

Azodicarbonamide $NH_2CON{=}NCONH_2$ is one of the most effective exothermic chemical blowing agents. The decomposition temperature of ADCA is 210°C. The decomposition products of ADCA at temperatures up to 220°C are nitrogen, carbon monoxide, and urea. The decomposition temperature of ADCA should be in the temperature range of 130–150°C, which ensures the necessary viscosity of pitch-thermoplastic for its foaming. To decrease the ADCA decomposition temperature to 130–150°C, zinc stearate was used as an initiator [9].

For effective pitch-thermoplastic foaming, in addition to the blowing agent, nucleating agents, and surfactants are required.

Nucleating agents serve as nuclei for the formation of cells and act as activators for the development of a fine cellular structure. ADCA, a fine powder, can act as hot spot nucleators. When ADCA decomposes, heat is released [10], which creates hot spots in the volume – local places with elevated temperatures. At these points, the viscosity of the matrix is lower due to the increased temperature, and they will serve as centers for the generation of gas bubbles. Thus, ADCA simultaneously performs the function of a dynamic nucleating agent during decomposition. In addition, the pitch-thermoplastic contains solid particles ranging in size from 1 to 3 μm, which can act as nucleating agents.

Surfactants are introduced into the composition of foamed materials to increase the dimensional stability of cellular structures. Anionic surfactants, calcium and zinc stearates, have been widely used in the production of foam materials based on PVC due to their environmental safety and the manifestation of a synergistic effect when used together [11, 12]. Calcium and zinc stearates are able to adsorb on the phase interface and lower the surface tension, which favorably affects the foaming process. In addition, calcium and zinc stearates function as heat stabilizers for pitch-thermoplastic [13, 14].

The article presents the results of the influence of the PTP composition and the amount of surfactants on the rheological properties of the pitch-polymer matrix. The kinetics of pitch-thermoplastic foaming under pressure have been studied as well as the properties of the obtained solid pitch-thermoplastic foams.

1.2 EXPERIMENTAL METHODS AND MATERIALS

1.2.1 PITCH-THERMOPLASTIC

Thermoplastic material is a product of low-temperature (170°C) modification of coal tar pitch with active polymer additives PVC 10–20 phr and PMMA 1–5 phr (phr is the amount of additive by weight per 100 parts by weight of coal tar pitch). Pitch-thermoplastic is characterized by rheological properties that allow it to be used as a precursor for gas-filled carbon materials [2, 6].

1.2.2 COMPLEX BLOWING AGENT

A mixture of azodicarbonamide (CAS: 123–77–3) and zinc stearate (CAS: 557–05–1) was used in the ratio ADCA: $ZnSt_2$ = 1:0.25. The addition of zinc stearate reduces the ADCA decomposition temperature to 130–150°C [9].

1.2.3 MIXTURE OF ZINC STEARATE AND CALCIUM STEARATE

The mixture of zinc stearate (CAS: 557–05–1) and calcium stearate (CAS: 1592–23–0) was used as a processing aid to control the rheological properties of the pitch-thermoplastic.

1.2.4 MELT FLOW INDEX (ASTM D1238; DIN ISO 1133)

The melt flow index (MFI) provides a correct assessment of the rheological characteristics of a viscous thermoplastic melt. MFI is the mass of material extruded through a capillary under given conditions (temperature and pressure) in 10 minutes and is measured in g/10 min. The higher MFI leads to the lower viscosity of the thermoplastic.

1.2.5 METHOD OF FOAMING UNDER PRESSURE

For modified coal tar pitch foaming under pressure, the extrusion method was chosen, for which a foaming chamber was used – the IIRT-AM device (Figure 1.1). The chamber is a cylindrical channel (diameter – 9.55 mm, height – 152 mm), into which a metal rod (2) (diameter – 9.475 mm) is inserted. In the installation, a steel plug (4) and sealing washers made of fluoroplastic/PTFE (3) were used to prevent material leakage. The extrusion chamber was heated by a heater (6) to a set temperature of 150°C.

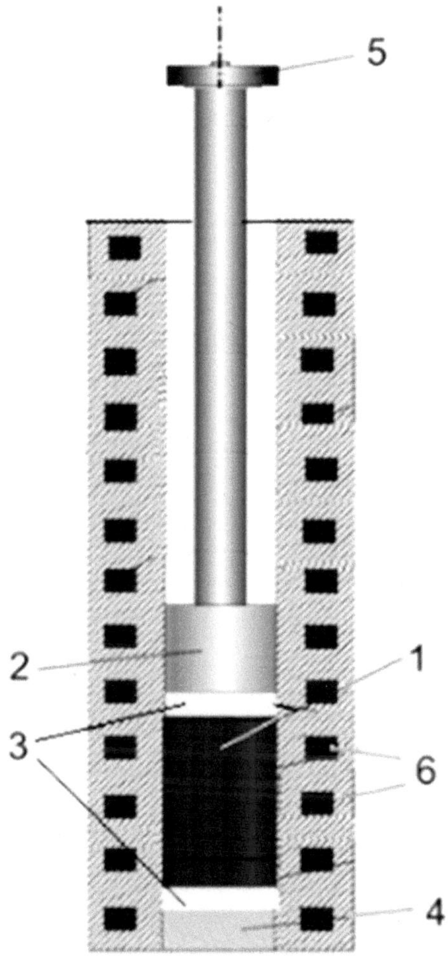

FIGURE 1.1 Scheme of foaming chamber.

Note: 1: sample of prepared material; 2: rod; 3: PTFE sealing washers; 4: steel plug; 5: load weight; and 6: electric chamber heating.

In the process of foaming, the studied material (1) was exposed to temperature during the action of a load (5) of a certain mass/weight on it, to create the necessary foaming pressure of 0.145 and 0.265 MPa.

The pressure in the working chamber (P, N/m^2) was determined by the formula:

$$P = P_0 + \frac{M}{S}$$

Where; S is the cross-section of the working cylindrical chamber of the IIRT-AM device, m^2; and M is the load weight, N.

The efficiency of foaming was characterized by the value of the foaming ratio – K_f:

$$K_f = \frac{V_f}{V_0} = \frac{\Delta H + H_0}{H_0} = \frac{\Delta H}{H_0} + 1$$

Where; V_f is the sample volume after foaming in the presence of a blowing agent, m^3; V_0 is the sample volume after complete melting, m^3; H_0 is the sample height after complete melting, m; and ΔH is the change in height of the sample due to gas release during decomposition of complex blowing agent under conditions of experiment, m.

In order to obtain results that are invariant concerning different temperatures and pressures, to calculate the foaming ratio of the material (K_f), the measured volumes of gases released during the foaming process were brought to conditions 0.1 MPa and 298 K:

$$K_f^{\circ} = \frac{V_0 + \nu_0}{V_0} = 1 + \frac{\Delta H T_0 P}{H_0 T P_0},$$

Where; V_0 is the increase of the sample volume due to the gas release during the decomposition of complex blowing agent, to conditions 0.1 MPa and 298 K, m^3; H_0 is the sample height after complete melting, m; and ΔH is the change in sample height due to gas release during decomposition of complex blowing agent under conditions of experiment, m.

$$T_0 = 298 \text{ K};$$

$$P_0 = 0.1 \text{ MPa};$$

Where; P is the pressure in the working chamber, MPa; and T is the temperature in the working chamber, K.

1.3 RESULTS AND DISCUSSION

1.3.1 INFLUENCE OF SURFACTANTS ON PITCH-THERMOPLASTIC VISCOSITY

It is known that additives of zinc and calcium stearates reduce internal and external friction during the foaming process [7]. Internal lubrication ensures a uniform and fast process of composition foaming due to a decrease in internal friction and heat generation as a result of the redistribution of mechanical energy. External lubrication increases surface slip and reduces adhesion to metal, allowing foam material to move and unload.

Figure 1.2 shows the effect of zinc and calcium stearates ($ZnSt_2$:$CaSt_2$=1:1 wt. part) in an amount of 0.5–1 phr (per 100 wt.parts of PTP) on the viscosity of PTP (modified with PVC 10–20 phr and PMMA 1–5 phr at a temperature of 170°C). The PTP samples had an MFI in the range of 1.5–9 g/10 min, depending on the time of modification.

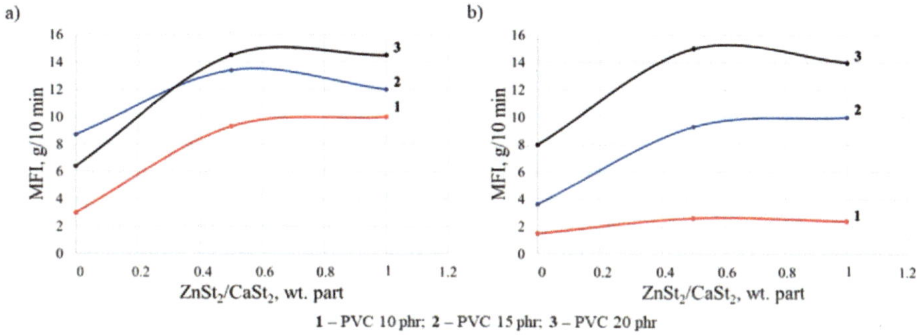

FIGURE 1.2 Dependence of pitch-thermoplastic MFI (load weight 5 kg at 150°C) on amount of Zn and Ca stearate mixture ($ZnSt_2$: $CaSt_2$ = 1: 1 wt. part) for various PTP compositions: a) PMMA 1 phr; b) PMMA 5 phr.

Figure 1.3 shows the effect of zinc and calcium stearates ($ZnSt_2$:$CaSt_2$ =1:1 wt. part) in an amount of 0.5–1 phr (per 100 wt.parts of PTP) on the viscosity of PTP (modified with PVC 10 phr and PMMA 1–3 phr at a temperature of 170°C). The PTP samples had an MFI in the range of 1.1–3.1 g/10 min.

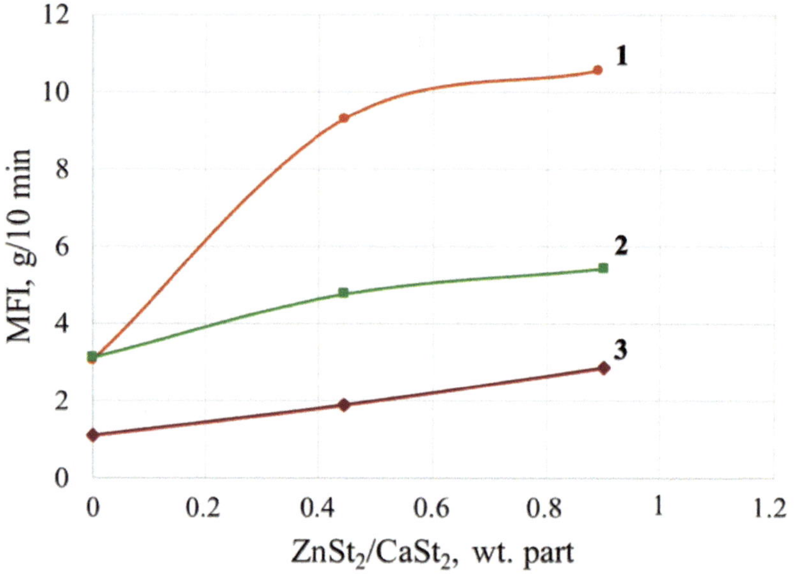

1 – PMMA 1 phr; **2** – PMMA 2 phr; **3** – PMMA 3 phr

FIGURE 1.3 Dependence of MFI (load weight 5 kg at 150°C) of PTP (PVC 10 phr, PMMA 1–3 phr) on amount of Zn and Ca stearate mixture ($ZnSt_2$: $CaSt_2$ = 1: 1 wt. part).

The lower viscosity of the pitch-thermoplastic led to the greater value of the MFI, which means the amount of mass of material flowed from the nozzle of the apparatus in 10 minutes.

Studies have shown that a mixture of zinc and calcium stearates ($ZnSt_2$:$CaSt_2$ = 1:1 wt. part) in an amount of 0.5–1 phr (per 100 wt.parts of PTP) reduced the viscosity of the composition, as evidenced by the increase in MFI, and regardless of the composition of the PTP. Surfactant molecules weaken the intermolecular interaction (dispersion, Van der Waals), thereby increasing the mobility of macromolecules and reducing the rigidity of the composition.

The addition of $ZnSt_2$/$CaSt_2$ in an amount of 0.5 phr had the maximum effect on the viscosity of PTP (increase in MFI by 2–3 times). With an increase in the amount of stearates to 1 phr, their influence on the PTP rheological properties was weakened.

The viscosity of the PTP samples (MFI 1.1–3.1 g/10 min), for which the amount of PVC was 10 phr, and PMMA 1–3 phr, decreased due to the addition of surface-active $ZnSt_2$/$CaSt_2$ in an amount of 0.5–1 phr, i.e., MFI increased by 1.7–3.5 times (Figure 1.3).

1.3.2 *KINETICS OF PTP FOAMING UNDER PRESSURE*

Free foaming at 150°C of coal tar pitch modified with PVC showed that its structure is not strong enough, which confirms the fragility of the resulting solid foams. Exceeding the pressure in gas bubbles over the pressure in the system and low viscosity led to the formation of an inhomogeneous large-celled structure. Therefore, it is necessary to change the component composition of the material and the process conditions in order to change the characteristics of the foamed structure and the properties of the solid foam.

As the most structured material, a pitch-thermoplastic obtained by modifying coal tar pitch with PVC 10 phr and PMMA (1–3 phr) was chosen as a matrix for foaming. Zinc and calcium stearates (1:1 wt. part) were added as a technological additive, which allows adjusting the viscosity, in the amount of 0.5–1.0 phr (per 100 wt.parts of PTP). For foaming, a complex blowing agent was investigated: a mixture of azodicarbonamide and zinc stearate (ADCA: $ZnSt_2$=1:0.25) in an amount of 1–3 phr (per 100 wt.parts of PTP).

The study of the kinetics of pitch-thermoplastic foaming was carried out according to the method described above (*see* section 1.2.5). After the zinc and calcium stearates addition, the studied PTP samples had MFI (g/10 min) 9.3–10.6 (PMMA 1 phr), 4.8–5.4 (PMMA 2 phr), 1.9–2.9 (PMMA 3 phr).

The kinetic curves shown in Figure 1.4 made it possible to determine the blowing agent and the amount of stearates mixture to ensure optimal foaming conditions (achieving the highest possible foaming ratio at the highest possible speed in a short time interval).

For pitch-thermoplastic (PVC 10 phr, PMMA 1 phr), the best foaming conditions were observed at an amount of stearates 1 phr (10.56 MFI) and CBA 1 and 2 phr (curves 5, 6 in Figure 1.4(a)). The foaming process proceeded at a fairly high speed and ended in the 15th minute with K_f = 2 (CBA=1 phr) and K_f =2.2 (CBA=2 phr). In addition, it should be noted that with an increase in the amount of CBA from 1 to 3 phr the kinetics of foaming changed dramatically, which ultimately led to a K_f decrease for stearates of 0.5 phr from 2.2 to 1.6 (curves 2, 4 in Figure 1.4(a)); for stearates 1 phr from 2 to 1.4 (curves 5, 7 in Figure 1.4(a)). Most likely, this occurs as a result of the initiation of reactions between CBA and PTP components or due to a large loss of gas through the walls of the matrix (due to high pressure in the bubbles).

For pitch-thermoplastic (PVC 10 phr, PMMA 2 phr) at the amount of stearates 1 phr (5.42 MFI) and CBA 2 phr, can be achieved a foaming ratio of 2.2 in 25 min (curve 6 in Figure 1.4(b)), and for stearates 0.5 phr (4.76 MFI)

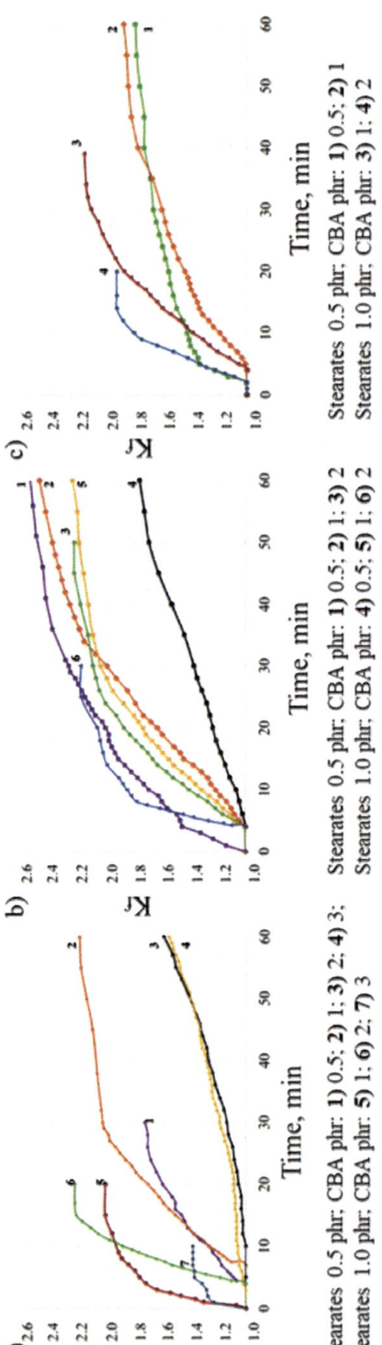

FIGURE 1.4 Kinetics of PTP foaming (P = 0.146 MPa, T = 150°C) for: a) PTP: PVC: PMMA = 100: 10: 1 phr; b) PTP: PVC: PMMA = 100: 10: 2 phr; c) PTP: PVC: PMMA = 100: 10: 3 phr.

and CBA 2 phr in 40 min (curve 3 in Figure 1.4(c)), i.e., the more stearates were, the less time required for foam formation. This is most likely due to the stronger plasticizing effect of stearates, which leads to a decrease in viscosity and facilitates the process. The maximum foaming ratio of 2.5–2.55 was typical for stearates 0.5 phr and CBA amount 0.5–1 phr (curves 1, 2 in Figure 1.4(c)), however, was fixed at the 60th min.

For pitch-thermoplastic (PVC 10 phr, PMMA 3 phr), the best foaming conditions were observed at an amount of stearates 1 phr (2.86 MFI) and CBA 1 and 2 phr (curves 3, 4 in Figure 1.4(c)). The foaming process proceeded at a fairly high speed and ended at the 15th minute with K_f=1.95 (CBA = 2 phr) and at 35 min K_f=2.15 (CBA=1 phr). There was a regularity in the decrease in K_f from 2.15 to 1.95 with an increase in the CBA amount from 1 to 2 phr (curves 3, 4 in Figure 1.4(c)). Most likely, this happens because of the reactions starting between CBA and PTP components or because a lot of gas is lost through the matrix walls (due to high pressure in the bubbles).

1.3.3 INFLUENCE OF BLOWING AGENT AND SURFACTANT ON MAXIMUM PTP FOAMING RATIO

The maximum foaming ratio of the investigated PTP samples (the viscosity of which is shown in Figure 1.3) at the blowing agent amount 1–3 phr was in the range of 1.5–2.5 (Figure 1.5). The higher viscosity of the PTP sample led to the lower K_f. It has been experimentally established that an increase in the CBA amount up to 3 phr was impractical since it either practically did not affect the K_f, or reduced it. The maximum foaming ratio of 2.2–2.4 was shown by the PTP samples (PMMA 2 phr), which had an MFI of 4.8–5.4 g/10 min at a CBA amount of 1–2 phr. The PTP samples (PMMA 1 phr, MFI 9.3–10.6 g/10 min) at CBA amount of 1–2 phr had K_f 2.0–2.2 (at a $ZnSt_2/CaSt_2$ amount of 1 phr).

As can be seen from Figure 1.5, the maximum foaming ratio K_{fmax} = 2.0–2.4 (P = 0.146 MPa, T = 150°C) was observed for PTP samples (PVC 10 phr, PMMA 1–2 phr: MFI 4.8–10.6 g/10 min) at CBA amount of 1–2 phr.

1.3.4 INFLUENCE OF PITCH-THERMOPLASTIC VISCOSITY ON DECOMPOSITION OF CBA

The determination of the gas release in the PTP foaming process showed that in a viscous system, complete decomposition of CBA did not occur. The gas release was less than calculated. The relative volume of the released gas (concerning the theoretical one) decreased with an increase in the amount of

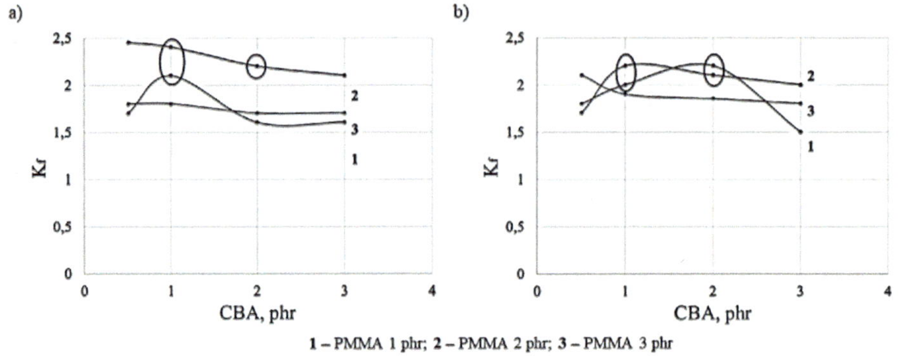

FIGURE 1.5 Dependence of maximum foaming ratio (P = 0.145 MPa, T = 150°C) of PTP (PVC 10 phr, PMMA 1–3 phr) on amount of CBA and Zn/Ca stearate mixture ($ZnSt_2$: $CaSt_2$ = 1: 1 wt. part): a) 0.5 phr; b) 1 phr.

CBA (Figure 1.6). At the CBA amount of 1 phr, the gas release was 60–68% of the theoretical for PTP samples with MFI 4.8–9.3 g/10 min; at the CBA amount of 2 phr – 30% of the theoretical for PTP samples with MFI 4.8, 5.4, 10.6 g/10 min. Reducing the CBA amount to 0.5 phr contributed to an increase in the CBA decomposition degree (the volume of released gas increased to 70–91% of the theoretical); however, the absolute volume of gas was small, which determined the low foaming ratio.

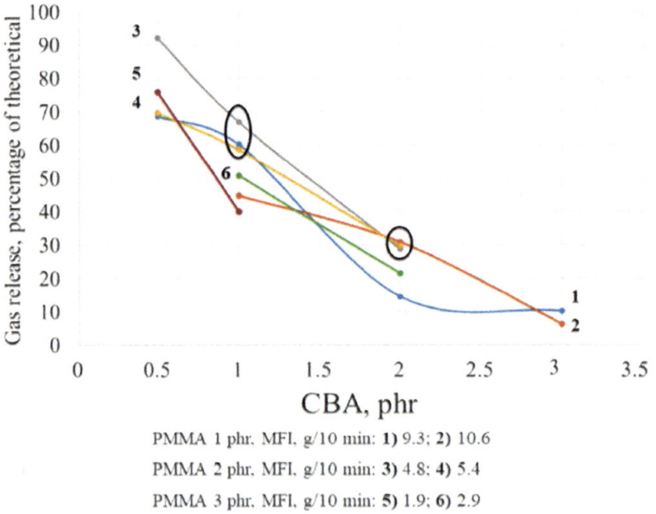

FIGURE 1.6 Gas release as a result of CBA decomposition during foaming of PTP of various viscosity (P = 0.145 MPa, T = 150°C).

Thus, experimental studies have shown that solid foams with K_f=2.0–2.4 can be obtained by foaming the PTP samples with an MFI of 4.8–10.6 g/10 min under a pressure of 0.146 MPa at T = 150°C using a chemical blowing agent in the amount of 1–2 phr and surfactants (zinc and calcium stearates 0.5–1.0 phr).

An analysis of the kinetic regularities has shown that PTP foaming is a rather complex process, which is characterized by the occurrence of many physical and chemical acts of interaction, such as the CBA decomposition, the foam formation, the stearates action, and the occurrence of chemical reactions between the PTP components cannot be ruled out. Therefore, conditions should be chosen for each PTP composition. Ultimately, the determining parameters will be the morphological characteristics of the obtained samples of solid foams.

1.3.5 INFLUENCE OF PRESSURE ON FOAMING RATIO

An increase in pressure from 0.145 MPa to 0.265 MPa during the PTP foaming reduced the foaming ratio (Figure 1.7). The main reason for such a decrease in K_f was that the volume of gas released in the form of bubbles decreased with increasing pressure (Table 1.1). This can probably be due to two reasons: a decrease in the rate of CBA decomposition and the dissolution of part of the released gas in the pitch-polymer matrix.

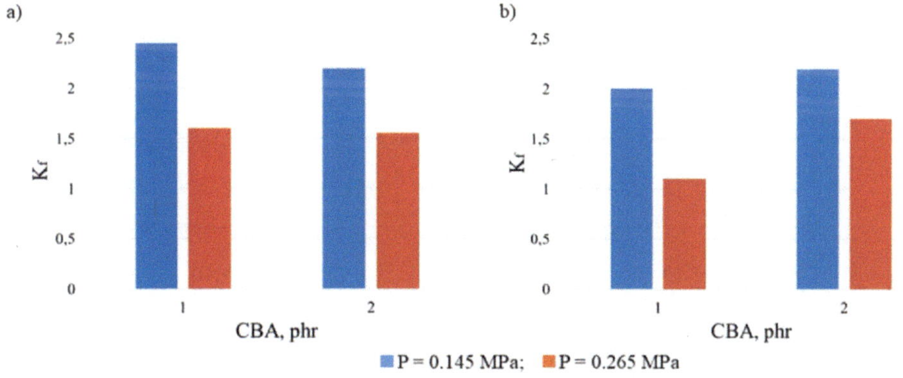

FIGURE 1.7 Influence of pressure on the foaming ratio of PTP (PVC 10 phr) depends on the amount of PMMA and Stearates at 150°C: a) PMMA 2 phr, $ZnSt_2/CaSt_2$ 0.5 phr; b) PMMA 1 phr, $ZnSt_2/CaSt_2$ 1 phr. MFI of samples, g/10 min: 1) 4.76; 2) 10.56.

TABLE 1.1 Influence of Pressure on Volume of Gas Release During PTP Foaming

Sample	PTP MFI, g/10 min	CBA, phr	Volume of Gas Release at T=150°C (cm³)	
			0.145 MPa	0.265 MPa
1	10.56	2	3.95	2.8
2	4.76	1	4.3	2.4

With an increase in the CBA amount from 1 to 2 phr, the volume of gas release increased, which led to an increase in the gas bubble pressure, which resisted external pressure. For this reason, with an increase in the CBA amount from 1 to 2 phr, the K_f difference for the two pressures decreased (Figure 1.7).

1.3.6 PROPERTIES OF PITH-THERMOPLASTIC SAMPLES FOAMED UNDER A PRESSURE

Table 1.2 shows the morphological properties of PTP samples foamed under a pressure of 0.145 MPa depending on their viscosity. Less viscous PTP samples, the MFI of which is in the range of 5.42–10.56 g/10 min had higher porosity (28–34%). With an increase in viscosity (MFI 1.88–4.76 g/10 min), the porosity of the samples decreased (17–22%).

TABLE 1.2 Morphological Properties of Pitch-Thermoplastic Foamed at P = 0.145 Mpa and 150°C (PVC 10 Phr, PMMA 1–2 Phr). Amount of $ZnSt_2/CaSt_2$ 0.5–1 phr

Sample	PTP MFI, g/10 min	Porosity (%)		Apparent Density (kg/m³)	
		CBA, phr		CBA, phr	
		1	2	1	2
1.	10.56	29	29	934	926
2.	9.3	34	29	870	928
3.	5.42	30	28	918	942
4.	4.76	21	20	1,030	1,050
5.	2.86	22	18	1,024	1,080
6.	1.88	20	17	1,051	1,089

It was noted that an increase in the CBA amount from 1 to 2 phr can slightly reduce the porosity and increase the apparent density of samples foamed at P = 0.145 MPa and 150°C.

For the studied PTP samples (Table 1.3), foamed at different pressures, it was found that the CBA amount affected the porosity of the samples at P = 0.265 MPa. The more CBA was used, the higher porosity was reached at P = 0.265 MPa. For the samples foamed at P = 0.145 MPa, the CBA amount had practically no effect on porosity.

TABLE 1.3 Dependence of Morphological Properties of Pitch-Thermoplastic Foamed at 150°C (PVC 10 phr, PMMA 1–2 phr) on Pressure. Amount of $ZnSt_2/CaSt_2$ 0.5–1 phr

Sample	PTP MFI, g/10 min	Pressure, MPa	Porosity (%)		Apparent Density (kg/m³)	
			CBA, phr		CBA, phr	
			1	2	1	2
1.	10.56	0.145	29	29	934	926
		0.265	25	29	986	938
2.	4.76	0.145	21	20	1,030	1,050
		0.265	23	33	1,000	878

It has been experimentally established that solid foam samples foamed at P = 0.265 MPa, obtained from a more viscous PTP (MFI 4.76 g/10 min), were characterized by the highest porosity of 33% and a low apparent density of 878 kg/m³ at a CBA amount of 2 phr (Table 1.3). This can be explained by the lower gas loss through the matrix walls as a result of diffusion processes for more viscous matrices compared to less viscous ones. However, at the CBA amount of 1 phr, the porosity of the PTP samples was lower. It can be assumed that the volume of gas released did not create sufficient pressure to overcome the external pressure P = 0.265 MPa. Therefore, part of the resulting gas remained dissolved in the matrix. To confirm this assumption, a series of experiments was carried out, in which PTP samples foamed at P = 0.265 MPa were kept at T = 120–140°C and P_{atm} for 26–30 hours. Analysis of the obtained samples showed an increase in their porosity by 1.4–1.8 times and a decrease in apparent density (Table 1.4), which may be a consequence of the diffusion of gas dissolved in the matrix into the gas phase.

TABLE 1.4 Characteristics of Solid Foam Obtained at T=150°C, P = 0.265 Mpa Before and After Holding at 130°C (P_{atm})

Sample	PTP MFI, g/10 min	Kf	Holding Time, Hours	Apparent Density, g/cm³		Porosity (%)	
				Before	After	Before	After
1	5.04	1.7	30	1,000	760	23	42
2	4.74	1.9	26	970	850	26	36

Solid foam (P = 0.265 MPa), depending on the PTP composition (PVC and PMMA modifiers), after additional holding at T = 120–140°C and P_{atm}, was characterized by a porosity of 36–42%, an apparent density of 760–850 kg/m³, a coefficient of thermal conductivity of 0.068–0.073 W/(m·K), and a compressive strength of 3.4–3.6 kgf/cm².

Solid foam, the structure of which was formed at a pressure of P = 0.145 MPa from PTP with MFI = 10.56 g/10 min, was characterized by ellipsoidal cells ranging in size from 20 μm to 200 μm, elongated in the direction of material movement from the extrusion chamber opening (Figure 1.8).

SEM of solid foam (Figure 1.8) obtained at P = 0.265 MPa from PTP with MFI = 4.76 g/10 min showed that the cells had an irregular shape, and elongated in the direction of foam movement from the installation chamber (Figure 1.1). The shape factor was approximately 4–5, and the cell size ranged from 25/100 μm to 50/250 μm.

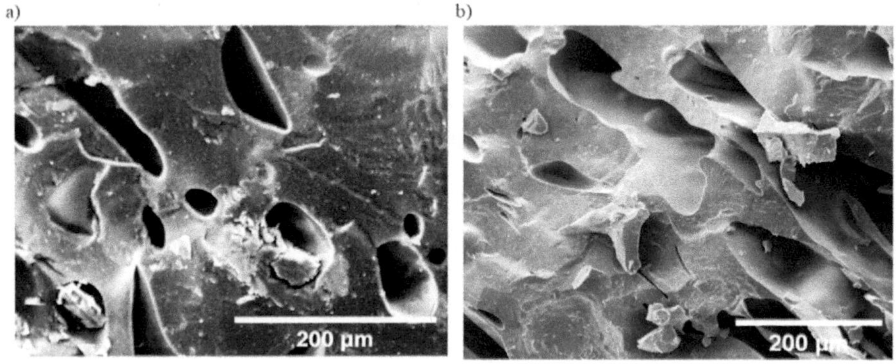

FIGURE 1.8 SEM of gas-filled PTP at 150°C: a) P = 0.145 MPa; MFI of PTP 10.6 g/10 min (PVC 10 phr, PMMA 1 phr), CBA 2 phr; b) P = 0.265 MPa; MFI of PTP 4.8 g/10 min (PVC 10 phr, PMMA 2 phr), CBA 1 phr.

A comparison of the porous structure of solid foams obtained at different pressures showed that with an increase in viscosity and pressure, the structure of the foams changed, and the cells acquired an irregular (torn) shape.

1.4 CONCLUSION

The study of the kinetics of pitch-thermoplastic foaming under pressure showed that the maximum foaming ratio of the studied PTP samples at a CBA amount of 1–3 phr reached 1.5–2.5. The higher viscosity of the PTP sample led to the lower K_f. The increase in pressure from 0.145 MPa to 0.265 MPa during the PTP foaming reduced the foaming ratio. Technological additive surfactant ($ZnSt_2/CaSt_2$) reduced the viscosity of the melt (an increase in MFI by 2–3 times), regardless of the pitch-thermoplastic composition.

In the process of PTP foaming, the complete decomposition of CBA did not occur. The relative volume of gas release (compared to the theoretical one) decreased with an increase in the CBA amount. Solid foams with K_f=2.0–2.4 were formed by foaming the pitch-thermoplastic (MFI 4.8–10.6 g/10 min) under a pressure of 0.146 MPa at T=150°C along with the action of CBA in the amount of 1–2 phr and the presence of 0.5–1.0 phr surfactants ($ZnSt_2/CaSt_2$). For solid foam samples foamed at P = 0.145 MPa, the CBA amount had practically no effect on porosity.

It was established that at P = 0.265 MPa, the CBA amount in the range of 1–2 phr affected the porosity of solid foam samples. The more CBA was used, the higher porosity was reached. The solid foam samples obtained by more viscous pitch-thermoplastic (MFI 4.76 g/10 min) foaming at P = 0.265 MPa and the CBA amount of 2 phr were characterized by the highest porosity of 33% and a low apparent density of 878 kg/m^3.

The porous structure of solid foams, depending on viscosity and pressure, was characterized by ellipsoidal cells ranging in size from 20 μm to 250 μm, elongated in the direction of material movement from the extrusion chamber opening. With an increase in viscosity and pressure, the cells acquired an irregular (torn) shape. The authors demonstrated the potential of using pitch-thermoplastic as a precursor for producing solid foams with the aid of chemical blowing agents at relatively low foaming temperatures (150°C). The solid foam obtained through pitch-thermoplastic (MFI = 4.76 g/10 min) foaming under a pressure of 0.265 MPa exhibited a porosity of 36–42%, an apparent density of 760–850 kg/m^3, a thermal conductivity coefficient of 0.068–0.073 W/(m·K), and compressive strength of 3.4–3.6 kgf/cm^2.

The received solid foams based on pitch-thermoplastic belong to non-carbonized materials, have heat-insulating properties, and are resistant to the action of aggressive environments.

KEYWORDS

- **azodicarbonamide**
- **blowing agent**
- **melt flow index**
- **pitch-thermoplastic**
- **solid foam**

REFERENCES

1. Krutko, I., Kaulin, V., & Satsyuk, K. (2015). Composite materials based on coal tar pitch. In *High Performance Polymers for Engineering-Based Composites* (pp. 265–277). Apple Academic Press, Inc. (Taylor & Francis Group).
2. Krutko, I., Danylo, I., & Kaulin, V. (2019). Carbon foams based on coal tar pitch. *Petroleum and Coal, 61*(5), 1102–1111.
3. Zhou, P., & Chen, Q. L. (2016). Preparation and characterization of carbon foam derived from coal pitch. *Journal of Analytical and Applied Pyrolysis, 122*, 370–376.
4. Wang, Y., He, Z., Zhan, L., & Liu, X. (2016). Coal tar pitch-based carbon foam for thermal insulating material. *Materials Letters, 169*, 95–98.
5. Xiang, L., Yanli, W., & Liang, Z. (2018). Carbon foams prepared from coal tar pitch for building thermal insulation material with low cost. *Chinese Journal of Chemical Engineering, 26*(2), 415–420.
6. Krutko, I., Kaulin, V., Danylo, I., Yavir, K., & Satsyuk, K. (2021). Change in Rheological Properties of Pitch-Thermoplastic under the Effect of Polar Polymers. In *Advanced Materials, Polymers, and Composites: New Research on Properties, Techniques, and Applications* (pp. 209–220). Apple Academic Press, Inc. (Taylor & Francis Group).
7. Krutko, I., Danylo, I., & Kaulin, V. (2019). Kinetics study of modified coal tar pitch foaming. *Petroleum and Coal, 61*(1), 150–159.
8. Krutko, I., Kaulin, V., Yavir, K., & Danylo, I. (2020). The Effect of Polyvinyl Chloride on the Properties of Coal Tar Pitch. *Petroleum and Coal, 62*(3), 888–898.
9. Krutko, I., Danylo, I., & Kaulin, V. (2019). Kinetics of azodicarbonamide decomposition in the presence of an initiator for obtaining solid foams. *Issues of Chemistry and Chemical Technology*, Ukrainian State University of Chemical Technology, *1*, 26–34.
10. Reyes-Labarta, J. A., & Marcilla, A. (2008). Kinetic study of the decompositions involved in the thermal degradation of commercial azodicarbonamide. *Journal of Applied Polymer Science, 107*(1), 339–346.
11. Vrandecic, N. S., Klaric, I., & Roje, U. (2001). Effect of Ca/Zn stabiliser on thermal degradation of poly (vinyl chloride)/chlorinated polyethylene blends. *Polymer Degradation and Stability, 74*(2), 203–212.
12. Balkose, D., Gokcel, H. I., & Goktepe, S. E. (2001). Synergism of Ca/Zn soaps in poly (vinyl chloride) thermal stability. *European Polymer Journal, 37*(6), 1191–1197.
13. Krutko, I., Yavir, K., & Kaulin, V. (2018). Effect of antioxidants on the stability of pitch-based polymer to thermo-oxidative action. *Chemistry and Chemical Technology, 12*(1), 109–113.
14. Krutko, I., Yavir, K., & Kaulin, V. (2018). The stabilizer nature influence on coal tar pitch degradation. *Issues of Chemistry and Chemical Technology*, Ukrainian State University of Chemical Technology, *4*, 97–102.

CHAPTER 2

Polymer Composites with AI in Environmental Monitoring: A Review

LELA MIRTSKHULAVA

Department of Computer Science, Ivane Javakhishvili Tbilisi State University, Tbilisi, Georgia

ABSTRACT

Polymer composites are widely used in environmental monitoring as sensors for detecting and quantifying specific contaminants in the environment. With the integration of artificial intelligence (AI) techniques, the analysis of polymer composites in environmental monitoring has become more sophisticated and accurate. AI can analyze large datasets of polymer properties and environmental parameters to predict the presence and concentration of contaminants in the environment. Machine learning algorithms have been used to analyze the electrical properties of polymer composite sensors and to identify specific contaminants. Additionally, AI models can be used to optimize the performance of polymer composite sensors by predicting the ideal polymer composition and sensor design for detecting specific contaminants. The integration of AI in polymer composite analysis has enabled real-time monitoring and early detection of pollution, thereby aiding in effective environmental management and protection. This abstract highlights the potential of AI in polymer composite analysis for environmental monitoring, offering a more comprehensive understanding of environmental contaminants and enabling efficient and cost-effective monitoring and management of environmental pollution.

Sustainability in Polymer Technology and Plastics Engineering: Concepts, Strategies, and Opportunities.
Tamara Tatrishvili, Neha Kanwar Rawat, Swati Gokul Talele, and A. K. Haghi (Eds.)

2.1 INTRODUCTION

Polymer composites are materials consisting of two or more different types of polymers that are combined to create a new material with unique properties. They are commonly used in various industries including aerospace, automotive, construction, and marine. Environmental monitoring is another application where polymer composites have found use due to their ability to sense changes in environmental conditions.

One example of the use of polymer composites in environmental monitoring is for detecting volatile organic compounds (VOCs) in the air. VOCs are a major contributor to air pollution and can have negative health effects on humans and wildlife. Polymer composites can be used as sensors to detect and quantify VOCs in the air. The polymer composite sensor is made up of a conductive polymer and a polymer that is selective to the target VOC. When the target VOC comes into contact with the polymer composite, it causes a change in the electrical properties of the sensor which can be detected and measured.

Another example is the detection of heavy metals in water. Heavy metals are toxic and can accumulate in aquatic organisms, causing harm to both the organisms and humans who consume them. Polymer composites can be used as sensors to detect and quantify heavy metals in water. The polymer composite sensor is made up of a polymer that is selective to the target heavy metal and a conductive polymer. When the target heavy metal comes into contact with the polymer composite, it causes a change in the electrical properties of the sensor which can be detected and measured.

2.2 RELATED WORKS ON ENVIRONMENTAL MONITORING

2.2.1 ICT TECHNIQUES IN ENVIRONMENTAL MONITORING

Environmental monitoring is a critical aspect of environmental management, which involves the collection and analysis of data to understand and manage the impact of human activities on the environment. With the increasing availability of Information and Communication Technology (ICT) tools, environmental monitoring has become more efficient, accurate, and cost-effective. In this section, we will discuss the role of ICT techniques in environmental monitoring such as:

- **Remote Sensing:** Remote sensing techniques involve the use of satellite and airborne sensors to collect data on environmental variables such as land cover, vegetation, water quality, and atmospheric conditions. This data can be used to assess environmental change over time, identify areas of concern, and support decision-making processes. Several studies have shown the effectiveness of remote sensing in environmental monitoring, including the use of satellite imagery to monitor water quality [1] and detect land-use changes [2].
- **Geographic Information Systems (GIS):** GIS techniques involve the use of spatial data to map and analyze environmental variables, such as land cover, topography, and water resources. GIS can be used to identify areas of high environmental risk, model environmental scenarios, and support decision-making processes. Several studies have demonstrated the effectiveness of GIS in environmental monitoring, including the use of GIS to map pollution sources [3] and model land-use changes [4].
- **Internet of Things (IoT):** IoT involves the use of sensors and other devices to collect and transmit data on environmental variables in real-time. IoT can be used to monitor air quality, water quality, noise levels, and other environmental variables, and provide real-time alerts to environmental risks. Several studies have shown the effectiveness of IoT in environmental monitoring, including the use of IoT sensors to monitor water quality in rivers [5], detect air pollutants [6], and detect real-time wildfires [7].

2.2.2 POLYMER-RELATED ISSUES

There are many scientific issues related to polymers, which are macromolecules composed of repeating units called monomers. Here are a few examples of polymer-related scientific issues:

1. **Polymer Synthesis and Processing:** One of the primary challenges in polymer science is designing and synthesizing polymers with specific properties, such as strength, flexibility, or heat resistance. This involves developing new methods for synthesizing polymers with controlled molecular weight, composition, and architecture. In addition, processing polymers into useful products often involves complex rheological phenomena, such as melt flow and deformation.

2. **Polymer Characterization:** Understanding the structure and properties of polymers is crucial for designing new materials and optimizing existing ones. Characterization techniques such as X-ray diffraction, NMR spectroscopy, and thermal analysis are used to determine the molecular structure, morphology, and thermal behavior of polymers.
3. **Polymer Degradation and Stability:** Polymers can degrade over time due to exposure to heat, light, oxygen, and other environmental factors. Understanding the mechanisms of polymer degradation and developing strategies to improve their stability is important for ensuring the long-term performance of polymer-based materials.
4. **Polymer Recycling and Sustainability:** The large-scale production and disposal of polymers have significant environmental impacts, including greenhouse gas emissions, land use, and waste generation. Developing new methods for recycling and upcycling polymers could help reduce these impacts and create more sustainable polymer-based products.
5. **Polymer-Based Materials for Advanced Applications:** Polymers have a wide range of applications in areas such as electronics, energy storage, biomedical devices, and advanced manufacturing. Developing new polymer-based materials with enhanced properties or novel functionalities is an important area of research for these applications.

2.3 EXPERIMENTAL METHODS OF POLYMER ANALYSIS IN ENVIRONMENTAL MONITORING USING AI

Experimental methods for polymer analysis in environmental monitoring using AI vary depending on the specific application and sample type. Here, we will discuss some examples of experimental methods used in recent studies.

Sample preparation typically involves the separation of microplastics from the sample matrix and the removal of organic matter and other contaminants [8]. Sample collection and preparation: For the analysis of microplastics in environmental samples, various collection methods have been utilized, such as filtration, sedimentation, and sieving [9].

2.3.1 ANALYSIS USING AI

Various AI techniques can be used for polymer analysis, including machine learning, deep learning, and neural network-based methods. Machine learning algorithms have been used to classify microplastics based on their physical and chemical properties, such as size, shape, color, and texture. Deep learning methods have been utilized to detect and quantify microplastics in water samples, using image analysis and object detection algorithms, and deep learning applications have been applied in predicting polymer properties [10]. Neural network-based methods have been developed for the analysis of the molecular weight distribution of polymers in complex mixtures [11].

Validation of results: Validation of results is critical for ensuring the accuracy and reliability of AI-based polymer analysis. Validation methods can include the use of reference standards, replicate analyses, and comparison with results obtained from traditional analytical methods. So, the experimental methods for polymer analysis using AI are constantly evolving, and new techniques and approaches are being developed to improve the accuracy and efficiency of the analysis [12–28].

2.4 POLYMER COMPOSITES AS SENSORS IN ENVIRONMENTAL MONITORING WITH AI

2.4.1 WAYS USING AI IN POLYMER COMPOSITE OPTIMIZATION

1. **Material Design and Optimization:** AI algorithms can be used to design and optimize the properties of polymer composite sensors for specific environmental monitoring applications. By learning the relationships between the material properties and the sensing performance, AI can generate new composite materials that are optimized for the desired sensing properties, such as sensitivity, selectivity, and response time.

2. **Data Analysis and Pattern Recognition:** AI techniques such as machine learning and pattern recognition can be used to analyze the data collected by polymer composite sensors in real time. By learning the patterns in the sensor response, AI algorithms can detect and identify specific environmental pollutants, such as volatile organic compounds, heavy metals, or gases, and provide real-time feedback on the environmental conditions.

3. **Sensor Network and Integration:** AI can be used to optimize the design and deployment of a sensor network for environmental monitoring applications. By learning the spatial distribution of environmental pollutants, AI algorithms can determine the optimal placement of the sensors and integrate the data from multiple sensors to provide a comprehensive view of the environmental conditions.
4. **Prediction and Prognostics:** AI techniques such as neural networks can be used to predict the behavior of polymer composite sensors under different environmental conditions. By learning the relationships between the input parameters, such as temperature, humidity, or pressure, and the sensor response, AI algorithms can provide real-time predictions of the sensor behavior and optimize the sensor performance for changing environmental conditions.

The combination of polymer composites and AI techniques can provide a powerful and versatile platform for environmental monitoring applications. By optimizing the design and performance of the sensors, and integrating the data from multiple sensors in real-time, polymer composite sensors can help monitor and protect the environment cost-effectively and efficiently.

There are several types of neural networks that can be used to optimize the performance of polymer composites by simulating their behavior under different environmental conditions. Here are a few examples:

i. **Multilayer Perceptron (MLP):** Multilayer perceptron is a type of feedforward neural network that consists of multiple layers of interconnected nodes. MLPs can be trained to predict the behavior of polymer composites under different environmental conditions by learning the relationships between the input parameters (such as the type and concentration of filler materials, processing parameters, and environmental conditions) and the output properties (such as mechanical strength, thermal stability, and corrosion resistance). MLPs have been used to optimize the mechanical properties of polymer composites for a variety of environmental monitoring applications.

ii. **Convolutional Neural Network (CNN):** Convolutional neural networks are a type of neural network that is commonly used for image and signal processing tasks. CNNs can be applied to polymer composites by treating the composite as an image or signal and learning the relationships between the input parameters and output properties. For example, CNNs can be used to optimize the design of a composite sensor by predicting the output signal based on the input parameters such as the type and concentration of the sensing materials, processing parameters, and environmental conditions.

iii. **Recurrent Neural Network (RNN):** Recurrent neural networks are a type of neural network that is commonly used for time-series and sequence-modeling tasks. RNNs can be applied to polymer composites by treating the environmental conditions as a time series and learning the relationships between the input parameters and output properties over time. For example, RNNs can be used to optimize the design of a composite material for monitoring environmental pollutants by predicting the changes in the material properties over time in response to different pollutants.

iv. **Autoencoder:** Autoencoders are a type of neural network that can be used for unsupervised learning tasks such as data compression and feature extraction. Autoencoders can be applied to polymer composites by learning the underlying patterns and features of the input data, such as the relationship between the filler materials and the resulting mechanical properties. Autoencoders can then be used to generate new composite materials that are optimized for specific environmental monitoring applications.

2.4.2 POLYMER PROPERTIES FOR ENVIRONMENTAL MONITORING

Polymers are widely used in environmental monitoring applications due to their unique properties, which make them well-suited for a range of sensing and detection applications. Some of the key properties of polymers that make them useful for environmental monitoring include:

- **Selectivity:** Polymers can be designed to selectively bind to specific analytes or environmental pollutants, allowing for the development of highly sensitive and specific sensors.
- **Stability:** Polymers are often resistant to environmental factors such as moisture, temperature, and pH changes, which allows them to maintain their properties and performance over time.

- **Sensitivity:** Polymers can be designed to be highly sensitive to small changes in environmental conditions, making them useful for monitoring parameters such as temperature, humidity, and pressure.
- **Durability:** Polymers are typically robust and resistant to wear and tear, which makes them suitable for use in harsh or remote environments.
- **Scalability:** Polymers can be manufactured in large quantities at low cost, making them ideal for use in large-scale environmental monitoring applications.

Some specific examples of polymers used in environmental monitoring applications include:

- **Molecularly Imprinted Polymers (MIPS):** These polymers are designed to selectively bind to target analytes with high affinity and specificity. MIPs have been used for the detection of a range of environmental pollutants, including heavy metals, pesticides, and organic compounds.
- **Conducting Polymers:** These polymers have electrical conductivity and can be used as sensing materials in electronic devices for environmental monitoring. They have been used for sensing parameters such as humidity, temperature, and gas concentrations.
- **Hydrogels:** These polymers have the ability to absorb and retain water, making them useful for monitoring environmental parameters such as soil moisture and humidity.
- **Fluorescent Polymers:** These polymers can be designed to emit light in response to specific environmental stimuli, making them useful for sensing applications such as pH and temperature monitoring.

2.4.3 ORGANIC POLYMERS

Organic polymers are large molecules composed of repeating units called monomers that are connected through covalent bonds. They are commonly referred to as macromolecules because of their large size, and they can be found in a variety of natural and synthetic forms.

Organic polymers are widely used in industry and everyday life because of their unique properties, such as flexibility, strength, and resistance to chemicals and heat. They are also generally lightweight and easy to process, which makes them attractive materials for a wide range of applications.

Some common types of organic polymers include:

- **Polyethylene:** Polyethylene is a synthetic polymer that is used in a wide range of applications, including packaging, toys, and construction materials. It is a highly flexible and durable material and is often used in its low-density form (LDPE) for applications requiring high flexibility and toughness.
- **Polypropylene:** Polypropylene is a thermoplastic polymer that is used in a variety of applications, including packaging, textiles, and automotive parts. It is highly resistant to chemicals and can be processed easily, making it a versatile material for a range of applications.
- **Polystyrene:** Polystyrene is a synthetic polymer that is used in applications such as packaging, insulation, and disposable utensils. It is a lightweight and rigid material and can be produced in a range of forms, including foam, sheets, and films.
- **Polyvinyl Chloride (PVC):** PVC is a widely used synthetic polymer that is used in applications such as pipes, electrical cables, and medical devices. It is highly resistant to chemicals and weathering and can be processed in a range of forms, including rigid and flexible.
- **Natural Rubber:** Natural rubber is a polymer derived from the sap of rubber trees. It is highly elastic and can be processed into a wide range of forms, including tires, hoses, and footwear.

Organic polymers are versatile and widely used materials that have transformed modern life in countless ways. They have applications in a range of industries, including packaging, automotive, construction, and healthcare, and are constantly being developed and refined to meet the evolving needs of society.

2.5 EXPERIMENTAL RESULTS

2.5.1 POLYMER PROPERTIES PREDICTION USING POLYMER GENOME

Predicting polymer properties using a database is an emerging area of research called "polymer genomics." The goal of polymer genomics is to identify genetic variations that affect polymer properties, such as strength, flexibility, and durability. This information can be used to design new polymers with improved properties or to optimize the manufacturing process to produce polymers with desired properties [29].

One approach to polymer genomics is to use high-throughput sequencing techniques to generate large amounts of genomic data from a population of organisms that exhibit a range of polymer properties. This data can then be analyzed to identify genetic variations that are associated with specific polymer properties.

Another approach is to use computational methods to predict the effects of specific genetic variations on polymer properties. For example, machine learning algorithms can be trained on large datasets of polymer properties and genetic variations to predict how specific genetic variations will affect polymer properties. This field of polymer genomics is still in its early stages, but it holds great promise for improving our understanding of polymer properties and for developing new polymers with improved performance characteristics.

2.5.2 POLYMER PREDICTION

In the Polymer Genome, we draw polymer (Figure 2.1).

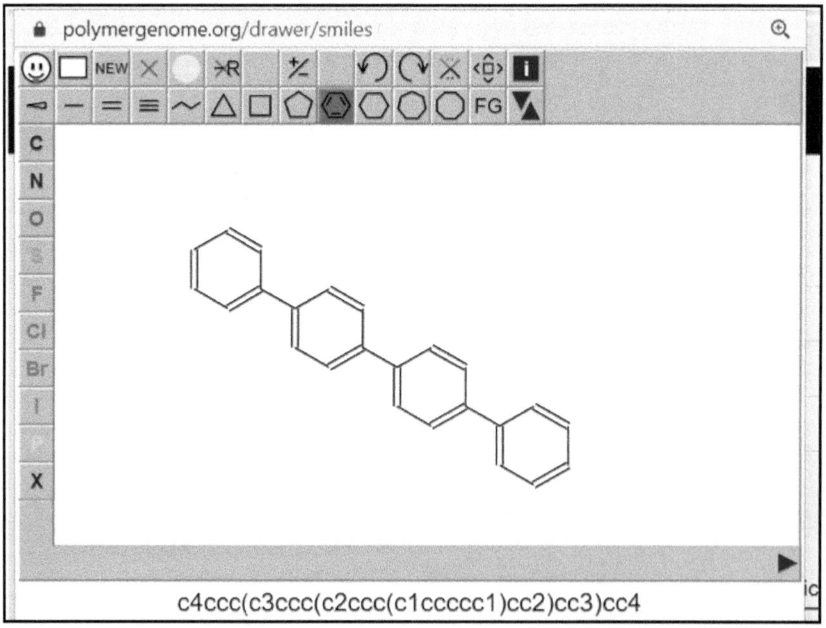

FIGURE 2.1 Polymer drawing.

Here is its [*]CC5CCC(C4CCC(C3CCC(c2ccc(c1ccccc1)cc2)C3)C4)C5[*] SMILE (simplified molecular-input line-entry system) formula which uses a short ASCII string to represent the structure of chemical species.

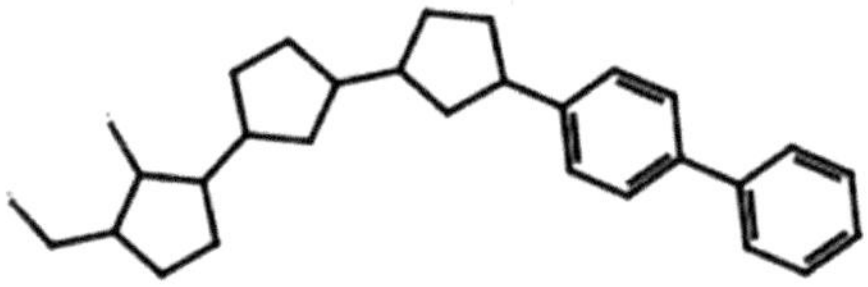

ELECTRONIC PROPERTIES

Bandgap (chain)	**4.5 eV**
Bandgap (bulk)	**4.3 eV**
Electron affinity	**2.3 eV**
Ionization energy	**6.2 eV**

DIELECTRIC & OPTICAL PROPERTIES

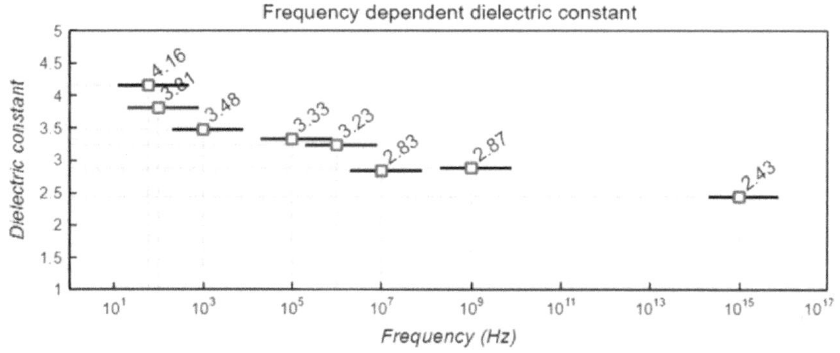

Dielectric constant (crystal)	**4.5**
Refractive index (bulk resin)	**1.6**
Refractive index (crystal)	**1.9**

PHYSICAL & THERMODYNAMIC PROPERTIES

Density	1.2 g/cc
Atomization energy	-5.94 eV/atom
Heat capacity	1 J/gK
Tendency to crystallize	36 %
Limiting oxygen index	34 %
Fractional free volume	0.18

THERMAL PROPERTIES

Glass transition temperature	400 K
Melting temperature	475 K
Thermal decomposition temperature	664 K
Thermal conductivity	0.21 W/mK

PHYSICAL & THERMODYNAMIC PROPERTIES

Density	1.2 g/cc
Atomization energy	-5.94 eV/atom
Heat capacity	1 J/gK
Tendency to crystallize	36 %
Limiting oxygen index	34 %
Fractional free volume	0.18

MECHANICAL PROPERTIES

Tensile strength	77 MPa
Young's modulus	2 MPa

PERMEABILITY PROPERTIES

He gas permeability	1.8 Barrer
H_2 gas permeability	2.0 Barrer
CO_2 gas permeability	1.7 Barrer
N_2 gas permeability	0.8 Barrer
O_2 gas permeability	1.1 Barrer
CH_4 gas permeability	0.8 Barrer

SOLUBILITY PROPERTIES

Hildebrand solubility parameter	19.7 $MPa^{1/2}$
Solvent	chlorobenzene, chloroform, 1,4-dioxane, THF, dichloromethane, acetic acid, nitrobenzene, benzene, NMP, m-cresol
Non-solvent	methanol, ethanol, isopropanol, water, DMSO, acetonitrile, acetone, diethyl ether, nhexane, DMF

2.6 CONCLUSION

The paper provides a comprehensive overview of the recent advances in the field of environmental monitoring using polymer composites and artificial intelligence (AI). We highlighted the advantages of using polymer composites in environmental monitoring, such as their high sensitivity and selectivity, and the potential for AI to improve the accuracy and efficiency of monitoring systems. The paper presents various examples of polymer composites used in environmental monitoring applications, such as detecting heavy metals, organic pollutants, and gases. We also discussed the use of AI in improving the detection and analysis of environmental data, such as machine learning algorithms for predicting pollution levels. The paper demonstrates the potential of polymer composites and AI in addressing environmental challenges such as pollution and climate change. The combination of these technologies offers a promising pathway for developing more efficient and accurate monitoring systems that can help protect the environment and human health. However, further research is needed to overcome the challenges associated with integrating AI into environmental monitoring systems and to develop more sustainable and cost-effective polymer composites for environmental applications.

KEYWORDS

- **artificial intelligence**
- **environmental monitoring**
- **geographic information systems**
- **information and communication technology**
- **Internet of things**
- **polymer informatics**
- **polymer-based sensors**
- **volatile organic compounds**

REFERENCES

1. Rocchini, D., Marcantonio, M., Delucchi, L., & Bacaro, G. (2019). Remote sensing for monitoring water quality in lakes and rivers: A review. *Science of the Total Environment, 653*, 890–906.
2. Fritz, S., See, L., Bayas, J. C. L., Waldner, F., Jacques, D. C., Becker-Reshef, I., Whitcraft, A. K., Baruth, B., Bonifacio, R., Crutchfield, J., et al. (2019). A comparison of global agricultural monitoring systems and current gaps. *Agricultural Systems, 168*, 258–272.

3. Plewe, B., Kelly, M., & Harrower, M. (2019). Mapping environmental injustice using GIS: The case of South Seattle, Washington. *Applied Geography, 108*, 75–87.
4. González-Tánago, I., Rojo-Alboreca, A., & Gómez-Limón, J. A. (2018). Land use changes, agricultural policies and farm structural adjustment in the European Union. *Land Use Policy, 78*, 8–19.
5. Jung, Y., Kim, J., & Choi, J. (2019). Development of a river water quality monitoring system using wireless sensor network and web-based decision support system. *Journal of Hydrology, 572*, 148–157.
6. Su, X., Huang, Y., Wang, X., & Gao, Z. (2019). A low-cost IoT-based air pollution monitoring system. *Journal of Cleaner Production, 218*, 300–308.
7. Mirtskhulava, L., Tvaliashvili, L., Bardavelidze, A., & Tvaliashvili, L. (2020). Environmental monitoring in flammable climate zones over IoT cloud. *International Journal of Simulation: Systems, Science & Technology*. DOI: 10.5013/IJSSST.a.21.02.34.
8. Deng, J., Guo, J., Xiong, X., Huang, S., & Zhang, W. (2021). Identification and classification of microplastics based on deep learning. *Environmental Science & Technology, 55*(4), 2545–2554.
9. Li, L., Gao, J., Zhao, Y., & Li, Y. (2021). Automated detection of microplastics in water samples based on deep learning. *Environmental Pollution, 287*, 117485.
10. Mirtskhulava, L. (2023). Deep learning applications in predicting polymer properties. In *Advanced Polymer Structures: Chemistry for Engineering Applications*. pp. 15–18. Apple Academy Press.
11. Wang, Y., Zhang, S., Wei, H., Wang, C., Cui, Z., & Chen, X. (2021). A neural network-based method for analyzing the molecular weight distribution of polymers in complex mixtures. *Analytical and Bioanalytical Chemistry, 413*(1), 195–202.
12. Hillmyer, M. A., Lodge, T. P., & Dorfman, K. D. (2016). Polymer informatics: An overview. *Macromolecules, 49*(16), 5767–5777. https://doi.org/10.1021/acs.macromol.6b01102.
13. Mannodi-Kanakkithodi, A., Pilania, G., & Huan, T. D. (2017). Computational polymer informatics: A review of tools and applications. *AIChE Journal, 63*(9), 3968–3984. https://doi.org/10.1002/aic.15788.
14. Gute, B. D., O'Brien, M., Reidy, M. P., & Elder, J. T. (2020). Polymer informatics: Current status and future opportunities. *Journal of Polymer Science Part B: Polymer Physics, 58*(19), 1309–1322. https://doi.org/10.1002/polb.25143.
15. Paul, A., Mishra, A. K., Mishra, R., Pandey, A. K., & Ramamurthy, P. C. (2021). Machine learning and polymer informatics. *Polymer Reviews, 61*(1), 64–101. https://doi.org/10.1080/15583724.2020.1869294.
16. Gute, B. D., O'Brien, M., & Elder, J. T. (2021). Advances in computational polymer informatics: Machine learning, data-driven approaches, and design principles. *Journal of Materials Chemistry A, 9*(9), 5009–5033. https://doi.org/10.1039/D0TA11881K.
17. Gao, S., Sun, J., & Huang, Y. (2022). Polymer informatics: A new approach for the development of polymer materials. *Polymer Chemistry, 13*(3), 271–285. https://doi.org/10.1039/D1PY01189J.
18. Gute, B. D., O'Brien, M., & Elder, J. T. (2022). Advances in polymer informatics: Applications in materials design and development. *Chemical Society Reviews, 51*(5), 2037–2059. https://doi.org/10.1039/D1CS00880H.

19. Kumar, A., Patel, B. B., Sharma, V., & Singh, B. (2021). Polymer artificial intelligence: A review of tools and applications. *Polymers, 13*(3), 319. https://doi.org/10.3390/polym13030319.
20. Chen, Q., Qian, L., Liu, Y., Jia, X., Wang, D., & Wang, W. (2021). Artificial intelligence for the design of polymer materials. *Journal of Materials Chemistry A, 9*(10), 5638–5661. https://doi.org/10.1039/D0TA11317A.
21. Li, Y., Zhang, J., Li, Z., Li, J., Li, L., & Yang, J. (2021). Polymer materials design using machine learning: A review. *AIChE Journal, 67*(4), e17212. https://doi.org/10.1002/aic.17212.
22. Liu, Y., Wang, D., Chen, Q., Qian, L., & Wang, W. (2021). Artificial intelligence-guided polymer design: Principles, applications, and future prospects. *Polymer, 233*, 123125. https://doi.org/10.1016/j.polymer.2021.123125.
23. Jha, A., Kumar, R., Kumar, A., Gupta, N., Bhanja, S., & Sharma, V. (2021). Artificial intelligence for polymer and materials design: Recent advances, challenges, and opportunities. *Journal of Polymer Science Part A: Polymer Chemistry, 59*(18), 2929–2959. https://doi.org/10.1002/pola.30830.
24. Pujari, S., Kalhapure, P., Ramalingam, B., Seidel, R., & Langer, R. (2021). Artificial intelligence in polymer chemistry and materials science: Principles, applications, and future prospects. *Chemical Reviews, 121*(20), 12035–12096. https://doi.org/10.1021/acs.chemrev.1c00457.
25. Chen, Q., Qian, L., Liu, Y., Jia, X., Wang, D., & Wang, W. (2022). Advances in polymer artificial intelligence: Principles, tools, and applications. *Accounts of Chemical Research, 55*(6), 1141–1153. https://doi.org/10.1021/acs.accounts.2c00007.
26. Nguyen, T. P., Nguyen, P. T., & Chae, K. H. (2015). Polymer composite sensors for environmental monitoring. *Sensors (Basel), 15*(7), 15736–15753. https://doi.org/10.3390/s150715736.
27. Singh, R., Gupta, V. K., Paul, A. K., & Kumar, S. (2020). Development of polymer composite-based sensors for detection of heavy metals in water: A review. *Critical Reviews in Analytical Chemistry, 50*(1), 69–83. https://doi.org/10.1080/10408347.2019.1685807.
28. Kumar, A., Kumar, A., & Bhatnagar, R. (2020). Polymer composites as sensors for environmental monitoring: A review. *Environmental Science and Pollution Research, 27*(12), 12644–12656. https://doi.org/10.1007/s11356-020-08275-4.
29. Polymer Genome Project. Retrieved from: https://www.polymergenome.org (accessed on 14 June 2024).

CHAPTER 3

Tensosensitivity Properties of Electrically Conductive Polymer Composites

L. SHAMANAURI,[1] NATIA SHENGELIA,[2] and E. CHKHAIDZE[3]

[1]*R. Dvali Institute of Machine Mechanics, Tbilisi, Georgia*

[2]*Sokhumi State University, Tbilisi, Georgia*

[3]*Assistant Professor, Faculty of Chemical Technology and Metallurgy, Georgian Technical University, Tbilisi, Georgia*

ABSTRACT

In this chapter, the tensosensitive properties of the composite materials observed based on the epoxy glue and silicon rubbers that contain graphite and carbon black have been studied. Based on this investigation, spatially prepared devices were demonstrated successfully: (1) tensosensitivity of observed composites enhanced by reduction of concentration of electric conducting filler and vice versa; (2) tensosensitivity of these materials enhanced by enhancing their elasticity; (3) the phenomena of relaxation in the polymer matrixes restrict the frequency of impulse deformations; (4) tensosensitivity of observed materials could be regulated by considering some amount of the low molecular plasticizer to the polymer matrix.

3.1 INTRODUCTION

It is widely acknowledged that polymers and composite materials are currently in a leading position among materials. This is explained by the fact that throughout their existence, they have been considered as materials of the future with limitless possibilities for practical application. This is also explained by the fact that from the very beginning, they have been

Sustainability in Polymer Technology and Plastics Engineering: Concepts, Strategies, and Opportunities. Tamara Tatrishvili, Neha Kanwar Rawat, Swati Gokul Talele, and A. K. Haghi (Eds.)

competing with traditional materials such as metals, ceramics, wood pulp, leather, and more. Currently, this competition continues with considerable success. What makes polymer materials competitive? First, their lightness, weight, high corrosion resistance, and relatively simple technology, which in turn reduces their cost. While in the past it was almost unthinkable to use polymers as elements in electronics, there is now increasing information about some radio-electronic elements or devices that are virtually entirely made of polymer material.

Despite a number of successes in the field of polymer electronics, there are still many unsolved problems, one of which is the stability of parameters. However, this Achilles heel of polymers, according to advanced scientists, is counting its last days. The science and technology of conductive polymer composites are relatively young but have high development potential. This is because while the creation of pure organic (in particular) polymer conductors is associated with some technological complexity, composite polymeric materials, or composites, are free of these complexities. They avoid the limited possibilities of synthesizing polymers and achieve this goal with much simpler technologies (diversity of fillers and their synergistic effect, creation of materials with the desired structure and properties with organic or inorganic modifiers).

Among the physical properties of polymer composites, the most important are their mechanical properties. However, in recent years, the diversity of functionality of these materials has been steadily increasing. So, for example, in the early times, in terms of electrical conductivity, polymers and composites based on them were typically dielectric, polymers and polymer composites with semiconducting and conductive properties are now increasingly being created.

In this case, the emphasis is on conductive composite materials, since converting dielectric polymers to chemically conductive electrodes is a much more complicated task than incorporating different typical conductive substances into the polymer and forming composite material.

With much simpler and cheaper technology, it is possible to obtain conductive materials whose range of conductivity can vary from dielectrics to narrow-banded conductors. The described capability makes the conductive polymer composites competitive materials for metals and semiconductors.

Thus, it is possible to consider composite materials, in addition to typical metals and semiconductors, among traditional conductors. Researchers and practitioners are currently focusing on the production of elements from polymer materials such as sensors for various physical parameters (pressure, temperature, electric and magnetic fields, photo and nuclear radiation).

There is an opportunity to expand the range of these measurement parameters, which is technically important. Sensor materials based on polymer materials have little history. There is a great deal of literature on polymer tensiometry and thermoelectricity, from which one of the most important sources of radiation is high sensitivity to measurement parameters.

The following works provide research on this topic. In the work [1], EPC-based tensografts have been developed, which have been specifically tested to measure the deformation of parachute troughs under different loading conditions. The sensor is based on a thermoplastic polymer and 35% technical carbon-based composite. The effects of sensor response on transmitters of different thickness rectangles (80 x 10 mm in surface dimensions) are tested for the effects of sensor thickness, aging, environmental humidity, and stretching speed on the electrical conductivity of the elements. They measured permeability, rate of change, and repetition after 10 cycles. The samples were allowed to relax for 1 min.

Based on ethylene rubber and technical carbon base, tensile-sensitive elements with a tensile susceptibility coefficient of 6–7 units have been prepared, although the mechanical strength is somewhat lower than its analogs [2]. The same is true of their stability. Similar work has been done on tensile-sensitive materials based on polyisoprene rubber and technical carbon [3]. To increase the strength of the materials [4], the authors of the work have created a tensile-sensitive element based on two basic polymers (polyester and epoxy resins) and carbon nanoparticles. Obviously, the authors used carbon nanotubes to enhance the mechanical strength of the composite. There is an interesting overview that discusses the technological methods of tensomechanical materials and possible regulatory factors for their properties [5].

In [6], the tensile sensitivity of EPC materials derived from isoprene and technical carbon based on fill concentration, ambient temperature, humidity, and ultraviolet radiation is studied. It is established that the sensitivity of the sensor is optimal when the concentration of the conductive component in the composite changes in the percolation threshold region, although in this transition area of conduction (dielectric–conductive) materials, it is highly volatile.

The development history of sensors built on polymeric conductive composites is provided in the work [7], which discusses the achievements in this field up to 2007. The facts presented here indicate the progress made in the field of polymer sensor technology.

To date, numerous theoretical or experimental works on the topic of polymer sensors have been undertaken, in which the authors attempt to delve deep into the mechanism of sensitivity regulation to physical parameters, but none of them have yet been fully substantiated (*see* [3, 8–13]).

3.2 EXPERIMENTAL SECTION

Electrically conducting rubbers were selected as polymeric composites based on the epoxy glue (ED-20 mark) and caoutchouc (SCTV mark – a product of Russia), in which the carbon fiber and synthetic graphite of P357 were filled. In some experiments, we used the low molecular weight caoutchouc of SCTN mark and glycerin (in epoxy glue-based composites) as a structure modifier.

Electrically conductive resins were obtained by the method commonly used in silicone rubber technology (peroxide vulcanization): mixing ingredients of a certain proportion on laboratory mills, removing the mixture from mills in plates, giving the desired shape to the conveyor (5 × 20….30 × 1–2 mm) and cracking of raw materials. Transfer to vulcanization camera, and vulcanization in two stages (first, during 20 min., 5 MPa, and 120C; second, at 180°C and at normal ambient pressure with subsequent spontaneous cooling). The modifier in the rubber mixture was introduced in the process of waltzing.

The epoxy glue-based composite was made by mechanical mixing of the permeable component in the initial mass and subsequently reinforced by the addition of an adhesive (polyethylene polyamine) of 1/10 by mass. The thin layer of the obtained mixture was fastened to the flat plate of Teflon by the copper electrodes placed therein. The cooling was carried out for one day at room temperature with further heat treatment at 50°C for 30 min. In some experiments, glycerol was used as a plasticizer in the amount of 5–10%, which was mixed with the mixture beforehand.

The obtained tensile-absorbing elements having the geometry shown in Figure 3.1 were coated with thin (0.5 mm) plates of textolite on a rotating disk-forming device (Figure 3.2) for experiments.

3.2.1 RESEARCH METHODOLOGY

The test specimens we received were tested on a self-propelled device that allowed the specimens to be tested for both bending and stretching deformations (Figure 3.2). The change of sample impedance from the device would go through what the Wheatstone bridge and digital oscilloscope were fixed on a personal computer.

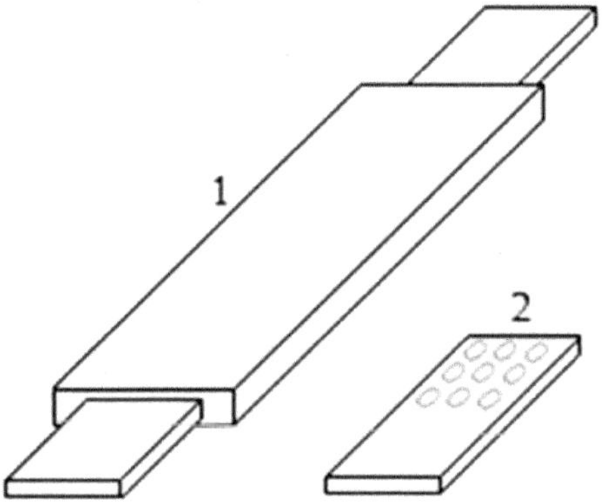

FIGURE 3.1 The geometric face of the test specimens.

Note: (1) The sample with electrodes made of perforated copper foil; (2) the electrode, the perforated part of which is inserted into the composite before being reinforced.

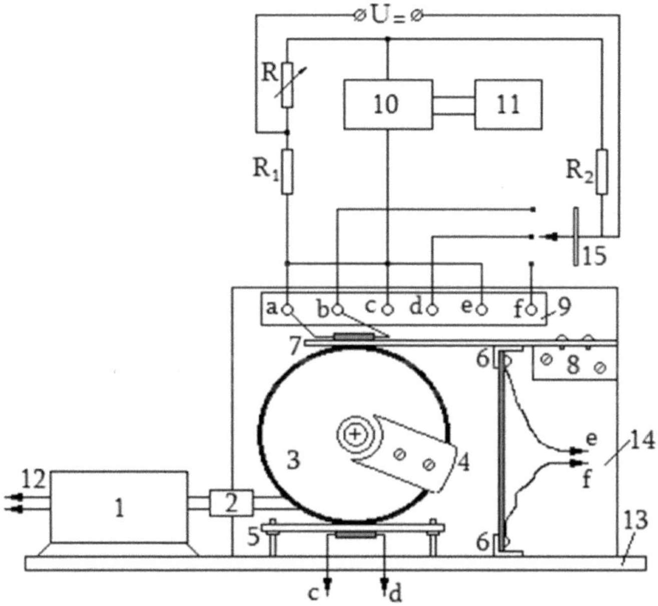

FIGURE 3.2 Sensor variable deformation device with a white stone boom.

Note: (1) Permanent power motor; (2) rubber pipe connecting motor; (3) plate attached to worm screw; (4) plate attached to plate, touching specimen (5) on second plate; (6) lower and upper straps of stretchable rubber specimen; (7) a specimen mounted on a plate; (8) clamping angle of plate 7; (9) clamp wires from samples; (10) oscilloscope; (11) computer; (12) power supply connectors for electric motors; (13) stand; (14) supporting board for plates and specimens; and (15) pattern switcher.

3.2.1.1 WORK ON THIS TOOL WAS CARRIED OUT IN THE FOLLOWING ORDER

1. First of all, in certain areas, samples based on conducting rubbers and epoxy glue are placed on the textolite plates 5 and 7 (Figure 3.2) to realize bending deformation. These plates are connected (by means of a switcher) to the Wheatstone Bridge. To realize stretching deformations (for rubbers), contacts shown in Figure 3.6 are used, which are also connected to the Wheatstone bridge;
2. The constant voltage engine 1 is switched with a constant voltage source using cables 12. Voltage changes (within a few volts) are regulated to control the number of rotations of the engine rotor. By means of a rubber tube connected to the engine, the disc 3 is rotated. The plate attached to the drive rotates the drive plate 4 with rotating patterns 5 and 7, starting with them and using contacts 6 for rubber stretching;
3. The mechanical deformation of the specimens periodically reflects the change in the electrical resistance of the deformable specimens, which is transferred to the oscilloscope embedded in the Wheatstone bridge and then to the personal computer for display on its screen. Switch 15 allows measurements to be carried out separately for each sample.

3.3 RESULTS AND DISCUSSION

It was experimentally shown that the tensile sensitivity of the samples depends on the concentration of the electrically conducting filler in them. Specimens based on rubber and epoxy glue were prepared to test this relationship. Figure 3.3 graphically illustrates the nature of this dependency under static loading.

It was also interesting to investigate how the tensile sensitivity of the composite was influenced by the hardness of the polymer matrix. Two substances were used for this purpose: one based on low-molecular-weight SCTV mark resin rubber and the other on glycerin-epoxy glue composites. The experiment aimed to determine the impact of polymer matrix softening on the tensile sensitivity of the composites.

In this study, the relationship between the tensile sensitivity coefficient of the composites and the filling concentration was examined, except that plasticizer (10–20 masses%) was added to the composites.

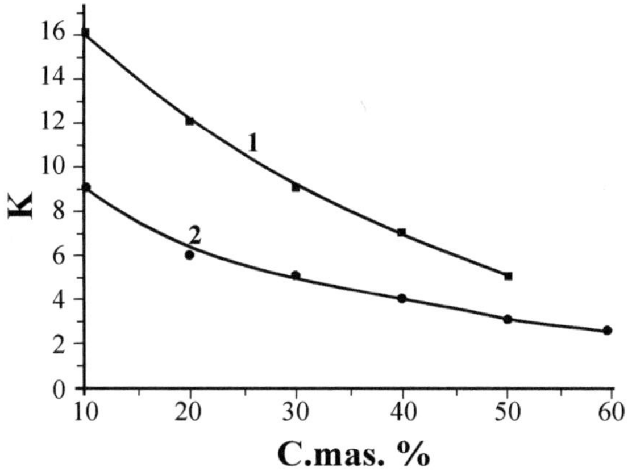

FIGURE 3.3 Dependence of the tensosensitivity coefficient of the composites on the permeability of the composites.

Note: (1) For carbon black (P357) filled rubber; and (2) for epoxy glue filled with the same filler.

It can be seen from Figure 3.4 that the values of the tensile susceptibility coefficients of composites based on both conductive rubber and epoxy glue are decreased with increasing concentration of the conductive component, i.e., with increasing conductivity. It can also be seen from the same figure that this coefficient of composites with the plasticizer is somewhat higher than the corresponding analogs that do not contain the plasticizer.

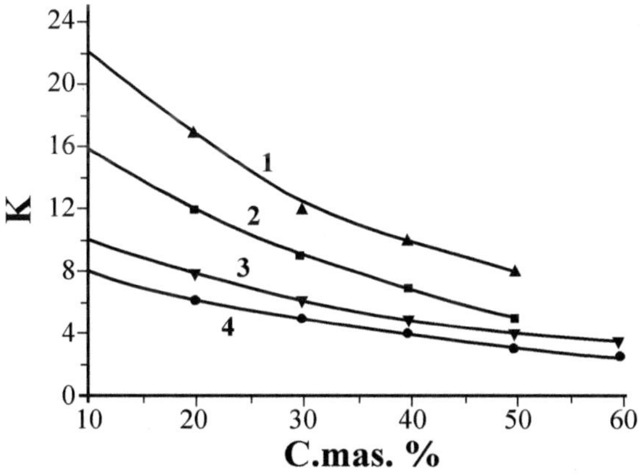

FIGURE 3.4 Dependence of tensosensitivity coefficient on conductive component concentration for composites containing plasticizer.

Note: (1) SCTV Rubber + SCTN Rubber (10 mass%) + P357 Tech. Carbon black; (2) without the same plasticizer; (3) Epoxy Glue + P357 Carbon black + glycerin (10 mass%); (4) without the same plasticizer.

This result indicates that the plasticizer promotes the movement of conductive particles in the polymer matrix, enhances the macromolecular activity, and at the same time prevents the coagulation of conductive particles and thus contributes to their maximum conductivity.

3.3.1 TENSOSENSITIVITY OF THE COMPOSITES TO IMPULSE DEFORMATION

Under variable deformation conditions, the exploitation of polymeric tensile elements greatly depends not only on the material content and the hardness of the material but also on the degree of repeatability of the deformation. If the intensity of the deformation repeatability for a metal-constructed sensor is practically negligible (the deformed body returns to its original state very quickly after removing the cause of deformation), we have a different situation for the tensile element made of polymeric material. Here, the relaxation phenomena characteristic of polymers come to the fore after removing deformation. It is a physical process that manifests itself in the initial transition of macromolecules from equilibrium to deformation. The viscosity of the macromolecular system, or polymer, is one of the determinants of relaxation. Clearly, the greater the degree of freedom the molecules have, the sooner they return to their original state in accordance with the Le Chatelier principle. Thus, when registering variable deformations, it is necessary to take into account the transition time to the initial state of the inevitable polymer matrix, that is, the relaxation time. To illustrate, Figure 3.5 shows the change in tensile element impedance over time – after unloading in the case of conductive polymer composites.

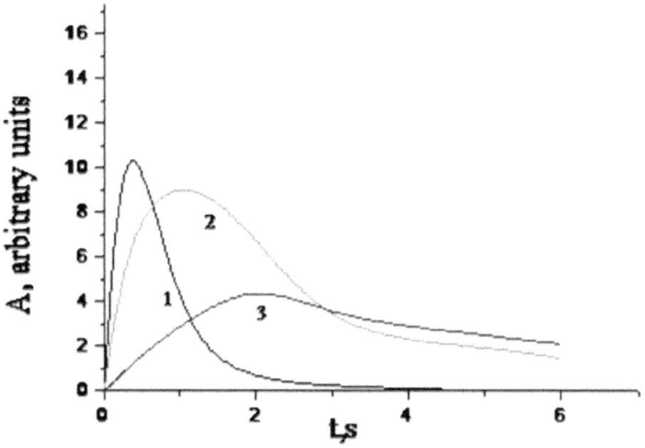

FIGURE 3.5 Dependence of the ohmic resistivity of composites after impulse deformation for 3 different samples with conductivities: (1) low; (2) middle; and (3) high.

The maximum curve in Figure 3.5 determines the maximum magnitude of the tensioelement impedance. Its subsequent period corresponds to the relaxation of the impedance of the element after lifting the load.

Similar curves are generally obtained in the case of conductive polymeric elements. Quantitatively different composites correspond to different curves. Specifically, for the composite where the impedance relaxation goes faster, the curve is narrower and the downward curve is slower than for the relatively large relaxation time; in other words, the curve is "narrower" for the relatively "soft" element than the rigid element.

Based on the analysis of the curves obtained by the uniform deformation of the tenso-elements, we have found that the voltage change on the tenso element is satisfactorily described by the formula:

$$R = R_1 - (R_1 - R_o) \exp(-t/\tau), \quad (1)$$

Where; Ro and R_1 are the initial and final values of the impedance in the relaxation process, and τ is the relaxation time.

Using the formula given above, we get:

$$\ln (R - R_1)/(Ro - R_1) = - t/\tau \quad (2)$$

The equilibrium of Formula (1) for relaxation kinetics is confirmed by the fact that we obtain a linear dependence on the coordinates of Formula (2) (Figure 3.6), from which we can determine the relaxation time.

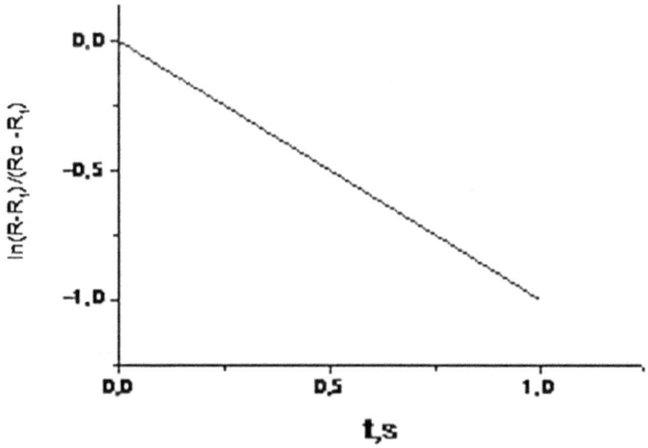

FIGURE 3.6 Electric conductance relaxation kinetics in the coordinates of formula (2) in conductive polymer composites.

The next step of the study was to test polymeric tensile elements to record impulse loads. The composite's specific volumetric electric resistivity, conductivity, and their tensosensitivity coefficients are listed in Tables 3.1 and 3.2.

TABLE 3.1 Specific Volumetric Electric Resistivity, Conductivity, and tensosensitivity Coefficient of Electric Conducting Rubbers

#	Ingredients (wt.%)	ρ_v, kOhm.cm	σ, MPa	K
1	СКТВ + П357 (20)	15.3	1.4	12.5
2	СКТВ + П357 (30)	3.2	2.3	9.2
3	СКТВ + П357 (40)	0.5	4.1	8.4
4	СКТВ + СКТН (10) + П357 (20)	8.4	1.2	18.1
5	СКТВ + СКТН (10) + П357 (30)	0.9	1.5	11.3
6	СКТВ + СКТН (20) + П357 (40)	0.1	1.4	12.5
7	СКТВ + П357 (20) + graphite (20)	0.2	3.2	10.5

TABLE 3.2 Specific Volumetric Electric Resistivity, Conductivity, and tensosensitivity Coefficient of Electric Conducting Composites Based on Epoxy Resin

#	Ingredients (wt.%)	Ingredients (wt.%)	Ingredients (wt.%)	Ingredients (wt.%)
1	ЭД–20 + П357 (40)	16.7	5.3	5.2
2	ЭД–20 + П357 (50)	2.2	6.5	4.8
3	ЭД–20 + П357 (60)	0.7	6.7	2.3
4	ЭД–20 + graphite (50)	5.4	7.6	3.5
5	ЭД–20 + graphite (60)	3.1	8.5	1.8
6	ЭД–20 + graphite (70)	1.4	9.3	1.2
7	ЭД–20 + glycerin (5) + graphite (50)	4.3	4.4	3.9
8	ЭД–20 + glycerin (10) + graphite (50)	2.1	3.5	4.3
9	ЭД–20 + П357 (30) + graphite (10)	7.5	5.1	6.4

The data in Tables 3.1 and 3.2 are partially illustrated in the dependence curves of the composite tensile sensitivity coefficients. From these tables, as well as from the diagrams (Figures 3.3 and 3.4), there is a regularity that establishes a certain correlation between the electrical conductivity of the composites and the tensile sensitivity coefficient. The positive effect of the plasticizer in enhancing tensile sensitivity is also noteworthy.

Among the data in Tables 3.1 and 3.2, the composites containing two different types of conductors at the same time (Table 3.1, position 3.7 and Table 3.2, position 3.9) also deserve attention, namely technical carbon P357 and graphite in the proportions indicated in the tables. These composites exhibit very useful properties – they increase the tensile sensitivity coefficient to some extent compared to composites containing only one (of these two) conductive components, the concentration of which is equal to the total concentration of both of these conductive components. This effect is known in the technical literature as a synergistic effect. Unfortunately, this effect still has no theoretical justification, but it is possible to obtain a synergistic mixture of ingredients through a purely experimental method (by varying the proportion of conductive components).

3.4 CONCLUSION

1. Electrolytic polymer composites based on silicone rubber (SCTV) and epoxy adhesive (ЭD–20) with carbon black (P357) carbon (20–60 wt.%), graphite (10–70 wt.%), low-molecular-weight SCTN silicone rubber (10–20 wt.%), and glycerol (5–10 wt.%) have been obtained;
2. Specimens equipped with electrodes made of perforated copper foil ($1 \times 5 \times 30$ mm in dimensions) have been obtained;
3. Special equipment has been developed for the tensile sensitivity testing of specimens in bending and stretching modes;
4. It is established that the performance of tensile elements in the dynamic mode is highly dependent on the frequency and amplitude of significant deformations;
5. It has been established that the mechanical relaxation phenomena characteristic in polymers are a major impediment to the measurement of variable deformations by polymeric tensile elements and reduce the range of tensometric measurements;
6. Specimens with a relatively small modulus compared to specimens with a high modulus show the ability to operate normally in a wider range of frequency deformations and amplitudes;
7. The use of low-molecular-weight rubber and glycerin in epoxy-based composites increases their elongation (adapts the module) while increasing the frequency and amplitude of the registration deformations;

8. With the use of the so-called synergistic effect of the binary fillers, enhancement of the technical characteristics of tensile elements has been developed.
9. The obtained tensile-sensitive polymer composites can be used in construction, mechanical engineering, and anywhere where tensometry of mechanical deformations is required.

ACKNOWLEDGMENTS

This work was supported by Shota Rustaveli National Science Foundation of Georgia (SRNSFG) under the GENIE project; grant number: YS-21-687.

KEYWORDS

- **coefficient of tensosensitivity**
- **electric conducting polymer composite**
- **mechanical deformation**
- **resistor**
- **tensosensitive**

REFERENCES

1. Cochrane, C., Lewandowski, M., & Koncar, V. (2010). A flexible strain sensor based on a conductive polymer composite for in situ measurements of parachute canopy deformation. *Sensors, 10*(9), 8291–8303.
2. Martinez, F., Obieta, G., Urible, I., Sikora, T., & Ochoteco, E. (2009). Polymer-based flexible strain sensor. *Procedia Chemistry, 1*(1), 915–918.
3. Kweon, Y., Lee, M., Park, T., Jang, H., & Jeong, A. (2019). Highly flexible chemical sensors based on polymer nanofiber field effect transistors. *Journal of Materials Chemistry C, 6*(1), 84–91.
4. Kozlowski, M., & Frackowiak, S. (2005). Chemical sensors based on polymer composites. *Sensors and Actuators B: Chemical, 109*(1), 141–145.
5. Huang, Y., Wang, W., Sun, Z., & Wang, Y. (2015). Multilayered flexible sensor for wide range pressure measurement based on CNTs/SB/SR composite. *Journal of Materials Research, 30*(12), 1865–1875.

6. Gupta, A., Sharma, S., Ahmad Mir, I., & Kumar, A. (2006). Advances in sensors based on conducting polymers. *Journal of Scientific & Industrial Research, 65*(1), 549–557.
7. Adhikari, B., & Majumdar, S. (2004). Polymer in sensor applications. *Progress in Polymer Science, 29*(7), 699–766.
8. Hu, Y., Zhao, T., Zhu, P., Zhang, Y., Liang, X., Sun, R., & Wong, C. (2012). A low cost, printable and stretchable strain sensor based on highly conductive elastic composites with tunable sensitivity for human monitoring. *Nano Research, 11*(4), 1938–1955.
9. Shi, G. D., Li, X. M., Chang, H. Y., Lin, Z. J., & Zhao, L. F. (2016). Graphene reinforced carbon nanotube network for strain sensors. *Advanced Functional Materials, 26*(1), 2078–2084.
10. Martinez, R. V., Branch, J. L., Fish, C. R., & Jin, L. H. (2013). Sensors with three-dimensional mobility based on flexible elastomers. *Advanced Materials, 25*(1), 205–212.
11. Matsuhisa, N., & Kaltenbrunner, M. (2015). Printable elastic conductors with a high conductivity for electronic textile applications. *Nature Communications, 6*(1), 7461.
12. Zhu, M., Xie, M., Xu, W., & Cheng, L. K. (2016). Nanocomposite-based stretchable deformation sensor matrix for a soft-bodied swallowing robot. *IEEE Sensors Journal, 16*(10), 3848–3855.
13. Millett, J. C. F., Bourne, N. K., & Rosenberg, Z. (2012). Ultra-sensitive strain sensors based on piezoresistive nanographene films. *Journal of Physics D: Applied Physics, 45*(6), 063112.

CHAPTER 4

Adsorption Properties of Silica Gels, In Situ, Modified with Nitrogen-Containing Polymers with Complexing Properties for Ions Cu(II), Fe(III), Pb(II)

I. SAVCHENKO,[1] E. YANOVSKA,[1] L. VRETIK,[1] D. STERNIK,[2] and O. KYCHKYRUK[3]

[1]*National Taras Shevchenko University of Kyiv, Kyiv, Ukraine*

[2]*Maria Curie-Skłodowska University, Lublin, Poland*

[3]*Ivan Franko Zhytomyr State University, Zhytomyr, Ukraine*

ABSTRACT

This review article introduces the latest achievements in the research of new polymer-inorganic composites based on silica gel and polymers of different chemical structures for the extraction of toxic metal ions from wastewater. This paper presents an overview of polymer-modified silica gels and reviews several of our recent studies using these composites as pollutant adsorbents, outlining the main characteristics of the various adsorption systems, demonstrating the specifics of each process, and identifying patterns to inform future research. Data are given on the possible practical use of the obtained hybrid materials as adsorbents of toxic metal ions.

4.1 INTRODUCTION

Immobilization of multifunctional polymers with complexing and ion-exchange properties on the porous surface of inorganic matrices increases the efficiency of such matrices as sorbents for ions of transition toxic metals.

Sustainability in Polymer Technology and Plastics Engineering: Concepts, Strategies, and Opportunities. Tamara Tatrishvili, Neha Kanwar Rawat, Swati Gokul Talele, and A. K. Haghi (Eds.)

Such multifunctional polymers include polyelectrolytes with a quaternary nitrogen atom in a side chain (polycations) or the main chain (polyionenes), e.g., polyhexamethylene guanidine, its derivatives, or polyaniline [1–9]. However, the ion-exchange properties of such polymers are significantly greater than the complexing ones. Sorbents based on them can remove mainly anionic forms of multivalent metals (molybdates, tungstates, chromates, dichromates, arsenates, etc.) or acidic remnants of mineral acids (in particular, nitrates, phosphates, arsenates) and have low sorption capacity toward cations of transition metals. Therefore, the search for organo-mineral composite materials bearing polymers with only complexing properties due to active (amino-, azo-, oxoquinoline, pyridine, mercapto-, etc.) groups which are able to remove cationic forms of toxic metals by their complexation is still relevant.

Undoubtedly, from an economic and ecological point of view, the best inorganic matrices for such composite materials are natural porous minerals currently being industrially developed from deposits [10–16]. However, it is necessary to establish a correlation between the chemical nature of the immobilized polymer and its adsorption capacity towards certain cationic forms of metals in terms of the complex process. It is also important to establish correlations between the method of polymer immobilization on a mineral carrier and the geometric shape of its placement on the inorganic surface. The shape of the macromolecule (coil, globule, aggregates of coils, etc.) significantly affects the accessibility of its complexing centers. All of the mentioned studies are important for clarifying the effectiveness of such sorbents for separating cations of transition toxic metals. That is why in the initial stages, it is necessary to use as mineral carriers for polymers with complexing properties porous inorganic materials with well-defined surface parameters, and minimal sorption capacity towards cations of transition metals. Such matrices include, in particular, mesoporous silica gels [17–21].

The main methods of polymer immobilization on the inorganic carrier are physical and chemical immobilization, sol-gel method, and *in situ* immobilization. Each of these methods has advantages and drawbacks.

Precipitation and further adsorption of previously synthesized polymer from a solution onto the inorganic carrier leads to the physical fixation of macromolecules on the mineral surface. Polymer immobilization on the solid surface of the carrier occurs due to the formation of hydrogen or π-π bonds, electrostatic forces, van der Waals forces, etc. It is advisable to use such a method to fix positively charged polymers on the negatively charged surface, e.g., silica matrix. In such a case, the polymer is held on a solid surface due to electrostatic forces.

The formation of covalent bonds between macromolecules and the inorganic carrier surface leads to a chemical fixation of the polymer. Synthesis of organo-mineral composites by the sol-gel method accompanies the formation of the inorganic component of the sol-gel composite around the polymer particles. As a result, areas of different nano-sized phases are formed, which can affect the desired properties of the composite. The method of *in situ* polymer immobilization on the surface realizes the direct formation of a polymer layer during the polymerization process in the presence of an inorganic matrix. This immobilization method is well-studied for polyaniline. A polymerization of aniline on the silica gel surface carried out with ammonium peroxodisulfate is described in [22]. It has been proven that the *in situ* polymerization of aniline allows the formation of a chemically and physically stable layer of polymer on a solid surface. It is a result of the simultaneous passage of interconnected processes: chemical – growth of macromolecules and physical–self-assembly of chains into supramolecular structures. Therefore, during the *in situ* polymerization, a special spatial structure of the polymer is irreversibly formed, which is firmly immobilized on the carrier surface [23]. It is noted that *in situ* polymerization of aniline is the only way to obtain polyaniline layers on porous and powdery nano-sized carriers. *In situ* polymerization of aniline is a versatile method concerning the chemical nature of the carrier. The only limitation is the acidity of the reaction medium, which must be in the interval pH=1–5.

The presented paper is devoted to the sorption properties of Cu (II), Fe(III), and Pb(II) ions with polymer-inorganic composites based on mesoporous silica gels with nitrogen-containing polymers of different chemical nature immobilized on the surface by *in situ* polymerization. The paper describes composites containing organic polymers bearing such groups as:

- **NH_2:** poly[2-amino-5-hydroxy-1,4-phenylene-bis(2-dimethacrylate)];
- **NH-C=O:** poly[N-(4-carboxyphenyl)methacrylamide];
- **N=N-:** poly[(4-methacryloyloxy-(4′-carboxy)azobenzene], poly[(4-methacryloyloxy-2-chloro-(4′-carboxy)azobenzene]; poly[4-methacryloyloxy-(4′-carboxy-2′-nitro)azobenzene]; and
- **Oxyqiunoline:** poly[8-oxyqiunoline methacrylate].

4.2 EXPERIMENTAL METHODS AND MATERIALS

4.2.1 METHODS

FTIR spectra of the samples of composite and the original mineral were recorded using an IR spectrometer with Fourier transformation (Thermo Nicolet Nexus FT-IR, USA). For this purpose, the samples were ground in an agate mortar and pressed with KBr. The FTIR spectra were recorded in the spectral range of 500–4000 cm^{-1} with 16 scans per spectrum at a resolution of 4 cm^{-1}.

- **Thermal Analysis:** The amount of immobilized polymer on the surface of the modified saponite was evaluated by thermogravimetric analysis, which was obtained on a synchronous TG/DTA analyzer "Shimadzu DTG-60 H" (Shimadzu, Japan) in the temperature range 15–1000°C. The heating rate of the samples was 10 deg/min. Differential scanning calorimetry was performed on an instrument "STA 449 Jupiter F1" (Netzsch, Germany) with a mass spectroscopic attachment "QMS 403C" (Germany).
- **Low-Temperature Adsorption-Desorption of Nitrogen:** The values of the specific surface area and the average pore diameter were calculated from the isotherms of low-temperature adsorption-desorption of nitrogen using sorbometer software "ASAP 2420 V1.01" (Micromeritics, USA). Before measurement, the samples were degassed at 100°C for 24 hours.
- **Surface Morphology Analysis:** The surface morphology of the composite was observed by using a scanning electron microscope (SEM, LEO 1430VP, Carl Zeiss, Germany).

4.2.1.1 DESCRIPTION OF ADSORPTION PROCESSES

Properties of the obtained composite to adsorb Cu (II), Pb (II), Cd(II), and Fe(III) were studied in static mode. Working solutions of nitrates of the corresponding metals were prepared in volumetric flasks of 25 ml, 50 ml, or 100 ml, diluting the solutions to the mark with water, and then adding the required volume to flat-bottomed flasks containing 0.1 g of adsorbent. The reaction proceeded while the flasks were shaken mechanically. Equilibrium concentrations of ions were measured using atomic absorption methods.

Working nitrate solutions of Cu(II), Pb(II), and Fe(III) are prepared with the sets of "standard sample solutions" of these salts on 1 M HNO_3 background (produced by A.V. Bogatsky FHI in Odesa) with concentrations of 1 and 10 mg/ml.

4.2.1.2 ANALYSIS OF ION CONCENTRATIONS

Absorption capacity (A) was calculated by formula (1):

$$A = (c_o - [M])\ V/m \quad (1)$$

Where; c_o is the concentration of the metal in the starting solution; $[M]$ is the concentration of the metal at equilibrium; V is the volume of the starting solution; and m is the mass of the composite.

Atomic Absorption Spectrophotometer "C115-M" (Ukraine) in the flame of a mixture of "acetylene–air" was used to measure equilibrium concentrations of copper (II), cadmium (II), lead (II), and iron (III). Equilibrium concentrations in solutions were calculated by comparing the intensities of spectral lines with those for standard solutions. The maximum wavelengths were: 324.7 nm for Cu (II), 228.8 nm for Cd (II), 283.3 nm for Pb (II), and 248.3 nm for Fe (III), with an aperture width of 0.5 cm.

Silica gel manufactured by Merck (fraction with a particle diameter of 0.1–0.2 mm) was used as an inorganic carrier of composites. The surface parameters of this original SiO_2 are given in Table 4.3. Synthesis of polymers was carried out in the presence of original silica particles by initiated radical polymerization of the corresponding monomers using α,α,′-azoisobutyronitrile (AIBN) as a radical initiator. Methods of composite synthesis are described in papers [24–30]. Table 4.1 shows the full and abbreviated names of the synthesized composites and the chemical structures of the immobilized polymers.

4.3 RESULTS AND DISCUSSION

Polymers *in situ* immobilization on the silica gel surface were confirmed by IR spectra comparative analysis of the composites and original silica gel. Examples of IR spectra of Composite 1 and Composite 2 are shown in Figures 4.1 and 4.2, respectively.

TABLE 4.1 Full and Abbreviated Names of Synthesized Composites and Structures of Nitrogen-Containing Composites

Full Name of Composite	Abbreviated Name of Composite	Schematic Representation of Composite Structure
Silica gel *in situ* modified with poly[2-amino-5-hydroxy-1,4-phenylene-bis(2-dimethacrylate)]	Composite 1	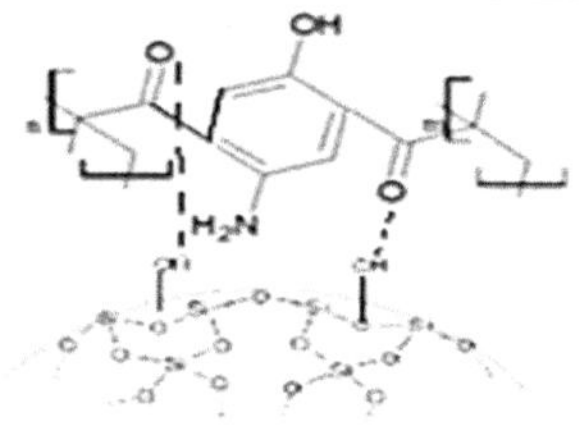
Silica gel *in situ* modified with poly[N-(4-carboxyphenyl) meth-acrylamide]	Composite 2	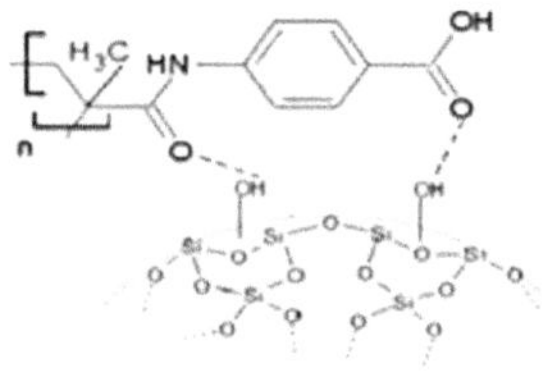
Silica gel *in situ* modified with poly[(4-methacryloyloxy-(4`-carboxy) azobenzene]	Composite 3	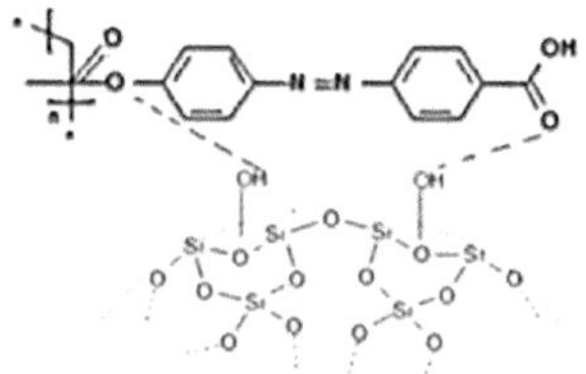
Silica gel *in situ* modified with poly[(4-methacryloyloxy-2-chloro-(4`-carboxy) azobenzene]	Composite 4	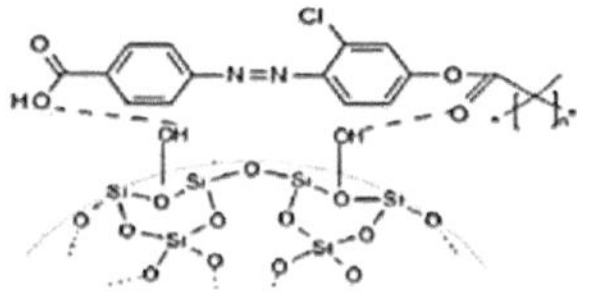
Silica gel *in situ* modified with poly[4-methacryloyl-oxy-(4`-carboxy-2`-nitro) azobenzene]	Composite 5	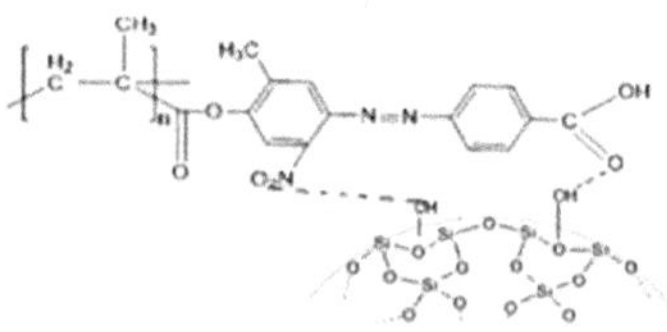
Silica gel *in situ* modified with poly[8-oxyqiunoline methacrylate]	Composite 6	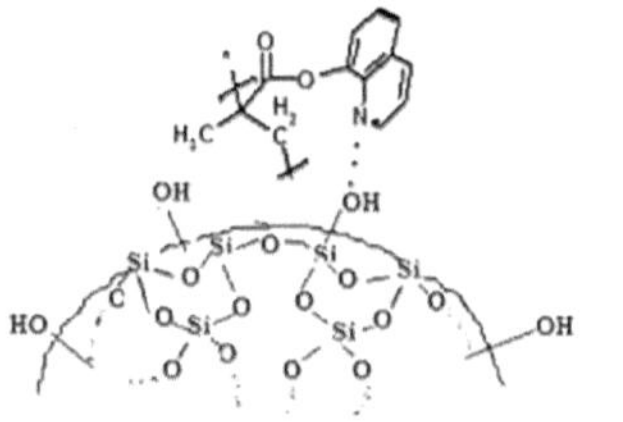

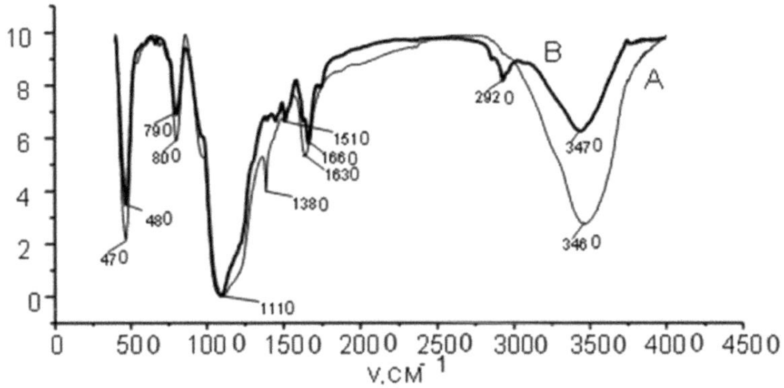

FIGURE 4.1 IR spectra of original silica gel (curveA) and composite 1 (curve B).

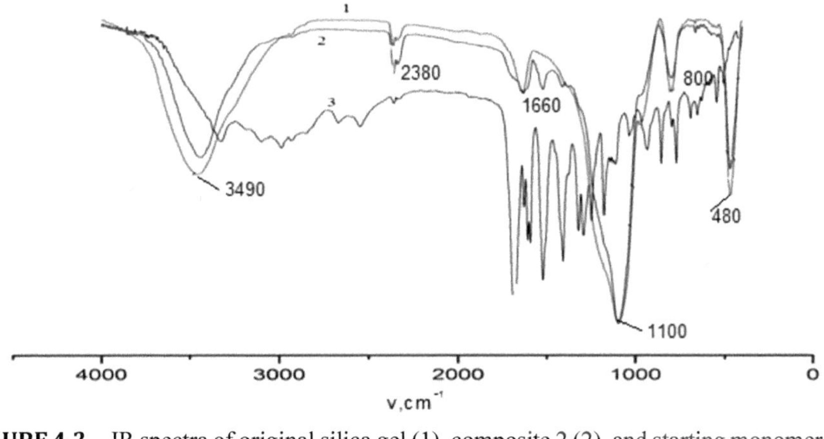

FIGURE 4.2 IR spectra of original silica gel (1), composite 2 (2), and starting monomer (3).

The obtained IR spectra of all synthesized composites contain an absorption band at 2,930 cm^{-1} and a number of bands in the region 1,700–1,200 cm^{-1}. The first band is attributed to the ν(C-H) vibration of sp^3-hybridized carbon. This band confirms the fact that polymerization has taken place on the surface of silica gel.

Thermogravimetric analysis of the synthesized composites was carried out to determine the weight fraction of immobilized polymers in the composites and to establish the temperature range of thermal decomposition of immobilized polymers. The results of the thermogravimetric analysis of the synthesized composites carried out under the same conditions are shown in Table 4.2.

TABLE 4.2 Thermogravimetric Analysis Data of the Synthesized Composites [24–30]

Composite	Temperature Range of Exothermic Effect	Weight Fraction of the Immobilized Polymer (± 0.1), wt.%
Composite 1	300–600°C	22.1
Composite 2	420–700°C	9.3
Composite 3	400–650°C	10.6
Composite 4	400–650°C	13.3
Composite 5	400–650°C	10.1
Composite 6	390–700°C	3.9

As one can see from the data in Table 4.2, the weight fractions of immobilized polymers in the composites are approximately the same and are ~10 wt.%. The exceptions are 22 wt.% for poly[2-amino-5-hydroxy-1,4-phenylene-bis(2-dimethacrylate)] in composite 1 and 4 wt.% for poly[8-oxyquinoline methacrylate] in composite 6. The temperature range of their final thermal decomposition with an exothermic effect is 400–650°C.

Thermograms combined with mass spectra in 3D format were obtained for all synthesized composites as well as mass spectra in 2D format at different temperatures. Examples of TG-MS-3D data of composites 2 and 4 are shown in Figures 4.3 and 4.4.

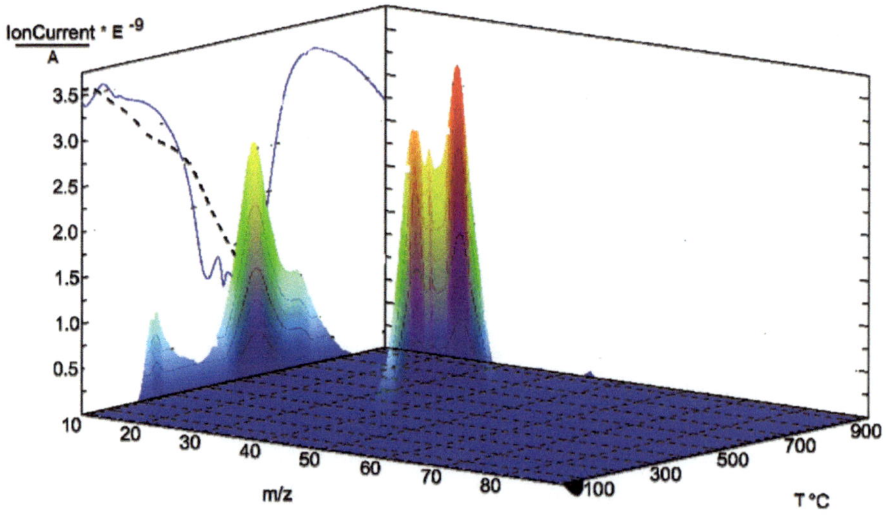

FIGURE 4.3 TG-MS-3D data of composite 2.

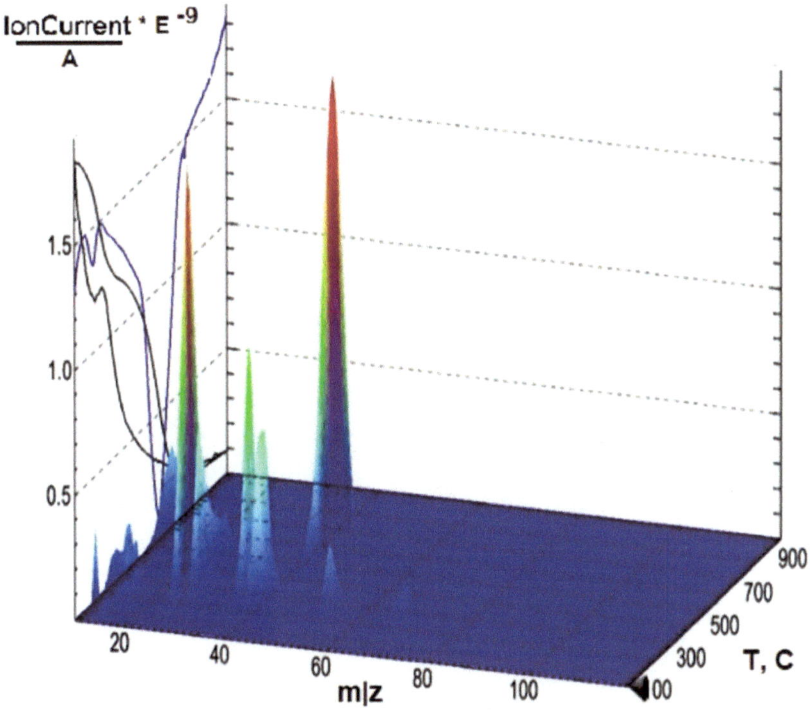

FIGURE 4.4 TG-MS-3D data of composite 4.

The products of thermal destruction of polymers immobilized on the silica gel surface were analyzed using mass spectra data. As could be expected, the final products of the thermal decomposition of all immobilized polymers are water and CO_2, and nitrogen is released in the form of N_2 or N_2O. Thermal destruction of the chlorine-containing polymer in composite 4 also resulted in Cl_2 formation.

Changes in the silica gel surface morphology after modification with polymers were recognized with the method of low-temperature adsorption-desorption of nitrogen. The obtained nitrogen adsorption-desorption isotherms of the synthesized composites and the original silica gel are similar. This observation indicates that the structure of the silica gel surface does not change after *in situ* immobilization of the polymers.

The values of the specific surface area of the composites and original silica gel were obtained by computer processing of the obtained nitrogen adsorption-desorption isotherms using the BET and Langmuir methods (*see* Table 4.3).

It was found that after surface modification by *in situ* immobilization of polymers, the specific surface area decreases by about 33%. The average diameter of silica gel pores calculated using the BJH method increases insignificantly.

By means of computer processing of the obtained nitrogen adsorption-desorption isotherms using the BET and Langmuir methods, the values of the specific surface area of the synthesized composites and the original silica gel were obtained and compared (Table 4.3). It was found that after surface modification by *in situ* immobilization of the corresponding polymers, the specific surface area decreases by about a third. At the same time, the average diameter of silica gel pores, calculated using the BJH method, increases insignificantly as a result of the *in situ* immobilization of polymers.

A comparison of the data in Tables 4.2 and 4.3 allows us to trace a clear correlation between the mass fraction of the immobilized polymer in the composition of the composite and the amount of reduction in the specific surface area and the average pore size of the original silica gel, namely: the greater the mass fraction of the polymer in the composition of the composite, the greater the reduction in the specific surface area of silica gel and the average volume of its pores.

TABLE 4.3 Comparison of Surface Parameters of Original and *in situ* Polymer-Modified Silica Gel [24–30]

Sample	$S_{пит}$ (m²/g)			Average Pore Volume (cm³/g)	Average Pore Size (nm)
	Calculated from Isotherm	BJH	Langmuir		
Original silica gel	447	464	687	0.68	5.9
Composite 1	345	363	593	0.41	5.5
Composite 2	252	264	396	0.42	6.3
Composite 3	386	402	600	0.61	6.1
Composite 4	309	327	494	0.51	6.2
Composite 5	341	358	539	0.57	6.4
Composite 6	390	407	610	0.62	6.1

The geometric shapes of the arrangement of polymers on the surface of silica gel were established using scanning electron microscopy. SEM photos of the surfaces of all composites are shown in Figures 4.5–4.10. As follows from the given photos, for polymers in the composites Composite 1, Composite 2, Composite 5, and Composite 6, an "island" or "globular" arrangement of the polymer on the silica gel surface is observed, and in the composites Composite 3 and Composite 4, the characteristic arrangement of polymers is mainly in the form of linear fibers.

Summarizing the obtained results, it can be stated that *in situ* immobilization of uncharged polymers on the surface of silica gels leads mainly to their "island" or "globular" form of placement on the surface. This leads to less availability of the active centers of the polymer for complex formation. This factor significantly reduces the potential sorption capacity of composites based on silica gel for cationic forms of transition metals and thus prevents the realization of the main advantage of polymers compared to immobilized low molecular weight compounds.

However, sometimes the chemical nature of the polymer makes it possible to form horizontal thread-like forms of the polymer on the surface of silica gel during *in situ* immobilization, which makes it available for participation in the complex formation of a much larger number of potentially active centers of polymer macromolecules. An example of this phenomenon is azopolymers in the composition of Composites 3 and 4. Most likely, their chemical structure made it possible to obtain similar chain forms of these polymers on the surface of silica gel (*see* Figures 4.7 and 4.8).

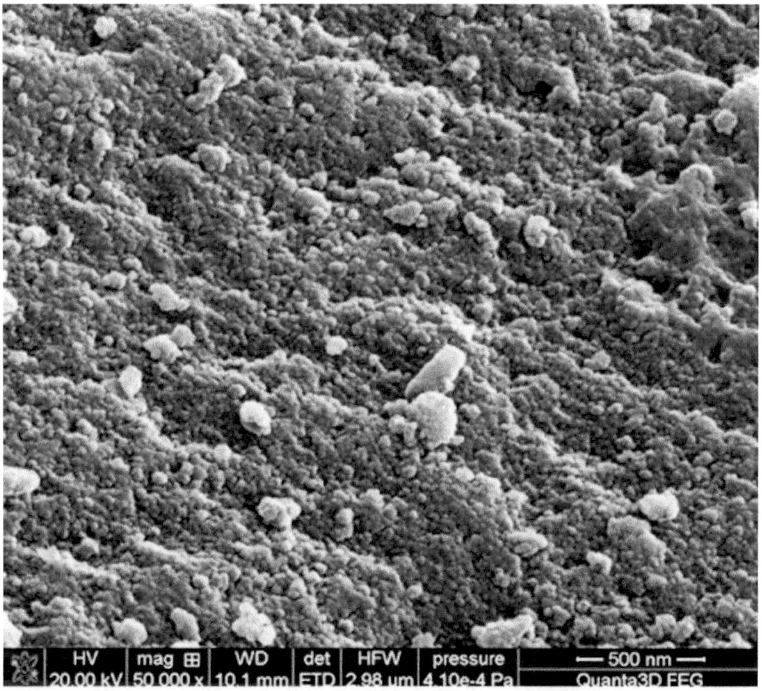

FIGURE 4.5 SEM photo of the surface of the composite 1 (x50000).

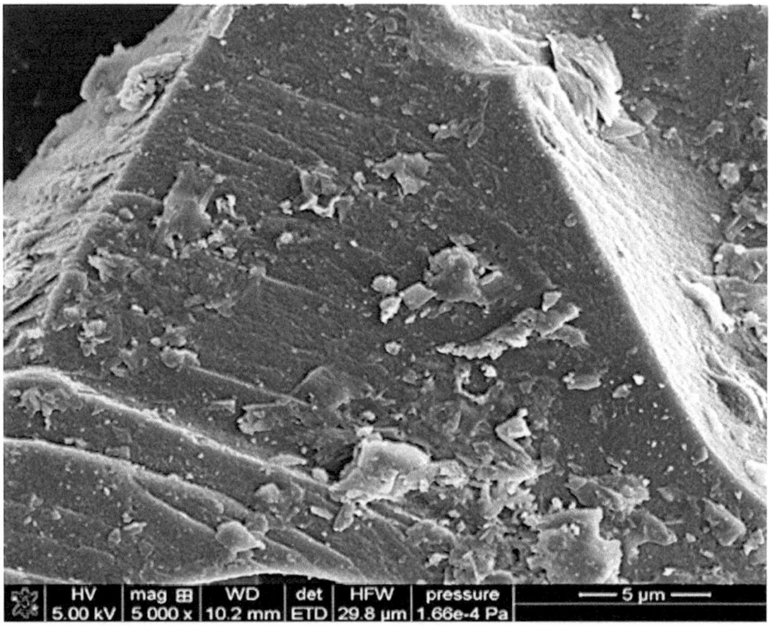

FIGURE 4.6 SEM photo of the surface of the composite 2 (x5000).

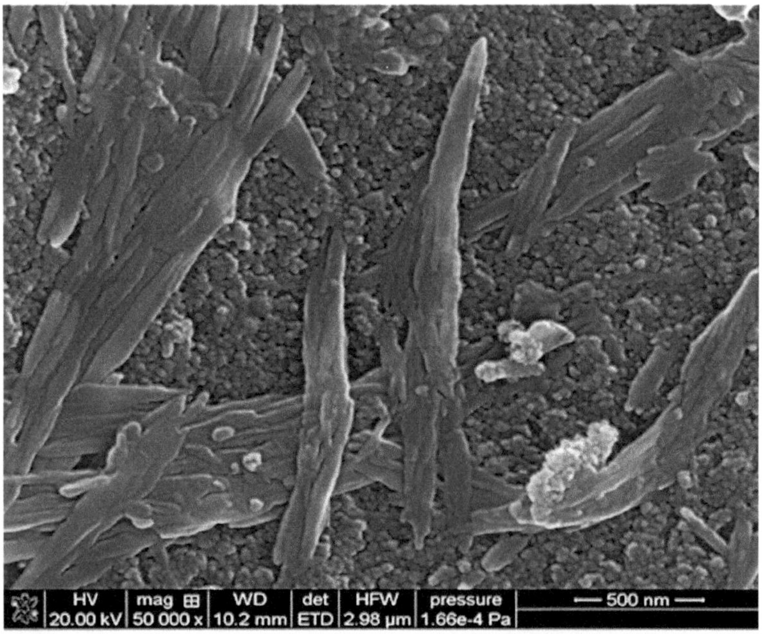

FIGURE 4.7 SEM photo of the surface of the composite 3 (x50000).

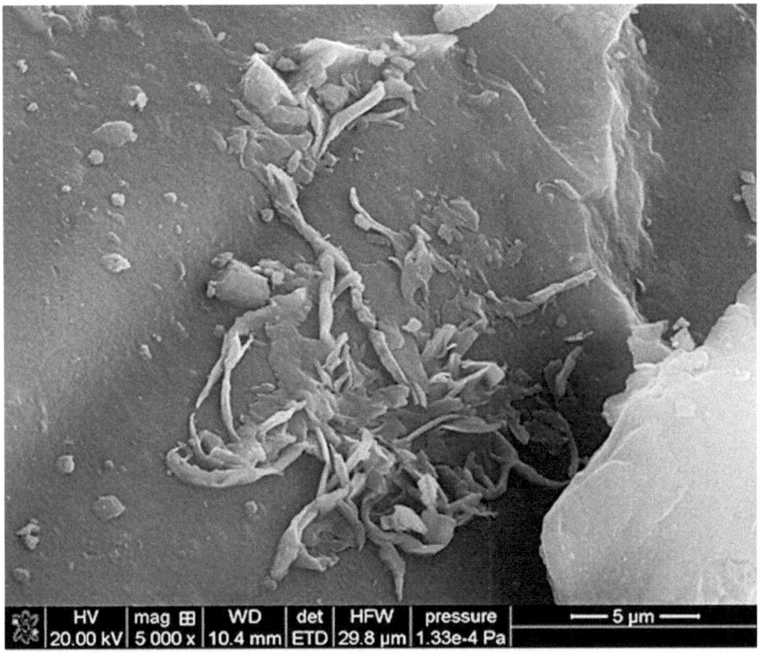

FIGURE 4.8 SEM photo of the surface of the composite 4 (x50000).

FIGURE 4.9 SEM photo of the surface of the composite 5 (x10000).

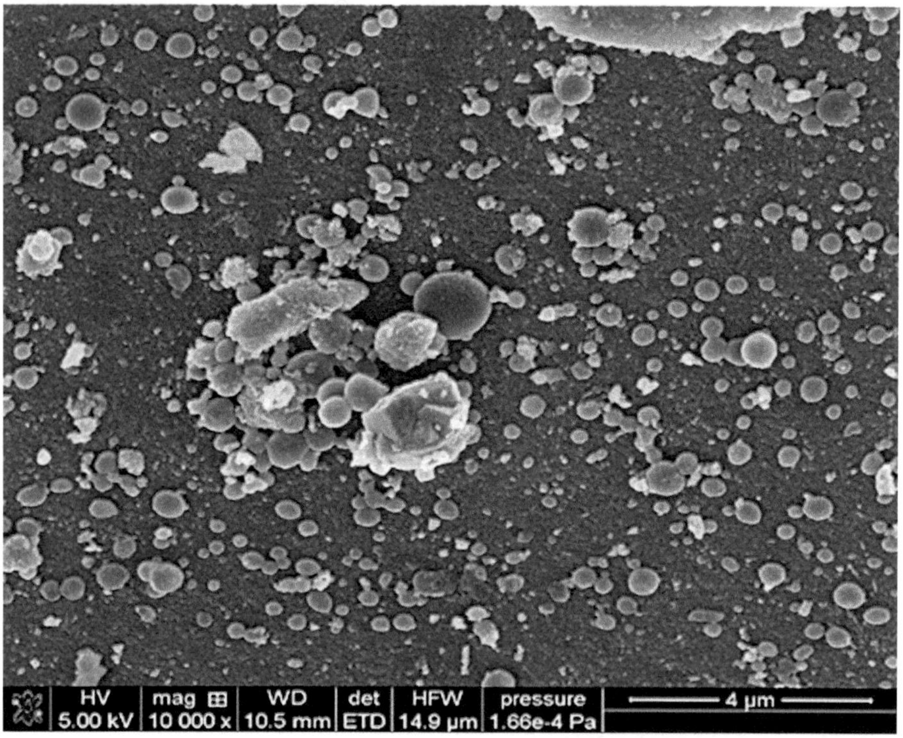

FIGURE 4.10 SEM photo of the surface of the composite 6 (x10000).

Since the immobilized polymer on the silica gel surface contains nitrogen-containing (amino-, azo-, oxyquinoline) groups, the synthesized composites should show sorption activity for those metal ions that have a high affinity for nitrogen-containing ligands. To confirm these considerations, we investigated the sorption of ions Cu (II), Fe (III), and Pb (II).

Studies of the sorption capacity of synthesized composites for Cu (II), Pb (II), and Fe (III) ions are described in detail in works [24–30]. These studies included the following: determination of the optimal pH range of the sorption medium; establishing the necessary contact time of phases to achieve sorption equilibrium in static mode; construction of sorption isotherms of the corresponding metal ions on the surface of the synthesized composites; establishing the sorption capacity for the listed metal ions and comparing it for the original silica gel and each of the synthesized composites. In particular, it was found that all synthesized composites show the maximum sorption capacity for selected ions in an aqueous environment without the additional addition of any buffer solutions to the initial solutions of these metal salts.

Therefore, the best sorption occurs with the original aqua complexes. And, in our opinion, it effectively occurs due to the formation of various ligand complexes on the surface of silica gel due to the replacement of oxygen atoms in the original aqua complexes with nitrogen atoms in the composition of immobilized polymers. Examples of schemes of such reactions are shown in Figures 4.11 and 4.12.

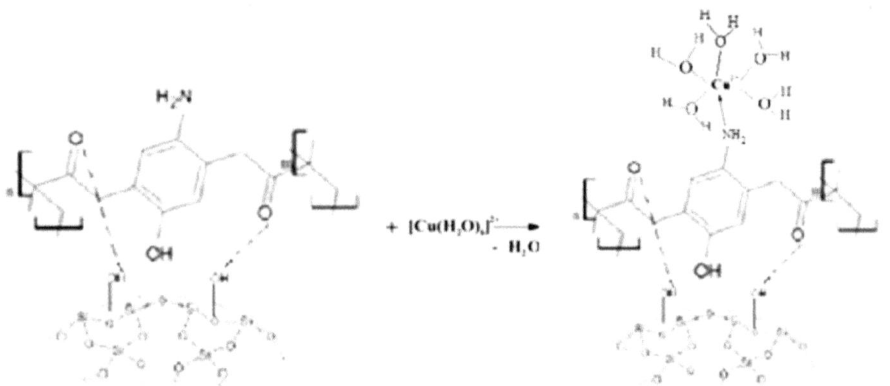

FIGURE 4.11 Scheme of complex formation of Cu (II) ions with composite 1 through the nitrogen atom of the amino group.

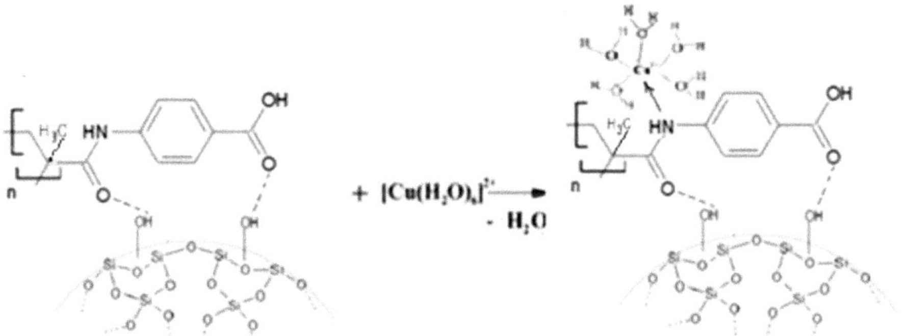

FIGURE 4.12 Scheme of complexation of Cu (II) ions with composite 2 through the nitrogen atom of the amide group.

Comparative characteristics of the sorption capacity of the synthesized composites and the original silica gel with respect to Cu (II), Pb (II), and Fe (III) ions are shown in Table 4.4.

TABLE 4.4 Comparison of The Sorption Capacity of the Original Silica Gel and Synthesized Composites with Respect to Cu (II), Pb (II) and Fe (III) Ions

Sample	Sorption Capacity, mmol/g		
	Cu (II)	**Fe (III)**	**Pb (II)**
Original silica gel	0.006	0.007	0.002
Composite 1	0.021	0.010	0.010
Composite 2	0.010	–	0.004
Composite 3	0.012	0.016	0.005
Composite 4	0.009	0.016	0.007
Composite 5	0.086	0.016	0.010
Composite 6	0.009	0.012	0.002

From the data in Table 4.4, it follows that the greatest sorption capacity for all studied ions is characteristic of silica gel *in situ* modified with poly[4-methacryloyloxy-(4′-carboxy-2′-nitro) azobenzene]. This polymer has two nitrogen atoms as part of its structural links, which increases the probability of complexation processes with selected transition metal ions. In favor of the fact that the chemical nature of this polymer plays a decisive role in this case, the fact that this polymer does not have a large (compared to other synthesized composites) concentration on the surface, because its mass fraction in the composition of the composite is 10%, nor the best (linear) geometric arrangement of macromolecules of the immobilized polymer on the surface of silica gel.

It also follows from the data of Table 4.4 that;

- The sorption capacity of composites with immobilized azopolymers is higher than that of composites with immobilized amino-, amido-, and oxyquinoline groups for Fe(III) ions.

  The sorption capacity of the composite with immobilized poly[2-amino-5-hydroxy-1,4-phenylene-bis(2-dimethacrylate)] is equal to the composite with immobilized poly[4-methacryloyloxy-(4′-carboxy-2′-nitro) azobenzene] relative to Pb(II) ions and the second largest relative to Cu(II) ions.

  These facts can be explained by the high concentration of this immobilized polymer compared to other synthesized composites (the mass fraction of poly[2-amino-5-hydroxy-1,4-phenylene-bis(2-dimethacrylate)] in the composition of the composite is 22%, which is twice the mass fractions of other polymers).

4.4 CONCLUSION

1. By *in situ* immobilization of poly[2-amino-5-hydroxy-1,4-phenylene-bis(2-dimethacrylate)], poly[N-(4-carboxyphenyl) methacrylamide], poly[(4-methacryloyloxy-(4′-carboxy)-azobenzene], poly[4-methacryloyloxy-2-chloro-(4′-carboxy)azobenzene], poly[4-methacryloyloxy-(4′-carboxy-2′-nitro)azobenzene] and poly[8-oxyquinoline methacrylate] on the surface of silica gel, 6 new composite materials were obtained.
2. The fact of the *in situ* immobilization of polymers on the surface of silica gel was confirmed by the methods of IR spectroscopy and thermogravimetric analysis combined with mass spectrometry. Thus, the possibility of using this method of surface modification with selected nitrogen-containing polymers was proven for the first time.
3. It was found that the mass fractions of *in situ* immobilized polymers in the composition of composites are in the range of 4–22%. The final thermal destruction of all immobilized polymers on the surface of silica gel occurs in the temperature range of 400–650°C with the predominant formation of carbon dioxide, water, and N_2 or N_2O.
4. It was established that *in situ* immobilized nitrogen-containing polymers have different locations on the silica gel surface: in the form of loops of macrochains, adsorbed tangles, or linear fibers. This modification almost does not affect the structure of its surface; the surface area is slightly reduced only.
5. It was found that all the synthesized composites show better sorption properties for Cu(II), Pb(II), and Fe(III) ions in a neutral aqueous environment compared to the original silica gel, with the exception of *in situ* immobilized poly[8-oxyquinoline methacrylate] for Pb ions (II).
6. The effect of the chemical nature of the immobilized polymer on the sorption properties of the obtained materials is shown: the composite with *in situ* immobilized poly[4-methacryloyloxy-(4′-carboxy-2′-nitro) azobenzene] was the best sorbent for Cu (II) and Pb (II) ions. Azopolymers fixed on the surface of silica gel are superior to polymers that contain nitrogen in the composition of amino-, oxyquinoline, and amide groups in terms of the sorption capacity for Fe (III) ions.

KEYWORDS

- **adsorption**
- ***in situ* immobilization**
- **polymer**
- **silica gel**
- **α,α,′-azoisobutyronitrile**

REFERENCES

1. Tertykh, V. A., Polishchuk, L. M., Yanovska, E. S., & Dadashev, A. D. (2008). Concentration of anions by silica adsorbents with immobilized nitrogen-containing polymers. *Adsorption Science & Technology, 26*(1-2), 59–68.
2. Polishchuk, L. N., Yanovskaya, E. S., & Yanishpolsky, V. V. (2007). Determination of anionic forms of Mo (VI), W (VI), Cr (VI), As (V), V (V) and P (V) using a silica gel using in situ immobilized polyonene. *Voprosy khimii i khim. tekhnologii, 6*, 119–123.
3. Polishchuk, L. N., Yanovskaya, E. S., & Yanishpolsky, V. V. (2007). Silica gel with immobilized polyonene in the type of anionic complexes Au (III). *Dopovidi NAN Ukrainy, 11*, 143–146.
4. Trofimchuk, A. K., Maglyovana, T. V., & Leshchenko, V. N. (2008). Development of the analytic form of reagents on the basis of silica gel impregnated with polyhexamethylenguanidine chloride. *Polish Journal of Chemistry, 82*, 453.
5. Ansari, R. (2006). Application of polyaniline and its composites for adsorption/recovery of chromium (VI) from aqueous solutions. *Acta Chimica Slovenica, 53*, 88–94.
6. Yanovska, E., Savchenko, I., Petrenko, O., & Davydov, V. (2022). Adsorption of some toxic metal ions on pine sawdust in situ immobilized by polyaniline. *Applied Nanoscience, 12*(4), 861–868.
7. Yanovska, E. S., Dadashev, A. D., & Tertykh, V. A. (2009). Inorganic anion–exchanger based on silica with grafted polyhexamethyleneguanidine hydrochloride. *Functional Materials, 16*(1), 105–109.
8. Dadashev, A. D., Tertykh, V. A., Yanovska, E. S., & Yanova, K. V. (2012). Ion–exchange properties of modified silicas with bound amides of polyhexamethyleneguanidine and maleic or o-phthalic acids with respect to metal-containing anions. *Chemistry, Physics and Technology of Surface, 3*(4), 419–428.
9. Dadashev, A. D., Tertykh, V. A., Yanovska, E. S., & Yanova, K. V. (2016). New approach to synthesis of silica with chemically bound guanidine hydrochloride for preconcentration of metal ions. *American Journal of Analytical Chemistry, 7*, 411–420.
10. Petrenko, O. V., Yanovska, E. S., Terebilenko, K. V., & Stus, N. V. (2020). *Green Chemistry*. Kyiv University, Ukraine, 239 pp. ISSN: 2079–1704 (print), 2518–1238 (online).

11. Jia, J., Liu, Y., & Sun, J. (2021). Preparation and characterization of chitosan/bentonite composites for Cr (VI) removal from aqueous solutions. *Adsorption Science & Technology*, Article ID 6681486, 15.
12. Vardy, S., Gruythuysen, J., & Baddiley, B. (2022). Kaolinitic clays as a potential source of dioxins in the Noosa River catchment, Queensland, Australia. *Environmental Chemistry, 19*(1), 1–12. https://doi.org/10.1071/EN21163.
13. Fernandes de Magalhães, S., Rodrigues da Silva, G., & Clark Peres, A. (2022). Zeolite application in wastewater treatment. *Adsorption Science & Technology*, Article ID 4544104, 26.
14. Vareda, J. P., Matos, D. P., Valente, A. J. M., & Durães, L. (2022). A new Schiff base organically modified silica aerogel-like material for metal ion adsorption with Ni selectivity. *Adsorption Science & Technology*, Article ID 8237403, 18. https://doi.org/10.1155/2022/8237403.
15. Srinivasan, R. (2011). Advances in application of natural clay and its composites in removal of biological, organic, and inorganic contaminants from drinking water. *Advances in Materials Science and Engineering*, Article ID 872531, 17. https://doi.org/10.1155/2011/872531.
16. Valpotic, H., & Gracner, D. (2017). Zeolite clinoptilolite nanoporous feed additive for animals of veterinary importance: Potentials and limitations. *Periodicum Biologorum, 119*, 159–172.
17. Yanovska, E. S., Dadashev, A. D., & Tertykh, V. A. (2009). Complexation of transition metal cations with polyhexamethylene guanidine chloride chemically fixed on the surface of silica gel. *Ukrainskiy Khimicheskiy Zhurnal, 75*(2), 69–74.
18. Budnyak, T. M., Tertykh, V. A., & Yanovska, E. S. (2013). Chitosan and its derivatives as effective sorbents for the extraction of metal ions. *Poverknost' Surface, 5*(20), 118–134.
19. Alsuhaimi, A. O., & Mccready, T. (2012). Rapid immobilisation of 8-hydroxyquinoline onto silica materials and its application for on-line solid-phase extraction of transition metals from environmental samples before ICP-OES determination. *Global NEST Journal, 14*, 55–65.
20. Kolodynska, D. (2012). Adsorption characteristics of chitosan modified by chelating agents of a new generation. *Chemical Engineering Journal, 179*, 33–43.
21. Budnyak, T., Tertykh, V., & Yanovska, E. (2014). Chitosan immobilized on the silica surface for the wastewater treatment. *Materials Science (Medžiagotyra), 20*(2), 177–182.
22. Stejskal, J., Trchova, M., Fedorova, S., Sapurina, I., & Zemek, J. (2003). Surface polymerization of aniline on silica gel. *Langmuir, 19*, 3013–3018.
23. Sapurina, I., & Stejskal, J. (2008). The mechanism of the oxidative polymerization of aniline and the formation of supramolecular polyaniline structures. *Polymer International, 57*, 295–1325.
24. Buriachenko, I., Yanovska, E., Vretik, L., Sternik, D., Kychkiruk, O., & Nikolaeva, O. (2016). In situ polymerization of 4-(methacryloylamino) phenyl-2-methylacrylate on the surface of silica gel. *French-Ukrainian Journal of Chemistry, 4*(2), 47–54.
25. Buriachenko, I., Yanovska, E., Savchenko, I., Sternik, D., Kychkiruk, O., & Ol'khovik, L. (2017). In situ immobilization on the silica gel surface and adsorption capacity of poly N-4-carboxyphenyl) methacrylamide on toxic metals ions. *Nanoscale Research Letters, 12*(1), 313(1–8).

26. Polonska, Y., Yanovska, E., Savchenko, I., Sternik, D., Kychkiruk, O., & Ol'khovik, L. (2019). In situ immobilization on the silica gel surface and adsorption capacity of polymer based azobenzene on toxic metal ions. *Applied Nanoscience, 9*(5), 657–664.
27. Polonska, Y., Yanovska, E., Savchenko, I., Sternik, D., Kychkiruk, O., & Ol'khovik, L. (2018). Adsorption capacity of poly[4-Methacroyloxy-(4'-carboxy-3-chloro)-azobenzene], in situ immobilized on the silica gel surface to toxic metals ions. *Molecular Crystals and Liquid Crystals, 672*(1), 92–103.
28. Polonska, Y., Yanovska, E., Savchenko, I., Sternik, D., Kychkiruk, O., & Ol'khovik, L. (2018). In situ immobilization on the silica gel surface and adsorption capacity of poly[4-methacroyloxy-(4′-carboxy-2′-nitro)-azobenzene] on toxic metals ions. *Molecular Crystals and Liquid Crystals, 671*(1), 164–174.
29. Polonska, Y., Yanovska, E., Savchenko, I., Sternik, D., Kychkiruk, O., & Ol'khovik, L. (2018). Azocontaining polymers, in situ immobilized on the silica gel surface. *Molecular Crystals and Liquid Crystals, 673*(1), 39–47.
30. Savchenko, I., Yanovska, E., Vretik, L., Sternik, D., & Kychkiruk, O. (2021). Synthesis, characterization and adsorption properties for metal ions of silica-gel functionalized by poly[8-methacroyloxy-quinoline]. *Molecular Crystals and Liquid Crystals, 719*, 103–115.

CHAPTER 5

Development of Environmentally Safe Oligomeric Nitrogen Fertilizer

ZAUR LOMTATIDZE, IA CHITREKASHVILI, GIVI PAPAVA, ETER GAVASHELIDZE, MARINA GURGENISHVILI, KETEVAN ARCHVADZE, KETEVAN PAPAVA, NANULI KHOTENASHVILI, NAZI GELASHVILI, NORA DOKHTURISHVILI, and RIVA LIPARTELIANI

TSU Petre Melikishvili Institute of Physical and Organic Chemistry, Tbilisi, Georgia

ABSTRACT

An environmentally safe and cost-effective technology for obtaining nitrogen fertilizers has been developed, which makes it possible to replace easily flushed nitrogen fertilizer with polymerized, which will not be easily washed off and provides dosed, regulated, controlled fertilizing of the plant and its harmonious development during the growing season. This makes it possible to prevent environmental pollution and practically prevent economic losses caused by the washing and evaporation of fertilizers. A targeted synthesis was carried out to obtain a new generation of fertilizers by polycondensation. The synthesis of oligomers is carried out in solution and the melt, as well as in the pores of a natural sorbent – zeolite. The process of polymer formation was studied and optimal reaction conditions were established. The process of microbial biodegradation of polymers and the mechanism of its occurrence has also been studied.

5.1 INTRODUCTION

Intensive population growth requires an increase in food production, especially grain crops. But agricultural land is gradually shrinking due to

Sustainability in Polymer Technology and Plastics Engineering: Concepts, Strategies, and Opportunities.
Tamara Tatrishvili, Neha Kanwar Rawat, Swati Gokul Talele, and A. K. Haghi (Eds.)

urbanization and intensive industrial development. One of the ways to solve the problem in these conditions is a significant increase in agrotechnical norms of mineral, especially nitrogen, fertilizers, but this leads to global environmental pollution. Nitrogen fertilizers are highly soluble in water and therefore easily washed out, which pollutes river waters and soil, and some evaporates, penetrates into higher layers of the atmosphere and destroys the protective ozone layer of the earth from harmful radiation, which is the cause of many serious diseases on earth. According to the UN, more than 200 million tons of nitrogen fertilizers are applied to the soil annually. Of these, for the reasons listed above, more than half are lost to no avail. As a result of these losses, there are large economic losses. The regular use of such fertilizers has an undesirable effect on the soil and the environment [1–3], leads to the accumulation of large amounts of nitrates and has a negative impact on human and animal health [4].

Therefore, the introduction of environmentally safe and cost-effective mineral fertilizers and the development of technologies for their production, as well as the protection of the environment from pollution, are currently urgent problems.

This technology allows us to replace easily flushed nitrogen fertilizers with polymerized ones that will not be washed off, and to provide dosed, regulated, and controlled fertilizing of the plant during the growing season.

These problems should be solved through the development of new technologies and their intensive application [5–12]. Studies and their results answer the question of solving this problem.

5.2 EXPERIMENTAL PART

Synthesis of nitrogen fertilizers of prolonged effect is realized in solution and melt, at the interaction of urea and formaldehyde. When the reaction is carried out in a solution, we use a 40% water solution of formaldehyde, while when it is carried out in a melt, we use paraformaldehyde. The urea/aldehyde molar ratio is 1:1–1.2, respectively. When the reaction is carried out in a solution, dehydration of the oligomer that is formed as a result of the reaction takes place under vacuum at 50–70°C. Then the oligomer is heated in a polymerization furnace at 100–110°C for 40 minutes [13].

The reaction process is continuous. Experiments are carried out for the first time in practice using a new method. The study of soil microflora and isolation of urea bacteria strains is achieved by the use of specific-selective nutrient media, while their cultivation is carried out on various liquid and solid nutrient media [14]. Quantitative determination of microorganisms of the I and II phases of nitrification is made in the medium.

Cultivation of a strain with urease activity was carried out in a glucose-peptone-containing medium (peptone – 10 g, glucose – 5 g, sodium chloride – 5 g, and tap water to make 1 L) at 28°C for 48 hours. The resulting culture suspension in sterile tap water or Zak's [5] solution was used as the inoculation material in all experimental variations.

In the cultivation dynamics (culture growth curve) for polymer degradation and ammonia release, the experiment was conducted as follows:

1. Inoculation material: the culture grown in flasks with the solid glucose-peptone medium was re-suspended in 5 ml of sterile water. The suspension from 2 flasks was transferred to a 1 L flask with 100 ml of liquid glucose-peptone medium. The cultivation lasted for 72 hours at 28°C on a shaking apparatus at 120–140 rev/min.
2. After cultivation, the biomass was collected by centrifugation (5000 rev/min, 10–15 min). The biomass was washed and re-suspended in Zak's B solution [5] without phenol red.
3. Microbial protein was determined in the biomass suspension according to the Bradford method. Inoculation of flasks was performed by calculating with respect to 1 mg/ml of microbial protein [15].
4. An experiment for the degradation of urea polymers and ammonia release was carried out in the Christensen glucose (10 g) modified medium [16], to which, alongside urea (control, 20 g/l), polymers were added according to the nitrogen content in them:

   i. Polymer #1–2.92 g (N – 31.8%);
   ii. Polymer #2–2.9 g (N – 32.02%);
   iii. Polymer #3–2.86 g (N – 32.5%);
   iv. Polymer #7–2.82 g (N – 33%);

Cultivation was performed in 1 L flasks containing 100 ml medium, on a shaking apparatus, at 28C, for 48 h.

Every 2 hours of cultivation, 3–3 ml suspension was taken from the flasks and ammonia was determined in them according to Belyaev's method [15].

For the incorporation of carbamide in zeolite pores needed for the preparation of structured fertilizer of prolonged effect, clinoptilolite-containing samples of natural, non-modified zeolite from various deposits were selected to carry out chemical, X-ray phase, and IR-spectral studies. The research included dispersion and fractionation according to grain size. Six fractions were isolated. The isolated clinoptilolite fractions were washed with distilled water at 100–120°C and dried in a thermostat. Spectroscopic studies of the samples were carried out to prove their identity to standard samples.

5.3 RESULTS AND DISCUSSION

The process of carbamide structuring in zeolite pores was studied. For this, natural, non-modified analcime was used. The grain size was 0.5–1 mm. Analcime was impregnated with a mix of carbamide and aldehyde components. Ammonium nitrate and ammonium chloride were used as catalysts.

Carbamide concentration in secondary pores of analcime equaled 10–15%. The structuring process was studied at 150°C. It was proved that during the process without a catalyst, low molecular oligomers are formed, which are completely washed off from analcime pores when treated with water. When catalysts are used, the structured product is formed. The reaction proceeds in two stages. In the first stage, oligomers are formed which later undergo structuring. As a result of the reaction, macromolecules are formed, which cannot be extracted from analcime pores. Thus, a chemical bond is created between analcime and structured carbamide molecules. When structured carbamide is used as chemical fertilizer, its degradation occurs gradually thanks to the action of microbacteria.

In the initial mixture, the carbamide/aldehyde ratio was 1:1.2 and 1:1.5, respectively. Temperature: 150°C, duration: 1.5 hours.

We studied the process of carbamide structuring in melt. The carbamide aldehyde molar ratio was 1:1 and 1:2, correspondingly. Temperature 135 and 150°C., reaction duration 2 h.

When the carbamide and aldehyde molar ratio is 1: 0.8–1 polymer of linear structure is formed. Simplified structure of linear macromolecules can be presented in the following form:

$$HNCONHCH_2[NHCONHCH_2]\ n{-}HNCONHCH_2$$

While introduced into soil, the polymer is degraded by the microorganisms effect. The structure of the polymer's degenerative ferment is as follows:

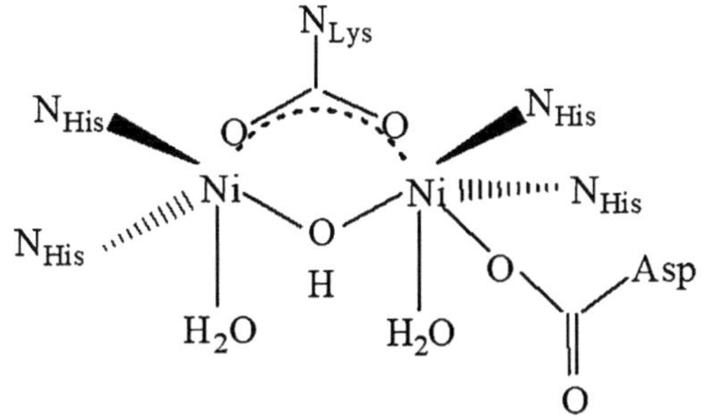

It contains various amino acids (histidine, lysine, asparagine) while in its active center, there are two Ni atoms.

Monoatomic nickel contributes to the polarization of urea carbonyl, activating it for nucleophilic attack. In this process, two water molecules bind to urea. The second nickel atom adds a hydroxide ion. This hydroxide ion attacks the positively charged carbon atom of the urea molecule. As a result, one of the two amino groups of urea becomes protonated. Schematically, this process can be presented as:

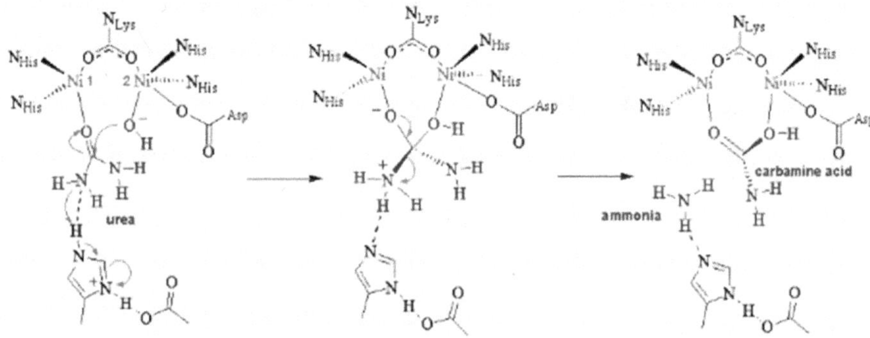

One ammonia molecule leaves urea and is transformed into carbamic acid, which is then broken down into ammonia and carbon dioxide. It should be noted that for urease activity to occur, an amino group must be bound to a mobile carbonyl group. If the transformation of these groups into carbamic acid proceeds through a fermentation reaction, a new amino group bound to a mobile carbamic group should be oriented from the polymer. With this mechanism in place, degradation will occur in stages. It is also possible that after each carbamic group is split off, the polymer will change its configuration, making it accessible for new groupings for fermentation. This would be an ideal model for a stage-wise release mechanism for ammonia.

When the ratio was 1:1, linear molecules were characterized by a prolongation effect. In this case, urease enzymes easily penetrate the polymer mass, and the degradation process proceeds quickly. When the ratio was 1:2, a polymer with a high crosslinking rate was formed. In this case, the penetration of enzymes into the polymer mass and the breaking of chemical bonds are difficult, leading to a longer biodegradation process.

Preliminary filling of secondary pores of zeolite (clinoptilolite) was achieved by a mechanical mix of carbamide and aldehyde components. At 130°C, the mix suffered melting and filled in the pores. A study of the structuring process was carried out at the following molar ratio of starting components: 1:1.5; 1:1.8; 1:2.0; 1:2.5; 1:3.0; 1:3.5; 1:4.0, correspondingly.

Temperature: 150, 155, 160, 170, and 180°C and process duration in minutes: 5, 10, 30, 60, 90, 120, 130, 140, 150, 180.

When we increased the amount of aldehyde, the speed of the created polymers increased and, accordingly, the solubility decreased. Optimal reaction conditions were established: the molar ratio of carbamide-aldehyde: 1:2.0, respectively; temperature: 150°C, duration of the process: 1.5 hours. As a result, a structured polymer was obtained.

The process of structuring in the melt was also studied. The influence of the molar ratio of the initial components, the duration of the process, and the temperature on the structuring process have been studied.

The molar ratio was changed in the following order: 1:2, 1:2.5, 1:3, 1:4; Temperature: 150, 160, 170, 180, 190, 200°C. Process duration, in minutes: 30, 60, 90, 120, 300, 600.

After 2 hours, a polymer was obtained that was practically insoluble in water. Cryogenic compositions based on carbamide structured with urease-producing bacteria were obtained, identified, and selected, taking into consideration the optimal cultivation medium for the bacteria's intense growth. Various concentrations of composites based on carbamide, which underwent degradation at various intensities due to the effect of urease-producing bacteria, were added to the cultivation medium as carbon and nitrogen sources. The influence of polymers on the growth and development of microorganisms was also observed, showing a significant intensification of their action. Experiments were conducted for urea, urea polymer degradation, and ammonia release using Christensen's glucose-modified medium (10 g) [13], in which polymers were introduced together with urea (control, 20 g/l) based on the nitrogen concentration in them:

1. Polymer #1–2.92 g (N – 31.8%);
2. Polymer #2–2.9 g (N – 32.02%);

3. Polymer #3–2.86 g (N – 32.5%);
4. Polymer #7–2.82 g (N – 33%); Cultivation was performed in 1 L flasks containing 100 ml medium, in shaking apparatus, at 28°C, for 48 h.

Every 2 h 3–3 ml suspension was taken from those flasks and ammonia was determined according to Belyaev's method [17].

The results of the experiments are offered in Figure 5.1.

Shows that in the control experiment urea utilization by the strain proceeds according to the classic scheme (according to the growth curve of culture). Namely, at the 4th hour of cultivation ammonia concentration in the medium is approximately 0.6 mg/ml, and by 10 hours of cultivation, it falls to zero, in the form of a descending exponent. The source of this ammonia is the peptone background of the Christensen medium. It is this source of peptone that the strain starts to utilize in the first place with proteolysis ferments, the activity of which is conditioned by the fact that inoculation material suffered an increase namely in the peptone-containing glucose-peptone medium. By the 10th hour of cultivation, the whole reserve of peptone is exhausted, and microorganisms begin to rearrange for the other source of nitrogen, urea, and urease release commences. As a result of the degradation of urea, thanks to urease, ammonia accumulation begins in the cultivation medium, the concentration of which from the 10th hour of cultivation begins to increase with ascending exponent. This is proved also by the appearance of a sharp odor of ammonia coming from the flask.

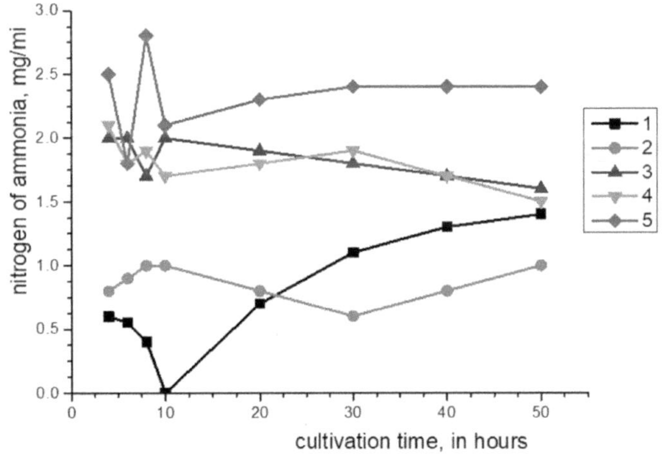

1. Series 1 test 2. Series 2 polymer #2 3. Series 3 polymer #3
4. Series 4 polymer #4 5. Series 5 polymer #7

FIGURE 5.1 Degradation of urea and its polymers with ammonia release. (1) control; (2) polymer #2; (3) polymer #3; (4) polymer #4; and (5) polymer #7; ammonia nitrogen mg/ml.

Based on investigations, Polymer #2 was considered the most perspective sample of degradation. At resuspending in water distillate and centrifuging (5000 rev/min, 10 min).

1 g Polymer #2 was washed 15 times in 100 ml water distillate. Simultaneously, ammonia was determined with Nessler's reagent in every 2 ml of washed supernatant. Formaldehyde was determined by silver mirror reaction. Results are given in Table 5.1.

TABLE 5.1 Absorption Data of Supernatants

#ofwashing	1	2	3	4	5	6	7	8	9	10	11	12	13	14	15
OD	2.6	2.5	2.5	2.0	1.6	1.6	1.2	1.2	0.5	0.2	0	0	0	0	0
Silver mirror reaction	+	+	–	–	–	–	–	–	–	–	–	–	–	–	–

OD – Absorption index of supernatant treated with Nessler's reagent on photoelectron colorimeter.
"+": reaction is positive, "–": reaction is negative.

Analysis of experimental results shows that we can remove admixtures from polymers by washing. At the same time, it was determined that alongside formaldehyde, they contained water-soluble admixtures of a nitrogenous character. Regardless of the negative silver mirror reaction for aldehydes, Nessler's reagent consistently indicated the presence of nitrogen-containing compounds in the supernatant.

In the next stage, washed Polymer #2 was dried and reduced to a constant weight at 90°C. After drying, 1 g of polymer was resuspended in 100 ml of distilled water. After centrifugation, the nitrogen concentration was measured using Nessler's reagent. The optical density equaled 1.2.

The transformation of polymer nitrogen into ammonia was achieved by using urease, where microbial ferment urease played the role of sulfuric acid in the cultivation medium.

Experiments were conducted to validate the method, where the reaction took place in a Christensen medium with an adequate quantity of urea polymers introduced as a nitrogen source. Inoculation was done using a 48-hour strain A flask culture. The same nutrient mediums served as controls, and their inoculation was done with sterile water. Cultivation lasted 72 hours on a shaking apparatus. After the cultivation was completed, the cultural liquid was centrifuged, and nitrogen was determined in the supernatant.

Logical results were received for Polymer #2 (based on previous experiments, where this polymer was considered as the prospective study object). In the control version, the mineral nitrogen background is slight, which might be the result of the degradation of peptone in the nutrient medium due to physical factors. The experiment shows that the activity of biodegradation increases almost three times. A new experiment was carried out with this in mind, in which Polymer #2 was used as a source of nitrogen. The experiment was conducted in three parallel versions:

1. Control #1: Christensen medium, with polymer, without inoculation.
2. Control #2: Christensen medium, with urea, inoculated.
3. Christensen medium, with polymer, inoculated.

Volume of nutrient medium – 100 ml, cultivation flask capacity 1 L, cultivation temperature 28°C, shaking apparatus 120–140 rev/min. Cultivation duration 172 h. Every 2 h, 2 ml of cultural liquid was taken from flasks, shifted to Vacutainer's flasks, and placed in an icebox. Then nitrogen was determined by the use of the modified Kjeldal method.

In the next stage, isolation of various strains of urobacteria, selection of active strains, and their propagation were performed. Strains of urobacteria were isolated from soils, and active strains were obtained by their selection, which were used for the degradation of structured carbamide, for its transformation into a form assimilated by plants, and then for the realization of ion exchange processes, with the purpose of transforming insoluble calcium phosphate into a soluble form. This process was realized in soil. Selected strains were tested with respect to polymer nitrogenous fertilizers, and the existence of their intermediary oligomers was checked. It is known that many synthesized compounds, including polymers, are characterized by biocide properties. Alongside this, a definite part possesses a selective toxic effect. When compounds with such properties are introduced into the soil, the property, as well as the quantitative composition of soil microflora, undergo changes, in some cases decreasing soil fertility rate.

Proceeding from the above statement, to determine the harmlessness of structured carbamide to soil microflora, we studied the effect of the listed polymers on the properties and quantitative composition of microflora in a certain type of soil (brown-carbonaceous, salty). Special attention was paid to the following physiological groups of soil microorganisms: cellulose-destroying, saprophytes, fungi, actinomycetes, free-living nitrogen fixers. Experiments were carried out in the field (Field Experiment I) as well as in laboratory conditions (*in vitro*).

Experiment I: Field experiment with brown-carbonaceous soil type. In the field (0.5 m^2), in cultivated soil, we introduced up to 1–5 g (according to variants) of structured carbamide compositions. The soil was watered once a day. On the third day of the experiment, a soil sample was taken (at 3 cm depth) and a soil suspension was prepared. The suspension was then diluted and plated on agar MPB (meat-peptone broth) in Petri dishes. The Petri dishes were placed in a thermostat at 28°C for 7 days. Simultaneously, the experiment was carried out with soil without fertilizer (control). After incubation, the following physiological groups were identified: cellulose-destroying bacteria, saprophytic bacteria, fungi, actinomycetes, freely living nitrogen-fixers, and urobacteria.

Analysis of results of the experiments shows that structured carbamide compositions don't make a marked effect on the property and quantitative composition of soil microflora. Experiment II (*in vitro*). Effect of structured carbamide compositions on some physiological groups isolated from various types of soils, on their growth and development – was studied. Analysis of experimental results proves that growth and development of microorganisms isolated from soil sown on agar, specific (for each physiological group) nutrient medium to which structured fertilizer was added, proceeded according to common dynamics, exceptions were freely living nitrogen-fixers (family *Azotobacter, Clostridium*) and urobacteria (family *Urobacillus*), where a marked excess of the biomass of the above-stated microorganisms compared to the control (medium without fertilizers) – was observed. Structured carbamide is not characterized by biocide properties. *In vitro* experiments enable us to suppose that it stimulates propagation of urobacteria in soil. Brown-carbonaceous and red soils were used in the experiment, which differ from each other by chemical composition and qualitative and quantitative composition of microflora.

Agarized medium (MPB) was prepared from suspensions of corresponding soil types (10 g soil sample per liter of distilled water), to which structured carbamide was added (experiment) or not added (control). Petri dishes were placed in a thermostat for 7 days; we incubated them and observed the formation of microorganisms and their growth and development.

Analysis of the experimental results shows (based on the total quantity of developed biomass) that soil type affects the quality of polymer degradation. The chemical composition of the soil determines the quantitative and qualitative composition of microflora, including the number of polymer destructors. It should be noted that polymers can also stimulate the abundance of urobacteria. Among the 4 structured carbamide-containing compositions, Polymer #2 was found to be the most promising. The degradation rate of this polymer by urobacteria is higher than that of the other studied polymers.

It was demonstrated that urobacteria are not adversely affected by the toxic effects of polymers and that microorganisms use these polymers as a nitrogen source, indicating the possibility of their degradation. The degradation of polymers in the soil occurs slowly.

Therefore, throughout the entire vegetation period, plants are supplied with nitrogen; nitrogen leaching and evaporation are minimized, plants assimilate nitrogen to the maximum, leading to increased productivity. This has a significant impact both economically and ecologically.

The obtained fertilizer of prolonged effect was used in field experiments under autumn wheat Bezostaya-1. Productivity indices show that the data suffered significant changes, similarly nitrogen norms per ha were changed. Productivity in the first control variant (2.4 t) is far lower, since nitrogen was not introduced there, in the second control variant (obtained according to agro rules) productivity equaled 4.1 tons per ha, while in the case of application of fertilizer with prolonged effect, the productivity in the fourth variant equaled 6.2 tons, which exceeds by 51.2% that of the control variant. This effect is explained by new technology developed by us. As for other variants, where nitrogen per ha norms were decreased from 80 to 30 kg, productivity indices were changed correspondingly compared to the control (obtained according to agro rules).

5.4 CONCLUSION

A significant economic and environmental effect was observed in those variants where nitrogen standards per hectare decreased to 70, 60, 50, and 40 kg, respectively. Productivity data of all these variants compared to the data of those obtained by the observance of agro norms are higher and exceed by 10.2–29.2%, which is explained by the application of the fertilizer of prolonged effect.

At the application of the fertilizer of prolonged effect, the hectare norm of nitrogen decreases by 50%, and alongside with the increase of productivity, it is of greatest economic and ecological significance, and can guarantee environment protection from pollution.

Technology of new generation, that was developed and preparation of multicomponent fertilizers of prolonged effect of focused designation for obtaining ecologically pure production and for environment protection is the newest trend. By development of the new technology, the new generation multi-component fertilizers of prolonged effect are obtained for the first time on the basis of urobacteria and polymerized nitrogenous fertilizers, which enable us to resolve practically definite ecological problems in agriculture.

Inculcation of new fertilizers will enable us to decrease hectare norms of nitrogenous fertilizers by 45–50%. Washing-off and evaporation of fertilizers, environment pollution, and economic losses conditioned by it will be brought to a minimum since the fertilizer is transformed into a hardly soluble state.

KEYWORDS

- **ammonia**
- **bacteria**
- **biodegradation**
- **fertilizer**
- **polycondensation**
- **polymer**
- **zeolite**

REFERENCES

1. U.S. Congress, Office of Technology Assessment. (1982). *Impacts of Technology on U.S. Cropland and Rangeland Productivity*. U.S. Government Printing Office, Washington, D.C.
2. U.S. Department of Agriculture. (1980). *Report and Recommendations on Organic Farming*. U.S. Government Printing Office, Washington, D.C.
3. Vogtmann, H. (1984). In Bezdicek, D. F., & Power, J. F. (Eds.), *Organic Farming: Current Technology and Its Role in a Sustainable Agriculture* (p. 19). American Society of Agronomy, Madison, WI.
4. Hodges, R. D., & Scofield, A. M. (1983). In Lockeretz, W. (Ed.), *Environmentally Sound Agriculture* (p. 3). Praeger, New York.
5. Papava, G., Dolidze, A., & Ebralidze, Q. (2000). Tiocarbamide-formaldehide matrix for obtaining the multicomponent polymer composition. *Georgian Engineering News, 2*, 131.
6. Ingram, C. W., Coyne, M. S., & Williams, D. W. (2005). Effects of commercial diazinon and imidacloprid on microbial urease activity in soil and sod. *Journal of Environmental Quality, 34*(5), 1573–1580. https://doi.org/10.2134/jeq2004.0433.

7. Papava, G., Lomtatidze Z., Dokhturishvili, N., Gugava, E., Gurgenishvili, M., Chitrekashvili, I., & Papava, K. (2019). Synthesis of nitrogen-containing biodegradable polymer fertilizers with the prolongation action. *Bulletin of the Georgian National Academy of Sciences, 13*(4), 80–84.
8. Papava, G., Gelashvili, N., Molodinashvili, Z., Gurgenishvili, M., & Chitrekashvili, I. (2011). Synthesis and study of phenol-formaldehyde type polymers on the basis of bisphenol with adamantane grouping. *Journal of the Balkan Tribological Association, 17*(3), 426–435.
9. Chitrekashvili, I.A., Gurgenishvli, M.B., Papava, G.Sh., Sherozia, V.A., Papava, K.R. Khotenashvili, N.Z., & Tabukashvili, Z.Sh. (2015). Zeolite Bazed Hybrrid Cationites. *High-Performance Polymers for Engineering-Based Composites,* 60–65.
10. Papava, G., Gelashvili, N., Molodinashvili, Z., Gurgenishvili, M., & Chitrekashvili, I. (2010). Synthesis and study of phenolformaldehyde type polymers on the base of bisphenol with adamantane grouping. Handbook of Condensed Phase Chemistry, 123–132.
11. Papava, G., Gurgenishvili, M., Chitrekashvili, I., Dokhturishvili, N., Gavashelidze, E., Gelashvil, N., & Archvadze, K. (2024). Synthesis Process Research of Urea-Formaldehyde Linear Oligomers when Carrying Out Polycondensation in a Solution to Obtain Biodegradable Nitrogen Fertilizers. *Environmental Technology and Sustainability: Physical, Chemical and Biological Technologies for Environmental Protection,* 139–144.
12. Papava, G. SH., Sarjveladze, I. V., Chitrekashvili, I. A., Medoidze, S. A., Gelashvili, N. S., Liparteliani, R. G., Papava, SH. R., & Khotenashvili, N. S. (2024). Study of the effect of polymerised urea on the yield of grass mixtures in the field. *Oxidation Communications, 47*(2), 384–392.
13. Pargham, W. E. (1984). Future perspectives for natural zeolites in agriculture. In *Use of Natural Zeolites in Culture and Agriculture* (pp. 283–265). Zeo Agriculture, Westview Press, Boulder, Colorado.
14. Zvyagintsev, D., Babeva, I., & Zenova, G. (2005). *Soil Biology*. Moscow State University.
15. Ramishvili, N., Gorozia, I., & Lomtatidze, Z. (2009). Study of nitrogen fixatives in acid-black earth soils of upper Abkhazia. *Bulletin of the Georgian National Academy of Sciences, 3*(1), 45–79.
16. Fillipov, Y., Egorov, T., & Sevastyanov, G. (1975). *Practicum in General Biochemistry*. Prosveshchenie, Moscow.
17. Tepper, E. H., & Shilnikova, V. K. (2004). *Workshop on Microbiology*. University of Bustard, Moscow.

CHAPTER 6

Green Plastics From Microorganisms: Current and Future Perspectives

G. VIDYA,[1] PRAVEEN C. RAMAMURTHY,[2] and K. N. MADHUSOODANAN[3]

[1]*Department of Chemistry, Dayananda Sagar College of Engineering, Bangalore, Karnataka, India*

[2]*Department of Materials Engineering, Indian Institute of Science, Bangalore, Karnataka, India*

[3]*Rubber Technology Division, Rubber Research Institute of India, Kottayam, Kerala, India*

ABSTRACT

The term green plastics comprises chemically divergent products that are produced from microorganisms under various environmental conditions. These are a better alternative to conventional petrochemical thermoplastics and currently gain more sensational public attention. Green plastics or bioplastics are a special type of biomaterial. The manufacture of microbial green plastics and their products could help to maintain the sustainability of the environment and also diminish the release of greenhouse gases. Green plastics are biodegradable and can be broken down fully into water, inorganic compounds, other organic materials, carbon dioxide, and compost by microorganisms under specific conditions, leaving no poisonous residue. Bioplastics are currently used in many disposable applications. They are polyesters consisting of dispersed nature and are present intracellularly in microbes as a form of storage granules having their chemical and physical characteristics like petrochemical plastics. Depending on the nature of microbes, green plastics differ in their monomer composition, physio-chemical properties,

Sustainability in Polymer Technology and Plastics Engineering: Concepts, Strategies, and Opportunities.
Tamara Tatrishvili, Neha Kanwar Rawat, Swati Gokul Talele, and A. K. Haghi (Eds.)

and macromolecular structure. This chapter gives more attention to the updated information on microbial green plastic manufacture, growth, use, bio-sustainability, commercial consistency, and industrial applications.

6.1 INTRODUCTION

During evolution, our world has gone through numerous momentous eons. Now we are in the middle of the "Plastic Age" Based on the various valuable applications of plastics, they are the most produced and used materials in our world because of their easy processability, plasticity, and constancy. Plastics have unconquered innumerable advantages over other materials used in earlier periods such as metals, minerals, ceramics, etc. However, the increasing consumption of petroleum-based thermoplastics leads to vital environmental grievances, and plastics that are deposited in landfills decompose gradually, which can persist in landfills for hundreds or even thousands of years. At present, people are very conscious of the hazardous effects of conventional petrochemical plastics that adversely affect the ecosystem. Therefore, these days extensive public interest is diminishing in petrochemical-derived plastics. Consequently, researchers are now making efforts to handle plastic waste on Earth by finding more sustainable and recyclable substitutes for plastic. Unlike petrochemical plastics, green plastics or bioplastics are plastics that are bio-based or biocompostable, at least moderately obtained from biomass, and that could decompose into natural constituents such as compost, water, and green gases by microorganisms with no artificial additives or catalysts. Hence, green plastics are obviously different from biodegradable plastics [1].

The first acknowledged bioplastic, polyhydroxybutyrate (PHB), was invented in 1926 by Maurice Lemoigne from his work with the *Bacillus megaterium* bacterium. In 1947, the first technical bio-plastic was presented on the world market that had excellent mechanical and chemical resistance properties. Followed by in the 1990s, more bio-plastics were introduced, such as polylactic acid (PLA) and derivatives of polyhydroxyalkanoates (PHA), which are based on different green plastics profiting from the fast technological advances in green chemistry and analytical chemistry in the regeneration of biomass such as cellulose, starch, sugars, etc. [2]. Green plastic refers to plastic made from biological materials such as wood, straw, vegetable oil and fats, corn starch, recyclable food waste, etc., apart from petroleum-derived plastics [3]. The manufacture of green plastics does not deplete fossil resources since they are produced from renewable resources. Bioplastics are made from a group called polyhydroxyalkanoate (PHA), which are polyesters naturally originated by bacterial fermentation upon lipids or sugar. The polyesters produced are deposited in the form of extremely refractive granules in the bacterial cell, and these bacteria can store carbon and energy. Based on microorganisms and their environmental conditions, homo-polyesters or co-polyesters with various hydroxyalkanoic acids are produced [4]. The green-plastic invention should be the fine answer for plastic-based pollution; hence, the production of bioplastics makes the greatest headway in the areas of agriculture and medicine. Therefore, due to their inherent biodegradability and biocompatibility, environmentally friendly manufacturing, and various applications including the medical sector, and consumable materials, green plastic has now become popular and emerged as a vital replacement for chemically derived plastics.

The principal focus of this chapter on microbial green plastics is to address the following questions and provide suitable explanations. The key questions are as follows:

1. What are the different types of green plastic manufacturing from microorganisms?
2. What is the current development of green plastics their commercial uses and their consistency?
3. Are these green plastics really good for our earth?

6.2 CLASSIFICATION OF GREEN PLASTICS AND THEIR PRODUCTION

The ecological issues from conventional plastics lead to a new way for the research of other alternatives. Green plastics are structurally comparable to conventional plastics, biologically renewable, and are glorified as a possible new material to reduce ecological problems. Green plastics or bioplastics are mostly divided into three types, which are

1. Natural polymers [Polymers from biomass]
2. Synthetic polymers [Polymers obtained from renewable resources]
3. Microbial polymers [Polymers produced by microbes]

The graphical representation based on the classification of green plastics is depicted in Figure 6.1. Additionally, biopolymers are made up of naturally non-toxic substituents and are considered inherently biocompatible. The basic idea for a cost-effective way of producing green polymers from biomass is grammatically shown in Figure 6.2 [6].

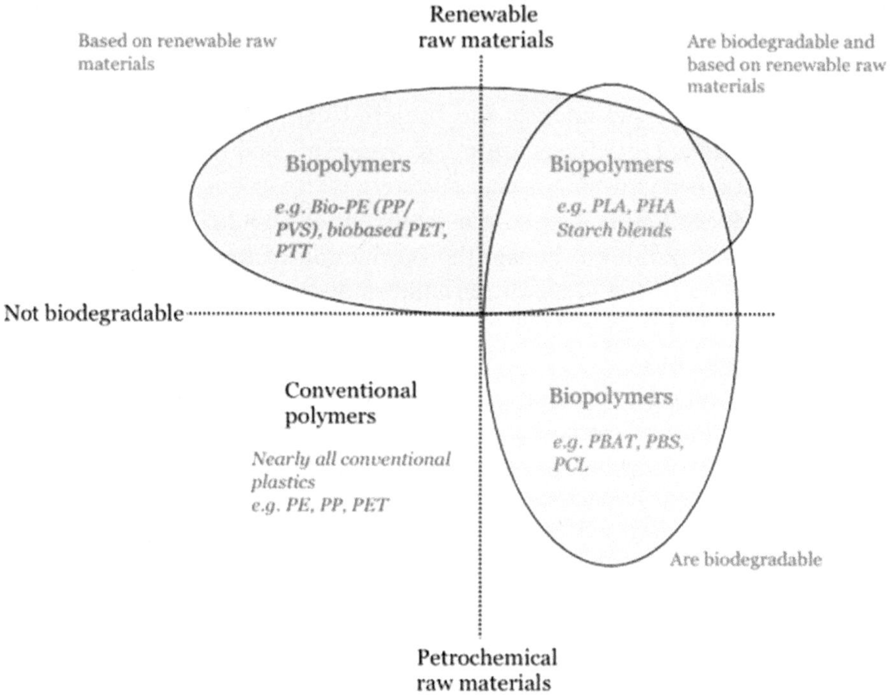

FIGURE 6.1 Understanding three different classes of bioplastics.

Source: Reprinted with permission from Ref. [5]. © 2012, Elsevier.

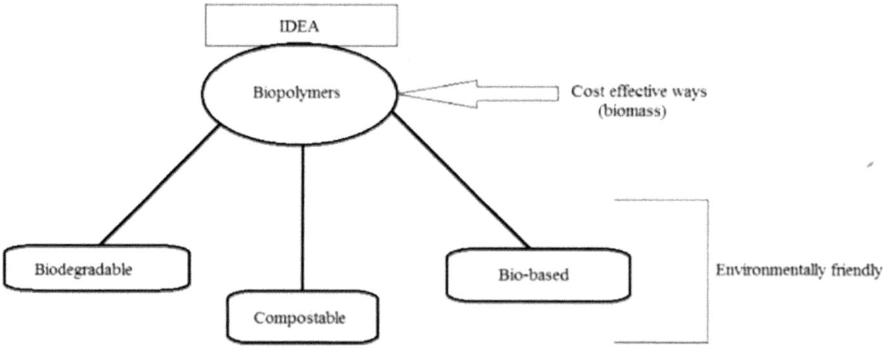

FIGURE 6.2 The fundamental idea of research of green plastics.

Source: Reprinted with permission from Ref. [6]. © 2017, Science PG.

6.3 GREEN PLASTICS FROM MICROORGANISMS

Microorganisms and algae are the main living organisms capable of degrading organic waste to form green plastics. Numerous bacteria species like *Azotobacter, Actinobacillus, Rhodobacterium, Agrobacterium*, etc., have the capacity to convert organic waste to bacterial PHB (polyhydroxybutyrate). Polyhydroxyalkanoates (PHA) are naturally occurring polyesters, and these are synthesized by bacteria and other microorganisms. The production of green plastics such as PHA and PHB from microorganisms is explained below, followed by a description of green plastics from algae. Figure 6.3 shows some important structures based on microbial green plastics.

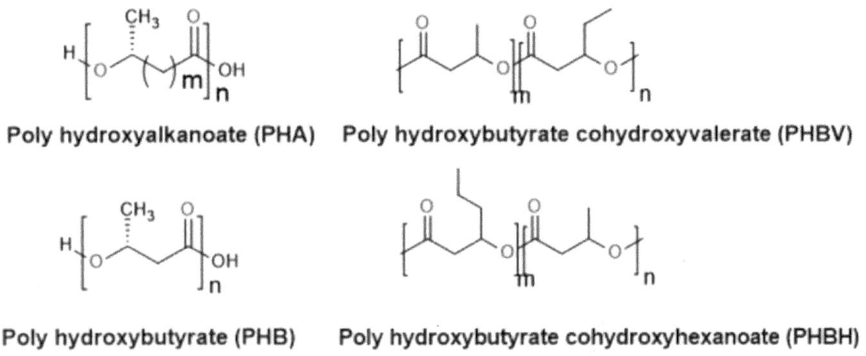

FIGURE 6.3 Structures based on some important microbial green plastics.

6.3.1 POLYHYDROXY ALKANOATES (PHA)

Polyhydroxyalkanoates (PHA) are linear polyesters naturally produced by microbes like bacteria due to the fermentation of sugar and lipids. They are able to store carbon and energy and are deposited in cells in the form of highly refractive granules. Depending on culture conditions, different types of bacteria could be involved in PHA production. They are categorized into two types: (i) one kind of bacteria that requires more carbon and controls other soil nutrients such as oxygen, phosphorus, calcium, and nitrogen for PHA synthesis. Examples include *Protomonas extorquens* and *P. oleovorans*; and (ii) bacteria that are not limited by nutrient restrictions can form PHA in the log phase. Examples include *Azotobacter vinelandii* and *Alcaligenes latus* [7]. The physical properties of PHA, such as biodegradability, renewability, and plasticity, depend on their synthesis pathways, chemical structure, monomer composition, etc. The production of PHA in the biosynthetic pathway depends on the particular substrate choice and the specificity of PHA synthase, which are critical for the production of short-chain-length and long-chain-length PHAs. For example, *Alcaligenes eutrophus* PHA synthase has a carbon chain length consisting of 3 to 5 carbon atoms, while bacteria like *P. oleovorans* provide polymerization to give 6 to 14 carbon atoms [8]. The granule size of PHA typically ranges from 0.2 to 0.5 µm and is accumulated in the cytoplasm. Figure 6.4 represents the synthesis and biodegradability of polyhydroxyalkanoates [9]. PHA hydrolases and depolymerases are the two main enzymes formed by bacteria that support the decomposition of PHA. The biocompatible nature of PHA shows very minimal toxic effects in living systems [9].

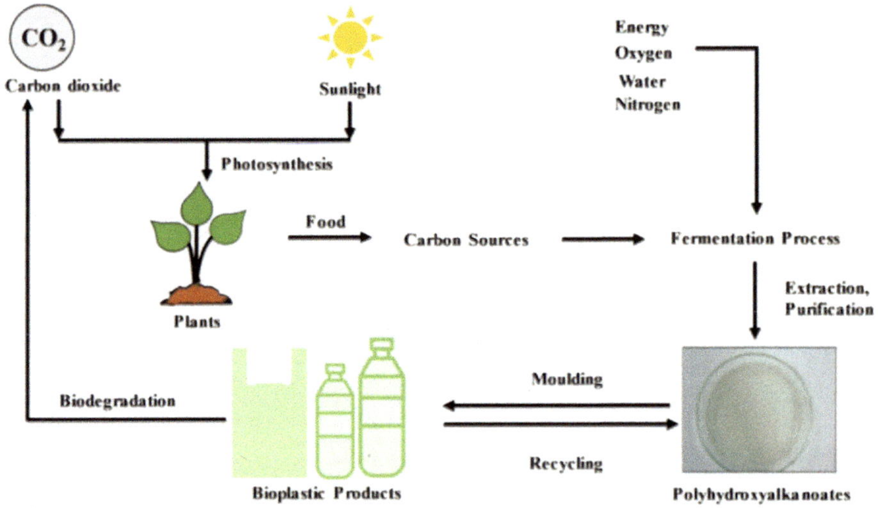

FIGURE 6.4 Synthesis and biodegradability of polyhydroxy alkenoates.
Source: Reprinted with permission from Ref. [9]. © 2020, Elsevier.

6.3.2 POLYHYDROXY BUTYRATE (PHB)

Bacterially synthesized renewable polyhydroxy butyrate (PHB) is fully biodegradable and hydrophobic thermoplastic polyester comprising a crystalline portion of almost 80%. The fundamental properties of PHB are impartially similar to those of some petrochemical synthetic polymers. Under optimal conditions, *Ralstonia eutropha* H16 produces 80% of dry cell weight, but the production cost of PHB fermentation processes carried out in bacteria is now very costly [10]. Nowadays, plant-related expression systems are quite interesting because no peripheral organic carbon source is essential, and that could be applied to large-scale production systems. Mainly three different bacterial enzymes are expressed in the cytosol or go to unlike parts of the plant cell, giving a good yield of PHB buildup in the plastids of *Arabidopsis thaliana* containing nearly 40% of dry cell weight. Unfortunately, due to some growth and infertility issues, plants are not suitable for high amounts of farming. Nowadays, a higher amount of PHB production in plants is achieved with the help of fertile offspring obtained from the plastid of *Nicotiana tabacum* with 18–19% PHB of their dry cell weight [11–13]. To produce novel PHB in a particular strain through enhancing PHB accumulation, metabolic engineering is a well-known tool to modify metabolic pathways. PHB biosynthesis with *Alcaligenes faecalis* was transferred to *E. coli* strains to get more PHB accumulation [14]. Now PHB is produced from different microbes such as yeast, recombinant bacteria, etc., and mostly PHB is soluble in most organic solvents, leading to the extraction of PHB from bacterial cells being more viable. Various types of microorganisms found in nature are able to decompose PHB under the influence of PHB depolymerase [15].

6.3.3 GREEN-PLASTICS PRODUCTION FROM ALGAE

The production of green plastics is not limited to microbes; algae can also produce green plastics. PHA, PLA, starch, some protein-related polymers, cellulose, and PVC have recently been reported to be produced from algal biomass. The main advantages of green plastics from algae are that they reduce the greenhouse effect by absorbing greenhouse gases, boost plastic quality, and have the capability to grow in large quantities in small spaces [16]. Bioplastic production from algae is done through various methods such as merging algae with other materials, using algae biomass directly, appropriate biorefinery processing, and the introduction of feasible genetic engineering for better green plastic manufacturing [17]. The key highlight of microalgae for the production of green plastics is its cost-effectiveness. Algae mainly use carbon dioxide and light for preparing food, so the cost of the substrate should be reduced during biosynthetic production.

6.4 CURRENT DEVELOPMENT AND INDUSTRIAL APPLICATIONS OF GREEN PLASTICS

Green plastics have emerged as one of the advanced and environmentally friendly materials developed in the past few years. Nowadays, this eco-friendly plastic is gaining attention due to its potential biodegradability and is also used as a promising material to replace conventional and chemical plastics. The current development for the research of green plastics mainly includes the upgrading of mechanical properties such as advancements in manufacturing technologies, establishment of standards, enhancing processability, progress in applicability, and reduction of manufacturing costs, etc. [18]. Current research on green plastics focuses on reducing these issues. A switch to green plastics would impact raw material prices. Another important factor is the fermentation carried out in bacteria, which is very costly and accompanies high manufacturing expenses, hindering the growth of green plastics. Improvements in bioplastics from microbes are achieved through genetic manipulation of cell factories, refining fermentation under different conditions, enzymatic alterations, and blending or mixing with other biopolymers, which are now being implemented [19]. All of these extensively modify the characteristics of green plastics such as elasticity, solubility, stability, crystallinity, glass transition temperature, and permeability. These properties must be enhanced for the efficacy of biomedical applications, including drug delivery, tissue engineering, and regenerative or reconstructive medicine [20]. For medicinal applications, biopolymers and the cell factory are certified as "generally recognized as safe" (GRAS), a title given by the FDA that introduces substances acknowledged as harmless or safe. This GRAS standard validates that polymers from microorganisms achieve purity standards and are harmless to apply in medical devices [21]. Compositional analysis of microbe-based polymers used for the manufacture of modern analytical instruments would help further improve their quality control analysis. Therefore, the progress of safe bacterial cell factories with the help of synthetic biology and bioengineering can lead to a plethora of new sustainable bioplastics. The efficacious industrial production of green plastics is related to different factors such as the price of microbial substrates, total weight of dry cell weight over substrate rate, efficiency, and the price of downstream processing based on the decontamination process. However, the bioengineering field helps to enlighten the upstream process via the use of low-cost substrates and improved productivity, and also their bioprocess optimization depends on upstream and downstream procedures that are essential for low-cost and scalable production [22]. However the manufacture of extracellular bioplastics is quite challenging due to the increased viscosity of the culture fluids, which affects the penetration of dissolved oxygen (DO), formation of ATP, and also improving the anaerobic environments or increasing energy-producing systems might boost throughput.

The other applications of green plastics are used for making disposable items for packaging, catering, and also for making vessels for daily household items, etc. Non-disposable applications include car interiors, mobile phone casings, carpet fibers, fuel lines, etc. Research is ongoing on the production of electroactive bioplastics for electrical uses because biodegradable products are very easy to recycle and are non-toxic. Green plastics degrade in soil, hence these plastics are now used in agriculture purposes such as being used as biodegradable carriers, plastic sheaths for crop protection, seedling containers, tubing for crop irrigation, etc. Green plastics are also applied in the construction and building industry for making various products such as fencing construction, decking, etc.

6.5 ARE GREEN PLASTICS BEING GOOD FOR THE ENVIRONMENT?

Green plastics are typically used as a sustainable substitute for conventional petrochemical plastics. Biodegradable green plastics can decompose into natural materials through bacterial action and merge into the soil without causing harm. This decay process is carried out by microorganisms in the presence of water and oxygen. The advantages of green plastics are that they are biodegradable, eco-friendly, derived from renewable resources, and have good processability. Most green plastics are 100% compostable and usually take 180 days or less to decompose in a composting facility. They are absorbed back into the earth and become nutrients for the soil. During the production process, green plastics emit two-thirds less harmful greenhouse gases compared to petrochemical plastics. The use of green plastics enables specific recycling approaches to protect the environment through microorganisms and biocatalysts [23, 24]. The developing microorganism-based recycling process used in a bioreactor setup has been established to allow upcycling and recovery of bioplastics into value-added chemicals and monomers [25]. Therefore, biotechnological waste treatment is also a very important criterion for the disposal of microbial plastics.

6.6 CONCLUSION

Microbial green plastics now garner more interest in scientific research and commerce due to their processing viability and environmental sustainability. The primary issue hindering the full-fledged commercialization of green plastics is the high manufacturing cost and lack of established production procedures. Scientists are still working on improving the biosynthetic pathways through genetic engineering to enhance the production of sustainable green plastics. Overall, plastics produced from microorganisms offer several advantages such as excellent biodegradability and biocompatibility, structural diversity based on microbes and their conditions, making them suitable for various applications including biomedical, pharmaceutical, tissue engineering, bio-implant, construction, packaging, disposable, and non-disposable applications. However, there are some shortcomings in this field such as the uneconomical production of microbial plastics, lack of cost-effective recycling procedures, and the need to achieve the right mechanical and physical properties for specific applications. Future research on green plastics should focus on achieving more economical production, exploring advanced applications, and enhancing overall environmental sustainability.

KEYWORDS

- **environment**
- **green-plastics**
- **microorganisms**
- **polyhydroxy butyrate**
- **polyhydroxyalkanoates**
- **sustainability**

REFERENCES

1. Park, Y. K., & Lee, J. (2022). Achievements in the production of bioplastics from microalgae. *Phytochemistry Reviews*. https://doi.org/10.1007/s11101-021-09788-8.
2. Geyer, R., Jambeck, J. R., & Law, K. L. (2017). Production, use, and fate of all plastics ever made. *Science Advances, 3*(7), e1700782. https://doi.org/10.1126/sciadv.1700782.
3. Moshood, T. D., Nawanir, G., Mahmud, F., Mohamad, F., Ahmad, M. H., & Ghani, A. A. (2022). Current research in green and sustainable chemistry. *Current Research in Green and Sustainable Chemistry,* 100273. https://doi.org/10.1016/j.crgsc.2022.100273.
4. Wani, S. J., Shaikh, S. S., & Sayyed, R. Z. (2015). Microbial biopolymers: An initiative step towards green plastic. *International Journal of Scientific & Engineering Research, 6*. ISSN 2229–5518.

5. Philp, J. C., Bartsev, A., Ritchie, R. J., Baucher, M. A., & Guy, K. (2013). Bioplastics science from a policy vantage point. *New Biotechnology, 30*(6), 635–646. https://doi.org/10.1016/j.nbt.2012.11.021.
6. Shamsuddin, I. M., Jafar, J. A., Shawai, A. S. A., Yusuf, S., Lateefah, M., & Aminu, I. (2017). Bioplastics as better alternative to petroplastics and their role in national sustainability: A review. *Advances in Bioscience and Bioengineering, 5*(4), 63–70. https://doi.org/10.11648/j.abb.20170504.13.
7. Guzik, M., Witko, T., Steinbüchel, A., Wojnarowska, M., Sołtysik, M., & Wawak, S. (2020). What has been trending in the research of polyhydroxyalkanoates? A systematic review. *Frontiers in Bioengineering and Biotechnology, 8*, 959. https://doi.org/10.3389/fbioe.2020.00959.
8. Dietrich, K., Dumont, M. J., Rio, L. F. D., & Orsat, V. (2017). Producing PHAs in the bioeconomy – Towards a sustainable bioplastic. *Sustainable Production and Consumption, 9*, 58–70. https://doi.org/10.1016/j.spc.2016.09.001.
9. Kumar, M., Rathour, R., Singh, R., Sun, Y., Pandey, A., Gnansounou, E., Lin, K. Y. A., Tsang, D. C. W., & Thakur, I. S. (2020). Bioplastics from waste biomass: Current trends and future prospects. *Journal of Cleaner Production, 263*, 121500. https://doi.org/10.1016/j.jclepro.2020.121500.
10. van Beilen, J. B., & Poirier, Y. (2008). Production of renewable polymers from crop plants. *The Plant Journal, 54*(4), 684–701. https://doi.org/10.1111/j.1365-313X.2008.03431.x.
11. Tatarev, K. B., McAvoy, S., Daughtry, S., Peoples, O. P., & Snell, K. D. (2011). High levels of bioplastic are produced in fertile transplastomic tobacco plants engineered with a synthetic operon for the production of polyhydroxy butyrate. *Plant Physiology, 155*(4), 1690–1708. https://doi.org/10.1104/pp.110.169581.
12. Nawrath, C., Poirier, Y., & Somerville, C. (1994). Targeting of the polyhydroxybutyrate biosynthetic pathway to the plastids of Arabidopsis thaliana results in high levels of polymer accumulation. *Proceedings of the National Academy of Sciences, 91*(26), 12760–12764. https://doi.org/10.1073/pnas.91.26.12760.
13. Goddijn, O. J. M., & Pen, J. (1995). Plants as bioreactors. *Trends in Biotechnology, 13*(9), 379–387. https://doi.org/10.1016/S0167-7799(00)88985-4.
14. Choi, J., & Lee, S. Y. (1997). Process analysis and economic evaluation for poly(3-hydroxybutyrate) production by fermentation. *Bioprocess Engineering, 17*(6), 335–342. https://doi.org/10.1007/s004490050394.
15. Jendrossek, D., & Handrick, R. (2002). Microbial degradation of polyhydroxyalkanoates. *Annual Review of Microbiology, 56*, 403–432. https://doi.org/10.1146/annurev.micro.56.012302.160838.
16. Zhang, C., Wang, C., Cao, G., Wang, D., & Ho, S. H. (2020). A sustainable solution to plastics pollution: An eco-friendly bioplastic film production from high-salt contained Spirulina sp. residues. *Journal of Hazardous Materials,* 121773. https://doi.org/10.1016/j.jhazmat.2019.121773.
17. Rahman, A., & Miller, C. D. (2017). Microalgae as a source of bioplastics. In A. Rahman & C. D. Miller (Eds.), *Algal Green Chemistry: Recent Progress in Biotechnology* (pp. 121–138). Elsevier. https://doi.org/10.1016/B978-0-444-63784-0.00006-0.
18. George, N., Debroy, A., Bhat, S., Singh, S., & Binda, S. (2022). Biowaste to bioplastics: An eco-friendly approach for a sustainable future. *Journal of Applied Biotechnology Reports, 8*(2), 221–233. https://doi.org/10.30491/JABR.2021.259403.1318.

19. Draper, J. (2013). *Bionanotechnology: Biological self-Assembly and Its Applications*. In B. H. A. Rehm (Ed.), Caister Academic Press.
20. Kai, D., & Loh, X. J. (2014). Polyhydroxyalkanoates: Chemical modifications toward biomedical applications. *ACS Sustainable Chemistry & Engineering, 2*(2), 106–119. https://doi.org/10.1021/sc400123z.
21. Moradali, M. F., & Rehm, B. H. A. (2020). Bacterial biopolymers: From pathogenesis to advanced materials. *Nature Reviews Microbiology, 18*(3), 195–210. https://doi.org/10.1038/s41579-019-0313-3.
22. Ko, S. H. (2017). Nanofluidic device for continuous multiparameter quality assurance of biologics. *Nature Nanotechnology, 12*(8), 804. https://doi.org/10.1038/nnano.2017.74.
23. Blank, L. M., Narancic, T., Mampel, J., Tiso, T., & O'Connor, K. (2020). Biotechnological upcycling of plastic waste and other non-conventional feedstocks in a circular economy. *Current Opinion in Biotechnology, 62*, 212–219. https://doi.org/10.1016/j.copbio.2019.11.011.
24. Batori, V., Åkesson, A., Zamani, A., Taherzadeh, M. J., & Horvath, I. S. (2018). Anaerobic degradation of bioplastics: A review. *Waste Management, 80*, 406–413. https://doi.org/10.1016/j.wasman.2018.09.040.
25. Bhatia, S. K., Otari, S. V., Jeon, J. M., Gurav, R., Choi, J. K., Bhatia, R. K., Pugazhendhi, A., Vinod, K., Banu, J. R., Yoon, J. J., Choi, K. Y., & Yang, Y. H. (2021). Biowaste-to-bioplastic (polyhydroxyalkanoates): Conversion technologies, strategies, challenges, and perspectives. *Bioresource Technology, 326*, 124733. https://doi.org/10.1016/j.biortech.2021.124733.

CHAPTER 7

Green Nanocomposites: Incorporating Biopolymers into Plastics for Enhanced Environmental Sustainability

DEEPA SHARMA, SMRITI BANSAL, and PURNIMA JAIN

Department of Chemistry, Netaji Subhas University of Technology (N.S.U.T), Azad Hind Fauj Marg, Dwarka, New Delhi, India

ABSTRACT

Advancements in the field of materials science and technological research have led to a significant increase in environmental concerns due to their large dependency on fossil fuel-based materials. This has caused a problem in limiting the natural resources for the future. Thus, highlighting a need for the development of sustainable and environmentally safe materials. Recent studies have focused on naturally derived biopolymers that show unique characteristics such as biodegradability and biocompatibility. In this regard, biopolymers have gained enormous attention from researchers as they can be synthesized using renewable resources and are abundantly available in nature. However, some modifications in the properties of biopolymers are essential for certain industrial uses such as environmental remediation and packaging applications. Thus, there is a necessity to explore the potential of biopolymers and their characteristic modifications through the incorporation of nanofillers or by producing nanocomposites with other polymeric plastics. Biopolymer-based nanocomposites can also be interchangeably termed green nanocomposites or biodegradable bioplastics depending on their specific properties and applications. This chapter specifically covers the importance of biodegradable nanocomposites/bioplastics and their synthesis techniques along with potential areas of applications. It also reviews the advantages

Sustainability in Polymer Technology and Plastics Engineering: Concepts, Strategies, and Opportunities. Tamara Tatrishvili, Neha Kanwar Rawat, Swati Gokul Talele, and A. K. Haghi (Eds.)

and limitations of different types of biopolymers as well as the strategies for the optimization of their nanocomposite properties. In conclusion, naturally derived biopolymers and their nanocomposites show the potential to lessen the environmental impact and contribute to sustainability.

7.1 INTRODUCTION

Polymeric materials, especially plastics, have become an inextricable part of our day-to-day life and are extensively used for various industrial applications, namely packaging, pharmaceuticals, biomedicine, and construction materials [1–4]. Plastics are fundamentally produced from fossil fuel-based polymers and are thus called petroleum-based plastics, including polyethylene terephthalate (PET), polypropylene (PP), polystyrene, polyvinyl chloride (PVC), high-density polyethylene (HDPE), and low-density polyethylene (LDPE), etc. [5–7]. They are lightweight synthetic polymers with a high strength-to-weight ratio, flexibility, chemical resistance, electrical insulation, longevity, low thermal conductivity, good moldability, and low production cost [5]. Despite these characteristics, the persistence of plastics in nature has become a vital concern for humans, wildlife, marine animals, and the sustainability of the environment due to their non-biodegradability, as represented in Figure 7.1 [8]. This results in the accumulation of enormous quantities of plastic waste products, especially plastic packaging, with a growth rate of 12% per year [1]. Geyer et al., in one of their studies, have estimated the accumulation of 12,000 Mt of plastic garbage (as shown in Figure 7.2) in landfills for disposal by the year 2050 [5, 7]. Thus, there is a need to address the usage and improper disposal of plastic waste in landfills and water bodies, which is leading to environmental deterioration.

Bioplastics are bioresource-based macromolecules that have quickly gained the keen interest of researchers. Essentially, bioplastic signifies an ensemble of materials that are biobased and biodegradable. The term bioplastic was primarily devised by the European Bioplastics, which is a European consortium employed as bioplastic suppliers [10]. They are believed to be ecologically and economically feasible to establish the bio-economy and shift reliance away from traditional petrochemical-based plastics [11]. In contrast to conventional plastics, bioplastics have several advantages, such as the capability to compost and degrade, resulting in a reduction in plastic waste generation and pollution [12]. The non-toxic and renewable characteristics of biopolymer-based plastics make them sustainable and safe for

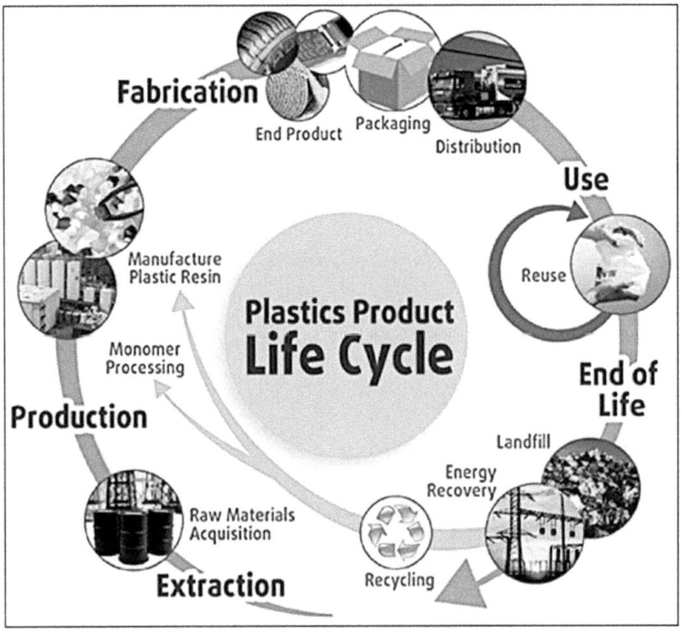

FIGURE 7.1 Common life cycle of plastic products both that can be recyclable and non-recyclable.

Source: Reprinted from Ref.[87]. https://creativecommons.org/licenses/by/

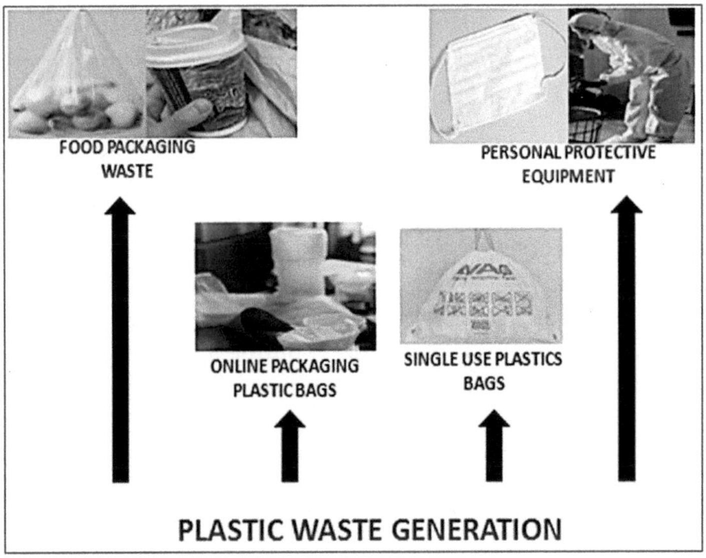

FIGURE 7.2 Sources of plastic waste generation.

both humans and the environment. Despite all these unique properties, they also have some limitations, which are shown in Figure 7.4. Bioplastics can be fabricated entirely or partially either from microbes like yeast or biomass sources like maize and sugarcane. Most bioplastics are biodegradable or compostable, but there can be bioplastics that are non-biodegradable in nature like petroleum-based plastics [13]. This chapter explicitly highlights the role of biodegradable bioplastics as a sustainable and ecological material for the development of green nanocomposites (Figure 7.3).

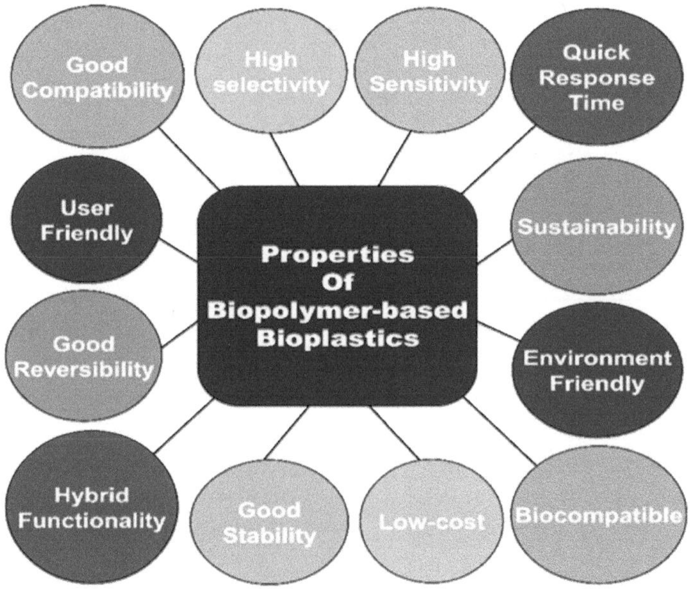

FIGURE 7.3 Properties of biopolymer-based bioplastics.

Biodegradable bioplastics are manufactured from natural or carbon-based materials such as proteins, fats, lipids, and polysaccharides [5]. According to Song et al. [15], some popular polysaccharides utilized to produce biopolymers are cellulose, starch, lignin, and chitin whereas, casein, and gluten as well as unsaturated fats and plant oils are natural lipids and proteins [15]. Biopolymer-based bioplastics possess unique properties such as biodegradability, biocompatibility, and sustainability over conventional fossil fuel-based plastics [16–18]. Cellulose is one of the most abundantly available plant-based biopolymers that originate in the cell walls of plants and are extracted using various chemical and mechanical methods [18]. Similarly, chitosan and alginate are biopolymers that can be derived from animal resources such as crab shells [19, 20]. Furthermore, the production of biopolymers from renewable sources such as rice husks, corn stalks, sugarcane, and rice stalks alters the dependency on fossil fuels and aids in the growth of a sustainable bio-economy. Thus, biopolymers are a great candidate to serve the sustainability of the environment as a greener solution to conventional plastics [21].

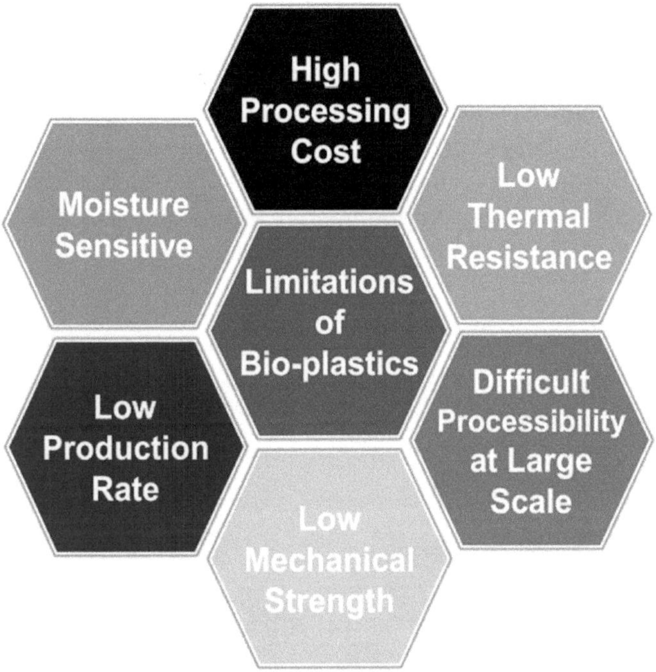

FIGURE 7.4 Some limitations associated with bioplastics.

The integration of biopolymers like cellulose, polyhydroxyalkanoates (PHA), chitosan, alginate, and polylactic acid (PLA) with nanoparticles produces environmentally friendly materials which are termed "green nanocomposites" [22]. It is a new area of materials that are extracted from renewable and sustainable resources including biopolymers, nanoparticles, and nano-fibers. Green nanocomposites comprise biopolymers like cellulose, starch, and chitin contributing to their renewability, biodegradability, and low toxicity [23]. Furthermore, green nanocomposites can be created using nanomaterials like nanoparticles, nanotubes, and graphene which are incorporated into a biopolymer matrix. Hence, biopolymer-based nanocomposites offer superior properties over conventional petroleum-based polymers having a high aspect ratio, increased surface area, improved mechanical strength, and special electrical and optical characteristics [24, 25]. Since they are synthesized from renewable and natural resources, they have lower carbon footprints. They also aid in diminishing the obstructive effects of conventional plastics on human health and the environment. Therefore, their manufacturing would not only help in decreasing the emission of greenhouse gases but also foster a circular economy as illustrated in Figure 7.5. Furthermore, the mechanical, optical, thermal, and barrier properties of biodegradable bioplastics can be tuned and enhanced by the reinforcement of nanoparticles or nano-fibers into the biopolymeric matrix as per desired applications [26]. Green nanocomposites have found a number of applications, including pharmaceuticals, electronics, packaging, energy storage products, and biomedical, contributing to a more sustainable future [27–29].

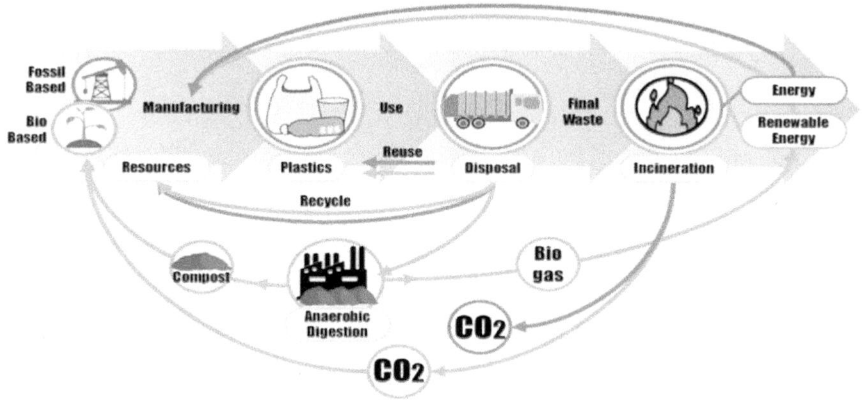

FIGURE 7.5 The main steps in the linear economy (straight arrows) and the additional steps introduced by the circular economy.

7.2 TYPES OF BIOPLASTICS AND THEIR CHARACTERISTICS

In this chapter, the main emphasis has been given to biodegradable bioplastics that are derived from naturally available plant and animal sources and are thus known as biobased and biodegradable bioplastics. Green nanocomposites comprise bio-resourced biopolymers which are complex substances with unique characteristic properties such as high mechanical strength, enhanced optical and electrical properties, biocompatibility, and biodegradability as a result of synergistic effects between the biopolymer matrix and the nanomaterials. The following section provides brief details about some of the popular biodegradable bioplastics namely, Polylactic acid (PLA), Thermoplastic starch (TPS), Polybutylene succinate (PBS), and Polyhydroxyalkanoates (PHA).

7.2.1 *POLYLACTIC ACID (PLA)*

PLA is an aliphatic polyester synthesized from lactic acid obtained from agricultural biomass, for example, cotton, maize, coffee waste, and food waste using polymerization techniques [30]. At present, PLA is the second most abundant manufactured bioplastic after starch-based bioplastics. It is an excellent substitute for traditional fossil fuel-based plastics because of its high abundance as a renewable resource, high mechanical strength, high transparency, and increased modulus with a percentage of elongation [31, 32]. Polylactic acid can be used in numerous applications such as agricultural films, packaging, and geotextiles, as the binding strands for cellulosic nonwovens, and lately, it is one of the primary thermoplastics for three-dimensional printing exploiting by the fused deposition modeling technique [31]. Tengrang et al. [33] and his coworkers compared the transparency of PLA by other bioplastics including Master-Bi, and PBS, and found that the translucency of PLA was the highest among all [33]. However, PLA has a strong flammability due to its characteristic structure and chemistry, with a 19–21% limiting oxygen index (LOI). Additionally, it readily starts to melt pouring during ignition which delimits its practical applications in various industries including textiles, electronics, packaging, and agriculture [34–37]. The mechanical strength of PLA is not much affected after recycling and therefore, recycled PLA could be easily used in the synthesis of PLA itself [38].

7.2.2 *THERMOPLASTIC STARCH (TPS)*

TPS is a renewable and biodegradable biopolymer derived from starch that is found in plants as a naturally occurring polysaccharide [39]. It is a type of bioplastic that is gaining demand as a greener alternative to petroleum-based plastics [40]. TPS is formed by plasticizing starch pellets with water and then heating the mixture to form a viscous melt. Plasticizers such as sorbitol or glycerol are often added to enhance the flexibility and processability of the material [41]. The resultant TPS can be processed using conventional polymer processing techniques such as extrusion, blow molding, and injection molding, allowing it to be used in a wide range of applications [42]. The major advantage of TPS is its biodegradability. Unlike traditional petroleum-based plastics, TPS can be broken down by microorganisms in the environment, leading to the production of harmless byproducts such as water, carbon dioxide, and biomass. This makes TPS an attractive option for applications where biodegradability is important, such as food packaging and agricultural films. In addition to its biodegradability, TPS has other favorable properties such as low toxicity, good barrier properties, and excellent film-forming ability. On the other hand, TPS does have significant disadvantages, such as limited water resistance, poor mechanical strength, and susceptibility to high temperatures and humidity [40]. Researchers are aiming to address these restrictions by adding reinforcing materials to TPS, such as natural fibers, nanoparticles, or biodegradable polymers. TPS has been exploited for different types of applications such as food packaging, disposable trash bags, eating utensils, compostable films, and bags for agriculture and retail [40]. Starch is a natural polymer that is commonly used in food packaging due to its availability, biodegradability, and low cost. However, it has poor mechanical features and moisture barrier compared to traditional non-recyclable plastic films. Nanocomposite concepts have emerged as a promising approach to improve the mechanical and barrier properties of starch-based materials, such as by incorporating nanofillers like waxy starch nanocrystals or kaolin, or by adding nanoparticles like ZnO or chitosan to the starch matrix [43].

7.2.3 POLYBUTYLENE SUCCINATE (PBS)

PBS is a biobased and biodegradable aliphatic polyester. It is a semi-crystalline polymer synthesized by the direct esterification reaction between succinic acid and 1,4-butanediol [44, 45]. It is the simplest and most direct method of synthesis for PBS. It has very good biodegradability, balanced mechanical properties, and thermoplastic processability. It is also a widely recognized biopolymer due to its excellent processing and superior chemical and thermal stability [46]. PBS can be produced using renewable resources such as corn, sugarcane, and cassava [47]. It is a white and crystalline thermoplastic with high biodegradability and exceptional mechanical properties [48]. However, there are substantial drawbacks to PBS that limit its practical pertinency, including its high cost, poor gas barrier properties, and smoothness [49]. The high cost of PBS can be reduced by reinforcing it with nanofibers obtained from natural resources [46]. It is a bioplastic polymer with a density of 1.25 g/cm^3, a glass transition temperature (Tg) range of about –45 to −10°C, and a melting temperature (Tm) in the range of 90–120°C [50]. Randomly oriented PBS samples have tensile strengths of 30–35 MPa, which is equivalent to polyolefins [51]. It is also a flexible polyester and depending on its crystallinity, it has a Young's modulus range between 300 and 500 MPa [51]. Aliotta et al. [47] and coworkers have reported the manufacturing of a highly biodegradable and bio-based PLA/PBS product designed for commercial use for thermoforming, injection molding, and calendering [47]. Thus, PBS could be a promising bioplastic that can be used for practical applications by chemically modifying its properties via the reinforcement of nanomaterials or other biopolymers.

7.2.4 POLYHYDROXYALKANOATES (PHA)

PHAs are biodegradable and optically active polymers produced by numerous bacteria in response to stress during a deficiency of inorganic nutrients, including nitrogen, phosphate, or oxygen deficiency, while carbon is abundant [53–56]. PHA is formed by microbial fermentation and then extracted through microbial cell lysis. PHAs are biocompatible, biodegradable, non-toxic polyesters made of (R)-3-hydroxy alkanoic acids [57] that have thermoplastic characteristics comparable to petrochemical plastics [54, 58]. PHAs may also be thermally transformed into PLA and are derived from renewable raw materials (such as fatty acids, maltose, and glucose) by biological conversion through the activity of various microorganisms [59]. They have various applications from packaging to textiles and medical implants [60]. Essentially, PHAs have not been widely used as bioplastics due to their high production and recovery costs [61].

7.3 SYNTHESIS TECHNIQUES FOR BIODEGRADABLE BIOPLASTICS

Biodegradable bioplastics can be produced from renewable sources like rice husk, corn stalks, starch, maize, and other biological resources. Thus, depending on their source and processibility with nanomaterials and nanofibers, they are defined as biopolymer-based nanocomposites or green nanocomposites. Various chemical and biological treatment methods are used for the extraction of raw materials from these renewable sources, which are further reinforced by the nanomaterials or blended with other polymers to form polymer composites. Subsequently, the resultant bioplastic can be processed into a number of products such as fibers and films. The inculcation of biodegradable bioplastics in place of standard plastics contributes to a sustainable environment. The characteristic properties of bioplastics also depend on the type of synthesis technique used during their manufacturing. Some of the common manufacturing processes of bioplastics have been discussed here.

7.3.1 MELT INTERCALATION METHOD

It is also called the melt mixing method. It is highly useful for creating thermoplastic bio-renewable nanocomposites. This approach can be used to synthesize polylactic acid (PLA), polyurethane, polyethylene terephthalate (PETP), isotactic polypropylene, styrene, and acrylonitrile (styrene-acrylonitrile), polybutadiene styrene, and polycarbonate polymers resulting in the intercalation of nanofillers within the layered matrices of polymers [62]. In this method, a high molecular weight polymeric matrix melt is created by applying high temperature, and then nanofillers are added and mixed homogeneously to produce a nanocomposite with a uniform structure [63]. However, the suitability of nanofillers with the polymeric matrix, surface functionalization, and processing conditions synergistically play a crucial role in producing a uniform dispersion. Depending on the degree of dispersion, and diffusion of the molten polymer matrix into the nanofiller, either an intercalated or non-intercalated layered nanocomposite is obtained [64]. The melt intercalation method is regarded as a modest and straightforward synthesis technique. Furthermore, this intercalation approach is considered to be versatile and ecologically feasible since it involves the melting of polymers which diminish the requirement of any solvent for the production of nanocomposite. Additionally, it is also found to be very economical and compatible with industrial processes such as injection molding, extrusion processing, and blending methods [65]. The key disadvantage of the melt intercalation method is its high processing temperature which may alter or impair the surface modifications of polymer and nanofillers.

*7.3.2 **IN SITU INTERCALATIVE POLYMERIZATION TECHNIQUE***

It is one of the most beneficial processing techniques for the manufacturing of bio-renewable bioplastics. Under this method, the nanofiller is dispersed in a monomer solution, followed by polymerization to form the nanocomposite. This approach can be employed for both thermoplastic and thermosetting polymers. In this process, the nanofillers are swelled within the low molecular weight monomers or liquid monomers contributing to the polymerization of polymer within the intercalated layer [66]. During the polymerization process, the initiation of polymeric reaction can be facilitated either by adding an organic initiator or by exposing the monomers to heat, which results in cationic substitution within the internal layers of nanofillers before the swelling of monomers occurs. The *in-situ* polymerization technique is found to be more operational than melt-intercalation or exfoliation adsorption processes. This approach also allows for the mixing of reactive and catalytically active organometallic species into some of the layers of nanofillers [67]. These active species might react with the monomer matrix, causing the polymerization process to occur inside the lamellar porosity. This technique was initially used to create a nylon-montmorillonite nanocomposite; afterward, it was utilized to create various thermoplastics and thermoset nanocomposites. Due to the lack of severe thermodynamic criteria, *in-situ* intercalative polymerization is a cost-effective, feasible, and acceptable process for thermoset-clay nanocomposites. However, this process has significant drawbacks, including a sluggish rate of reaction, and reliance on subsequent procedures such as nanofiller swelling and monomer diffusion rate in the nanofiller layer [68]. Sharma et al. [69] prepared titanium dioxide-encapsulated nano-cellulose grafted-polyaniline (TiO_2-NC-g-PANI) using the *in-situ* intercalation polymerization method. TiO_2-NC-g-PANI has been used as an electrochemical sensor for the detection of pesticides in food and water samples [69]. Wang et al. [70] synthesized a chitosan-g-poly(acrylic acid)/montmorillonite (CTS-g-PAA/MMT) nanocomposites using the same technique for the development of an effective adsorbent that can be used for the elimination of cationic dye-methylene blue from water samples [70].

7.3.3 EXFOLIATION ADSORPTION

In this processing technique, the nanofiller components are first disseminated into a solvent like water, toluene, or chloroform in which the polymeric or monomeric/prepolymer matrix is soluble and subsequently mixed with the polymeric solution [71]. The intercalation and displacement process takes place between the nanofiller interlayer and that of the polymeric matrix and the solvent [72]. Gradually the polymer matrix adsorbs the nanofillers into the interlayer resulting in the formation of layered nanocomposites which can be recovered by solvent evaporation or precipitation [73]. Although this method is easier and more modest than other techniques, there are some major disadvantages associated with this process. From an industrial perspective, this technique proves to be both extensive and expensive as it requires an appropriate solvent in which the polymer is soluble. Thus, this might be an expensive approach for large-scale production as it adds to the high cost of organic solvents and their cleanup. Furthermore, the significant capacity of solvents used and their subsequent disposal makes this approach hazardous to the environment [74]. Some of the biodegradable nanocomposites produced using exfoliation adsorption methods are poly (L-lactic acid)/layered silicate nanocomposite [75] and heat-sealable cellulose/polyvinyl alcohol-based packaging film [76].

7.3.4 TEMPLATE SYNTHESIS

Another name for template synthesis is the sol-gel technology. This technique is distinctive from other synthesis approaches involving the production of polymer in the presence of nanofillers whereas, in this approach, the nanofillers are synthesized with respect to the polymer matrix which acts as a template for nanomaterial or nanofiller production. Essentially, the polymeric matrix facilitates the nucleation and development of the host nanofillers and traps them exclusively within the layers as they propagate [77–79]. This method modifies the product shape by changing the nucleation and growth of the crystal during nano-layer production. This production route is most suitable for multilayer nanofiller dispersion in polymeric matrices at the nanoscale level [72]. Template synthesis is often employed to construct hydroxide-based double-layer nanocomposites, but this is less supple for the layered silicates [80] since the fabrication process typically involves high temperatures, which is unfavorable for the polymeric components. Sometimes, agglomeration might also occur during the development of nanofillers [81]. Afzal et al. [82] proposed a facile sol-gel process for the synthesis of highly active TiO_2-chitosan through controlled acid crystallization that was employed for photocatalytic elimination of methylene orange (99%) within 180 minutes under UV exposure [82].

7.4 APPLICATIONS OF BIODEGRADABLE BIOPLASTICS

Bioplastics are manufactured from bio-resourced macromolecules such as maize cellulose, sugarcane, and starch, which reduces dependency on non-renewable fossil fuel-based resources and also decreases the environmental impact of typical plastics. They can naturally degrade into water, carbon dioxide (CO_2), and other natural chemicals. They may be utilized in a number of applications, covering packaging, agriculture, and biomedicine. Biodegradable bioplastics have the potential (as shown in the Figure 7.5) to minimize plastic waste and environmental pollution, unlike conventional plastics which linger within the environment for hundreds of years. Following are some vital areas of applications involving biopolymer-based biodegradable bioplastics:

7.4.1 ENVIRONMENTAL REMEDIATION

Biopolymer-based bioplastics can be used for environmental remediation, such as removing pollutants from water and soil. Basically, they are used to develop adsorbents that can adsorb contaminants corresponding to heavy metals, organic dyes, and pollutants. Bioplastics can also be used for bioremediation, where microorganisms can break down pollutants into simpler by-products that are not harmful to the environment. In research conducted by Rahaman et al. [83] and his coworkers, they produced a composite adsorbent by making some chemical modifications in eco-friendly and biodegradable materials such as cellulose and chitosan and optimized its conditions for adsorption of heavy metal ions including Cr, Pb, and Cd obtained from polluted water wastes [83].

7.4.2 PACKAGING

Biodegradable bioplastics are derived from renewable bio-resourced macromolecules and are a potential and sustainable alternative to traditional plastics due to their biodegradable and compostable nature. They could be used for various packaging applications such as bags, wrappers, and other containers, reducing reliance on fossil fuels. The use of bioplastics in packaging applications aids in the reduction of plastic waste and pollution of the environment as they can be broken down into natural substances under specific environmental conditions or by the action of microbes. In a study, the functional properties of photochemically modified starch solutions as biodegradable food packaging have been studied. Microbial polymers such as PHA and PLA, plant-based polymers (cellulose, hemicellulose, starch, and lignin), and protein-based polymers like gelatin, keratin, and soy protein are some of the most commonly used components as smart packaging materials. Reinforcements of bio-based ingredients into the bioplastics such as plant essential oils and similar natural additives improve the antibacterial, oxygen-barrier, and anti-fungal properties of the packaging materials [61].

7.4.3 *BIOMEDICAL APPLICATIONS*

Biodegradable bioplastics are excessively exploited for biomedical applications. Drug delivery, tissue engineering, scaffolding, and medical implants are some examples of biomedical applications that involve the use of biodegradable bioplastics. Bio-resourced bioplastics can be designed to have specific properties, such as biodegradability, biocompatibility, more physiological repair, possible tissue growth, and mechanical strength, making them suitable for many medical applications [84]. They also aid in reducing the risk of infection and improve patient outcomes as they can be internally broken down by the body's natural processes, such as sutures which are used to hold the tissues after major injury or surgery. The PLA, poly (D, L-lactic-co-glycolic acid), and poly(glycolic acid) have been allowed to be used in health-related activities by the Food and Drug Administration. PLA is regarded as one of the strongest poly(a-hydroxyl acids) [85], and PLA-derived plates and screws have been observed to stimulate bone repair and become entirely absorbed after four years of implantation in dogs [86].

7.4.4 *SENSING APPLICATION*

Biopolymer-based plastics can be used for sensing applications, such as chemical sensors, and biosensors. Biosensors can be manufactured with specific properties, including high sensitivity, stability, and specificity/selectivity making them suitable for a number of sensing applications. Since biodegradable bioplastics are derived from biopolymers, they can be modified and integrated with electronic components for the development of wearable and smart sensing devices. Koelle et al. reported a DIY approach for developing functional electronic sensors and circuits that were produced using three conductive bioplastics and their formulations for smart electronics and sensors [52].

7.4.5 *AGRICULTURAL APPLICATIONS*

Agricultural applications of biopolymers include seed coatings, plant pots, and seed coatings [14]. Biodegradable bioplastics are derived from renewable sources such as cellulose, starch, and chitosan, reducing the reliance on non-renewable resources. Bioplastics can also facilitate improving soil health, preventing soil erosion, and reducing waste thereby contributing to sustainable agricultural practices. In a recent study, it was observed that, unlike traditional plastics, the accumulation of bioplastic in landfills does not distress the germination of seeds, however, it may cause a delay in the sprouting of seeds [9]. Thus, more research needs to be carried out to study the detrimental as well as essential effects of bioplastics on the environment and agriculture.

7.5 CONCLUSION AND FUTURE PERSPECTIVES

In conclusion, biodegradable bioplastics can be classified as green nanocomposites which present a promising solution to the rapidly growing problems associated with plastic waste management and contamination of the environment. Biodegradable bioplastics have great potential to substitute traditional plastic materials and contribute to environmental sustainability due to their biodegradable nature and renewable resources. However, a lot of research is still required to eliminate the shortcomings linked with biopolymers in order to ideally substitute petroleum-based plastics. The processability of biopolymers, their stability, large-scale production, and associated cost are some major shortcomings that need to be researched and monitored. Furthermore, the several characteristic properties of biopolymer-based plastics can be enhanced and tuned by the reinforcement of nanofillers such as metal nanoparticles or carbon-based nanofillers to achieve improved mechanical strength, optical, electrical, thermal, and other barrier properties in accordance with the desired applications. Thus, biopolymer-based nanocomposites being environment-friendly and sustainable show great potential to lessen the environmental consequences of numerous industries, including packaging, biomedical, and construction. Additionally, the advancement and application of biopolymers-based materials may open up new opportunities for adopting and using bio-based resources and reassure the circular economy.

Therefore, future studies based on bioplastics should focus on enhancing their performance and operationalization, emphasizing improvements in their mechanical strength, longevity, and stability. Additionally, optimizing the uniform dispersion of nanofillers into biopolymeric matrices is a crucial aspect. Furthermore, when designing synthesis methods for bioplastics, more effort should be focused on creating affordable and scalable synthesis techniques to accelerate their commercialization. Depending on the developed synthesis system and other crucial factors associated with their performance would make biopolymer-based nanocomposites competitive with traditional petroleum-based plastics in terms of performance, availability, and affordability for long-term success. However, considering the increasing demand and business of plastics, completely replacing them is a bit challenging. Yet, the growing awareness and strong desire for sustainable alternatives for the environment create a demanding need for accelerating research on green nanocomposites for their widespread adoption in the future.

ACKNOWLEDGMENT

The authors are wholeheartedly thankful to the Netaji Subhas University of Technology, for providing the lab facilities and financial aid.

KEYWORDS

- **bioplastics**
- **biopolymers**
- **environmental sustainability**
- **green nanocomposites**
- **low-density polyethylene**
- **polyhydroxyalkanoates**
- **polystyrene, polyvinylchloride**
- **polylactic acid**

REFERENCES

1. Muniyasamy, S., Seydibeyoğlu, Ö., Thulasinathan, B., & Arun, A. (2018). Biopolymer Synthesis and Biodegradation. In *Sustainable Biotechnology – Enzymatic Resources of Renewable Energy* (pp. 399–421). Springer International Publishing. https://doi.org/10.1007/978-3-319-95480-6_15.
2. Han, J. -W., Ruiz-Garcia, L., Qian, J. -P., & Yang, X. -T. (2018). Food Packaging: A Comprehensive Review and Future Trends. *Comprehensive Reviews in Food Science and Food Safety, 17*(4), 860–877. https://doi.org/10.1111/1541-4337.12343.

3. Babaremu, K., Oladijo, O. P., & Akinlabi, E. (2023). Biopolymers: A suitable replacement for plastics in product packaging. *Advanced Industrial and Engineering Polymer Research. 6*(4), 333–340. https://doi.org/10.1016/j.aiepr.2023.01.001.
4. Saygin, H., & Baysal, A. (2020). Similarities and Discrepancies Between Bio-Based and Conventional Submicron-Sized Plastics: In Relation to Clinically Important Bacteria. *Bulletin of Environmental Contamination and Toxicology, 105*(1), 26–35. https://doi.org/10.1007/s00128-020-02908-8.
5. Shafqat, A., Tahir, A., Mahmood, A., Tabinda, A. B., Yasar, A., & Pugazhendhi, A. (2020). A review on environmental significance carbon foot prints of starch-based bio-plastic: A substitute of conventional plastics. *Biocatalysis and Agricultural Biotechnology, 27*, 101540. https://doi.org/10.1016/j.bcab.2020.101540.
6. Ismail, N. A., (2016). Synthesis and characterization of biodegradable starch-based bioplastics. *Materials Science Forum, 846*, 673–678. https://doi.org/10.4028/www.scientific.net/MSF.846.673.
7. Geyer, R., Jambeck, J. R., & Law, K. L. (2017). Production, use, and fate of all plastics ever made. *Science Advances, 3*(7), e1700782. https://doi.org/10.1126/sciadv.1700782.
8. Worm, B., Lotze, H. K., Jubinville, I., Wilcox, C., & Jambeck, J. (2017). Plastic as a Persistent Marine Pollutant. *Annual Review of Environment and Resources, 42*(1), 1–26. https://doi.org/10.1146/annurev-environ-102016-060700.
9. Liwarska-Bizukojc, E. (2021). Effect of (bio)plastics on soil environment: A review. *Science of The Total Environment, 795*, 148889. https://doi.org/10.1016/j.scitotenv.2021.148889.
10. Sid, S., Mor, R. S., Kishore, A., & Sharanagat, V. S. (2021). Bio-sourced polymers as alternatives to conventional food packaging materials: A review. *Trends in Food Science & Technology, 115*, 87–104. https://doi.org/10.1016/j.tifs.2021.06.026.
11. Nanda, S., Patra, B. R., Patel, R., Bakos, J., & Dalai, A. K. (2022). Innovations in applications and prospects of bioplastics and biopolymers: A review. *Environmental Chemistry Letters, 20*(1), 379–395. https://doi.org/10.1007/s10311-021-01334-4.
12. Rendón-Villalobos, R., Ortíz-Sánchez, A., Tovar-Sánchez, E., & Flores-Huicochea, E. (2016). The Role of Biopolymers in Obtaining Environmentally Friendly Materials. In *Composites from Renewable and Sustainable Materials*. InTech. https://doi.org/10.5772/65265.
13. Ashter, S. A. (2016). New Developments. In *Introduction to Bioplastics Engineering* (pp. 251–274). Elsevier. https://doi.org/10.1016/B978-0-323-39396-6.00010-5.
14. Raj, S. N., Lavanya, S. N., Sudisha, J., & Shetty, H. S. (2011). Applications of Biopolymers in Agriculture with Special Reference to Role of Plant Derived Biopolymers in Crop Protection. In *Biopolymers* (pp. 459–481). John Wiley & Sons, Inc. https://doi.org/10.1002/9781118164792.ch16.
15. Song, J. H., Murphy, R. J., Narayan, R., & Davies, G. B. H. (2009). Biodegradable and compostable alternatives to conventional plastics. *Philosophical Transactions of the Royal Society B: Biological Sciences, 364*(1526), 2127–2139. https://doi.org/10.1098/rstb.2008.0289.
16. Thakur, S., Chaudhary, J., Sharma, B., Verma, A., Tamulevicius, S., & Thakur, V. K. (2018). Sustainability of bioplastics: Opportunities and challenges. *Current Opinion in Green and Sustainable Chemistry, 13*, 68–75. https://doi.org/10.1016/j.cogsc.2018.04.013.

17. Zhu, Y., Romain, C., & Williams, C. K. (2016). Sustainable polymers from renewable resources. *Nature, 540*(7633), 354–362. https://doi.org/10.1038/nature21001.
18. Pandey, J. K., Ahn, S. H., Lee, C. S., Mohanty, A. K., & Misra, M. (2010). Recent Advances in the Application of Natural Fiber Based Composites. *Macromolecular Materials and Engineering, 295*(11), 975–989. https://doi.org/10.1002/mame.201000095.
19. Wang, H., Liu, J., Fan, X., Ren, J., Liu, Q., & Kong, B. (2022). Fabrication, characterization, and application of green crosslinked sodium alginate hydrogel films by natural crab-shell powders to achieve drug sustained release. *LWT, 171*, 114147. https://doi.org/10.1016/J.LWT.2022.114147.
20. Hajiali, F., et al. (2022). Extraction of Chitin from Green Crab Shells by Mechanochemistry and Aging. *ACS Sustainable Chemistry & Engineering, 10*(34), 11348–11357. https://doi.org/10.1021/acssuschemeng.2c02966.
21. Moshood, T. D., Nawanir, G., Mahmud, F., Mohamad, F., Ahmad, M. H., & AbdulGhani, A. (2022). Sustainability of biodegradable plastics: New problem or solution to solve the global plastic pollution? *Current Research in Green and Sustainable Chemistry, 5*, 100273. https://doi.org/10.1016/j.crgsc.2022.100273.
22. El-Sayed, S. M., & Youssef, A. M. (2023). Eco-friendly biodegradable nanocomposite materials and their recent use in food packaging applications: a review. *Sustainable Food Technology, 1*(2), 215–227. https://doi.org/10.1039/D2FB00021K.
23. Dassanayake, R. S., Acharya, S., & Abidi, N. (2019). Biopolymer-Based Materials from Polysaccharides: Properties, Processing, Characterization and Sorption Applications. In *Advanced Sorption Process Applications*. IntechOpen. https://doi.org/10.5772/intechopen.80898.
24. Aldas, M., Ferri, J. M., Lopez-Martinez, J., Samper, M. D., & Arrieta, M. P. (2020). Effect of pine resin derivatives on the structural, thermal, and mechanical properties of Mater-Bi type bioplastic. *Journal of Applied Polymer Science, 137*(4), 48236. https://doi.org/10.1002/app.48236.
25. Xiong, R., Grant, A. M., Ma, R., Zhang, S., & Tsukruk, V. V. (2018). Naturally-derived biopolymer nanocomposites: Interfacial design, properties and emerging applications. *Materials Science and Engineering: R: Reports, 125*, 1–41. https://doi.org/10.1016/j.mser.2018.01.002.
26. Dantas de Oliveira, A., & Gonçalves Beatrice, C. A. (2019). Polymer Nanocomposites with Different Types of Nanofiller. In *Nanocomposites – Recent Evolutions*. IntechOpen. https://doi.org/10.5772/intechopen.81329.
27. Martău, G. A., Mihai, M., & Vodnar, D. C. (2019). The Use of Chitosan, Alginate, and Pectin in the Biomedical and Food Sector—Biocompatibility, Bioadhesiveness, and Biodegradability. *Polymers, 11*(11), 1837. https://doi.org/10.3390/polym11111837.
28. Gong, J., Chen, X., & Tang, T. (2019). Recent progress in controlled carbonization of (waste) polymers. *Progress in Polymer Science, 94*, 1–32. https://doi.org/10.1016/j.progpolymsci.2019.04.001.
29. González-Henríquez, C. M., Sarabia-Vallejos, M. A., & Rodriguez-Hernandez, J. (2019). Polymers for additive manufacturing and 4D-printing: Materials, methodologies, and biomedical applications. *Progress in Polymer Science, 94*, 57–116. https://doi.org/10.1016/j.progpolymsci.2019.03.001.
30. Nasrollahzadeh, M. (2021). *Biopolymer-Based Metal Nanoparticle Chemistry for Sustainable Applications*. Elsevier. 1st Edition – March 5, 2021; eBook ISBN: 9780128221235.

31. Brodin, M., Vallejos, M., Opedal, M. T., Area, M. C., & Chinga-Carrasco, G. (2017). Lignocellulosics as sustainable resources for production of bioplastics – A review. *Journal of Cleaner Production, 162*, 646–664. https://doi.org/10.1016/j.jclepro.2017.05.209.
32. Kalia, S. (2011). *Biopolymers: Biomedical and Environmental Applications*. John Wiley & Sons.
33. Tengrang, S., Leabwan, N., Loylerd, K., Enmak, P., Changprasert, S., & Korppiboon, A. (2017). Observations of the effect of different bioplastic packaging on prolonging the storage life of rambutan and biodegradability of the packaging. *Acta Horticulturae, 1186*, 177–184. https://doi.org/10.17660/ActaHortic.2017.1186.27.
34. Hamad, K., Kaseem, M., Yang, H. W., Deri, F., & Ko, Y. G. (2015). Properties and medical applications of polylactic acid: A review. *Express Polymer Letters, 9*(5), 435–455. https://doi.org/10.3144/cxprcsspolymlctt.2015.42.
35. Chow, W. S., Teoh, E. L., & Karger-Kocsis, J. (2018). Flame retarded poly (lactic acid): A review. *Express Polymer Letters, 12*(5), 396–417. https://doi.org/10.3144/expresspolymlett.2018.34.
36. Long, L., et al. (2017). Effects of bridged DOPO derivatives on the thermal stability and flame-retardant properties of poly (lactic acid). *Polymer Degradation and Stability, 139*, 55–66. https://doi.org/10.1016/j.polymdegradstab.2017.03.016.
37. Zhang, S., et al. (2018). Intercalation of phosphotungstic acid into layered double hydroxides by reconstruction method and its application in intumescent flame-retardant poly (lactic acid) composites. *Polymer Degradation and Stability, 147*, 142–150. https://doi.org/10.1016/j.polymdegradstab.2017.12.004.
38. Żenkiewicz, M., Richert, J., Rytlewski, P., Moraczewski, K., Stepczyńska, M., & Karasiewicz, T. (2009). Characterization of multi-extruded poly (lactic acid). *Polymer Testing, 28*(4), 412–418. https://doi.org/10.1016/j.polymertesting.2009.01.012.
39. Torres, F. G., Troncoso, O. P., Torres, C., Díaz, D. A., & Amaya, E. (2011). Biodegradability and mechanical properties of starch films from Andean crops. *International Journal of Biological Macromolecules, 48*(4), 603–606. https://doi.org/10.1016/j.ijbiomac.2011.01.026.
40. Jayarathna, S., Andersson, M., & Andersson, R. (2022). Recent Advances in Starch-Based Blends and Composites for Bioplastics Applications. *Polymers, 14*(21), 4557. https://doi.org/10.3390/polym14214557.
41. Nasseri, R., & Mohammadi, N. (2014). Starch-based nanocomposites: A comparative performance study of cellulose whiskers and starch nanoparticles. *Carbohydrate Polymers, 106*, 432–439. https://doi.org/10.1016/j.carbpol.2014.01.029.
42. Curvelo, A. (2001). Thermoplastic starch–cellulosic fibers composites: preliminary results. *Carbohydrate Polymers, 45*(2), 183–188. https://doi.org/10.1016/S0144-8617(00)00314-3.
43. Jafarzadeh, S., et al. (2023). Green synthesis of nanomaterials for smart biopolymer packaging: challenges and outlooks. *Journal of Nanostructure in Chemistry, 14*, 113–136. https://doi.org/10.1007/s40097-023-00527-3.
44. Rafiqah, S. A., et al. (2021). A Review on Properties and Application of Bio-Based Poly (Butylene Succinate). *Polymers, 13*(9), 1436. https://doi.org/10.3390/polym13091436.
45. Barrino, F., De La Rosa-Ramírez, H., Schiraldi, C., López-Martínez, J., & Samper, M. D. (2023). Preparation and Characterization of New Bioplastics Based on Polybutylene Succinate (PBS). *Polymers, 15*(5), 1212. https://doi.org/10.3390/polym15051212.

46. Mochane, M. J., Magagula, S. I., Sefadi, J. S., & Mokhena, T. C. (2021). A Review on Green Composites Based on Natural Fiber-Reinforced Polybutylene Succinate (PBS). *Polymers, 13*(8), 1200. https://doi.org/10.3390/polym13081200.
47. Aliotta, L., Seggiani, M., Lazzeri, A., Gigante, V., & Cinelli, P. (2022). A Brief Review of Poly (Butylene Succinate) (PBS) and Its Main Copolymers: Synthesis, Blends, Composites, Biodegradability, and Applications. *Polymers, 14*(4), 844. https://doi.org/10.3390/polym14040844.
48. Ali, F. B., & Mohan, R. (2009). Thermal, mechanical, and rheological properties of biodegradable polybutylene succinate/carbon nanotubes nanocomposites. *Polymer Composites, 525*(1–2), 141–149. https://doi.org/10.1002/pc.20913.
49. Bin, T., Qu, J., Liu, L., Feng, Y., Hu, S., & Yin, X. (2011). Non-isothermal crystallization kinetics and dynamic mechanical thermal properties of poly (butylene succinate) composites reinforced with cotton stalk bast fibers. *Thermochimica Acta, 525*(1–2), 141–149. https://doi.org/10.1016/j.tca.2011.08.003.
50. Debuissy, T., Pollet, E., & Avérous, L. (2017). Synthesis and characterization of biobased poly (butylene succinate-ran-butylene adipate). Analysis of the composition-dependent physicochemical properties. *European Polymer Journal, 87*, 84–98. https://doi.org/10.1016/j.eurpolymj.2016.12.012.
51. Xu, J., & Guo, B.-H. (2010). Poly (butylene succinate) and its copolymers: Research, development and industrialization. *Biotechnology Journal, 5*(11), 1149–1163. https://doi.org/10.1002/biot.201000136.
52. Koelle, M., Nicolae, M., Nittala, A. S., Teyssier, M., & Steimle, J. (2022). Prototyping Soft Devices with Interactive Bioplastics. In *The 35th Annual ACM Symposium on User Interface Software and Technology* (pp. 1–16). ACM. https://doi.org/10.1145/3526113.3545623.
53. Steinbüchel, A., & Lütke-Eversloh, T. (2003). Metabolic engineering and pathway construction for biotechnological production of relevant polyhydroxyalkanoates in microorganisms. *Biochemical Engineering Journal, 16*(2), 81–96. https://doi.org/10.1016/S1369-703X(03)00036-6.
54. Verlinden, R. A. J., Hill, D. J., Kenward, M. A., Williams, C. D., & Radecka, I. (2007). Bacterial synthesis of biodegradable polyhydroxyalkanoates. *Journal of Applied Microbiology, 102*(6), 1437–1449. https://doi.org/10.1111/j.1365-2672.2007.03335.x.
55. Sudesh, K., Abe, H., & Doi, Y. (2000). Synthesis, structure and properties of polyhydroxyalkanoates: biological polyesters. *Progress in Polymer Science, 25*(10), 1503–1555. https://doi.org/10.1016/S0079-6700(00)00035-6.
56. Ray, S., & Kalia, V. C. (2017). Polyhydroxyalkanoate Production and Degradation Patterns in Bacillus Species. *Indian Journal of Microbiology, 57*(4), 387–392. https://doi.org/10.1007/s12088-017-0676-y.
57. Kenny, S. T., et al. (2008). Up-Cycling of PET (Polyethylene Terephthalate) to the Biodegradable Plastic PHA (Polyhydroxyalkanoate). *Environmental Science & Technology, 42*(20), 7696–7701. https://doi.org/10.1021/es801010e.
58. Możejko-Ciesielska, J., & Kiewisz, R. (2016). Bacterial polyhydroxyalkanoates: Still fabulous? *Microbiological Research, 192*, 271–282. https://doi.org/10.1016/j.micres.2016.07.010.
59. Kawaguchi, H., Hasunuma, T., Ogino, C., & Kondo, A. (2016). Bioprocessing of bio-based chemicals produced from lignocellulosic feedstocks. *Current Opinion in Biotechnology, 42*, 30–39. https://doi.org/10.1016/j.copbio.2016.02.031.

60. Keenan, T. M., Nakas, J. P., & Tanenbaum, S. W. (2006). Polyhydroxyalkanoate copolymers from forest biomass. *Journal of Industrial Microbiology and Biotechnology, 33*(7), 616–626. https://doi.org/10.1007/s10295-006-0131-2.
61. Asgher, M., Qamar, S. A., Bilal, M., & Iqbal, H. M. N. (2020). Bio-based active food packaging materials: Sustainable alternative to conventional petrochemical-based packaging materials. *Food Research International, 137*, 109625. https://doi.org/10.1016/j.foodres.2020.109625.
62. Itapu, B., & Jayatissa, A. (2018). A Review in Graphene/Polymer Composites. *Chemical Science International Journal, 23*(3), 1–16. https://doi.org/10.9734/CSJI/2018/41031.
63. Dlamini, D. S., Li, J., & Mamba, B. B. (2019). Critical review of montmorillonite/polymer mixed-matrix filtration membranes: Possibilities and challenges. *Applied Clay Science, 168*, 21–30. https://doi.org/10.1016/j.clay.2018.10.016.
64. Mittal, V. (2009). Polymer Layered Silicate Nanocomposites: A Review. *Materials, 2*(3), 992–1057. https://doi.org/10.3390/ma2030992.
65. Balakrishnan, H., Hassan, A., Imran, M., & Wahit, M. U. (2012). Toughening of Polylactic Acid Nanocomposites: A Short Review. *Polymer-Plastics Technology and Engineering, 51*(2), 175–192. https://doi.org/10.1080/03602559.2011.618329.
66. Qin, X., Wu, Y., Wang, K., Tan, H., & Nie, J. (2009). In-situ synthesis of exfoliated nanocomposites by photopolymerization using a novel montmorillonite-anchored initiator. *Applied Clay Science, 45*(3), 133–138. https://doi.org/10.1016/j.clay.2009.04.014.
67. Namazi, H., Mosadegh, M., & Dadkhah, A. (2009). New intercalated layer silicate nanocomposites based on synthesized starch-g-PCL prepared via solution intercalation and in situ polymerization methods: As a comparative study. *Carbohydrate Polymers, 75*(4), 665–669. https://doi.org/10.1016/j.carbpol.2008.09.006.
68. Abedi, S., & Abdouss, M. (2014). A review of clay-supported Ziegler–Natta catalysts for production of polyolefin/clay nanocomposites through in situ polymerization. *Applied Catalysis A: General, 475*, 386–409. https://doi.org/10.1016/j.apcata.2014.01.028.
69. Sharma, D., Kushwaha, C. S., & Kumari, P. (2023). Preparation of Titanium Dioxide Encapsulated Rice Husk Derived Nanocellulose Grafted Polyaniline for Potentiometric Sensing of Residual Malathion. *Chemistry Africa, 6*(1), 233–243. https://doi.org/10.1007/s42250-022-00461-x.
70. Wang, L., Zhang, J., & Wang, A. (2008). Removal of methylene blue from aqueous solution using chitosan-g-poly (acrylic acid)/montmorillonite super adsorbent nanocomposite. *Colloids and Surfaces A: Physicochemical and Engineering Aspects, 322*(1–3), 47–53. https://doi.org/10.1016/j.colsurfa.2008.02.019.
71. Li, B. G., Xu, Y. Z., Bai, L., Dai, H., Xie, C. C., & Li, H. B. (2017). Ultrasonic effect on fabrication of intercalated MgAl-LDH/PVA nanocomposites via exfoliation-adsorption route. *Key Engineering Materials, 727*, 532–536. https://doi.org/10.4028/www.scientific.net/KEM.727.532.
72. Thakur, M., Chandel, M., Rani, A., & Sharma, A. (2022). Introduction to Bio renewable Nanocomposite Materials: Methods of Preparation, Current Developments, and Future Perspectives (pp. 1–24). https://doi.org/10.1021/bk-2022-1411.ch001.
73. Jacob, J., Robert, V., Valapa, R. B., Kuriakose, S., Thomas, S., & Loganathan, S. (2022). Poly (lactic acid)/Polyethyleneimine Functionalized Mesoporous Silica Bio composite Films for Food Packaging. *ACS Applied Polymer Materials, 4*(7), 4632–4642. https://doi.org/10.1021/acsapm.1c01551.

74. Saleh, T. A., Parthasarathy, P., & Irfan, M. (2019). Advanced functional polymer nanocomposites and their use in water ultra-purification. *Trends in Environmental Analytical Chemistry, 24*, e00067. https://doi.org/10.1016/j.teac.2019.e00067.
75. Alexandre, M., & Dubois, P. (2000). Polymer-layered silicate nanocomposites: Preparation, properties and uses of a new class of materials. *Materials Science and Engineering: R: Reports, 28*(1–2), 1–63. https://doi.org/10.1016/S0927-796X(00)00012-7.
76. Sharma, M. V. S. S. Deepa. (2021). Preparation and characterization of rice husk derived cellulose and polyvinyl alcohol blended heat sealable packaging film. *Indian Journal of Chemical Technology*. https://doi.org/10.56042/ijct.v28i4.50183.
77. Bhatia, P., Pachauri, A., Sood, A., & Gaur, M. K. (2019). A mini review: Polymer-matrix nanocomposites and its synthesis techniques (p. 150011). https://doi.org/10.1063/1.5122560.
78. Arya, A., & Sharma, A. L. (2020). Investigation on enhancement of electrical, dielectric and ion transport properties of nanoclay-based blend polymer nanocomposites. *Polymer Bulletin, 77*(6), 2965–2999. https://doi.org/10.1007/s00289-019-02893-x.
79. Xie, Y., Kocaefe, D., Chen, C., & Kocaefe, Y. (2016). Review of research on template methods in preparation of nanomaterials. *Journal of Nanomaterials, 2016*, 1–10. https://doi.org/10.1155/2016/2302595.
80. Rajani, A., Chauhan, P., & Dave, P. Y. (2021). Nanocomposites: A new tendency of structure in nanotechnology and material science. *Journal of Nanoscience and Technology, 7*(1), 937–941. https://doi.org/10.30799/jnst.315.21070103.
81. Poisson, C., et al. (2008). Mechanical, Optical and Barrier Properties of Pa6/Nanoclay-Based Single- and Multilayer Blown Films. *Polymers and Polymer Composites, 16*(6), 349–358. https://doi.org/10.1177/096739110801600601.
82. Afzal, S., Samsudin, E. M., Julkapli, N. M., & Hamid, S. B. A. (2016). Controlled acid catalyzed sol gel for the synthesis of highly active TiO_2-chitosan nanocomposite and its corresponding photocatalytic activity. *Environmental Science and Pollution Research, 23*(22), 23158–23168. https://doi.org/10.1007/s11356-016-7507-2.
83. Rahaman, Md. H., Islam, Md. A., Islam, Md. M., Rahman, Md. A., & Alam, S. M. N. (2021). Biodegradable composite adsorbent of modified cellulose and chitosan to remove heavy metal ions from aqueous solution. *Current Research in Green and Sustainable Chemistry, 4*, 100119. https://doi.org/10.1016/j.crgsc.2021.100119.
84. Jain, R., & Tiwari, A. (2014). Bioplastics for use in medical industry. *Asian Journal of Pharmaceutics, 8*(2), 139. https://doi.org/10.4103/0973-8398.134981.
85. Makadia, H. K., & Siegel, S. J. (2011). Poly Lactic-co-Glycolic Acid (PLGA) as Biodegradable Controlled Drug Delivery Carrier. *Polymers (Basel), 3*(3), 1377–1397. https://doi.org/10.3390/polym3031377.
86. Thian, E. S., & Best, S. M. (2008). Materials viewpoints in bone regenerative medicine: progress and prospects. *Materials Science and Technology, 24*(9), 1027–1030. https://doi.org/10.1179/174328408X341717.
87. Devasahayam, Sheila & Raman, R & Chennakesavulu, K. & Bhattacharya, Sankar. (2019). materials Plastics-Villain or Hero? Polymers and Recycled Polymers in Mineral and Metallurgical Processing-A Review. Materials. 9. 655. 10.3390/ma12040655.

CHAPTER 8

Polymer and Plastic Fundamentals and Its Connecting with Current Environmental, Biomedical, and Pharmaceutical Engineering

SAJID S. MANSURI,[1] HARSHAL S. PATIL,[1] DIKSHA K. NUNSE,[1] SHILPA M. JUMDE,[2] PRIYANKA R. GUPTA,[2] SWATI G. TALELE,[2] and LAXMIKANT B. BORSE[2]

[1]*Department of Pharmaceutics, Savitribai Phule Pune University, Sandip Institute of Pharmaceutical Sciences, Mahiravani, Nashik, Maharashtra, India*

[2]*Department of Pharmaceutics, Sandip Institute of Pharmaceutical Sciences, Mahiravani, Nashik, Maharashtra, India*

ABSTRACT

The current chapter focuses on the use of polymers in biomedical applications and pharmaceutical medication delivery of therapeutic medicines. As they regulate the medication's release from the device, polymers form the foundation of a pharmaceutical drug delivery system. Because they can be broken down into non-toxic monomers and, more importantly, because they can be used to build controlled release devices, biodegradable polymers are attractive for application. Natural polymers may be utilized to transport drugs at specified rates, and since they are readily available and have favorable physicochemical properties, they can be employed as a polymer for drug delivery systems. Because of their well-known biocompatibility and biodegradability, biodegradable polymers have been utilized extensively in biomedical applications. Polymers are typically employed as implants in the biomedical field and are

Sustainability in Polymer Technology and Plastics Engineering: Concepts, Strategies, and Opportunities. Tamara Tatrishvili, Neha Kanwar Rawat, Swati Gokul Talele, and A. K. Haghi (Eds.)

intended to provide long-term service. These advancements help increase the effectiveness of medical therapy and reduce unwanted side effects and other forms of patient annoyance. Polymers' primary function is to shield drugs from their physiological environment and delay their release to increase their stability. Based on their supply, structure, and processing techniques, polymers can be categorized. Thermoplastics, thermosetting polymers, elastomers, and biopolymers are examples of common categorization schemes. Each variety of polymer has special qualities that make it appropriate for particular purposes. Polymer characteristics are influenced by elements such as their chemical makeup, molecular weight, and manufacturing circumstances. Mechanical strength, thermal stability, chemical resistance, and electrical conductivity are a few of the important characteristics of polymers. Selecting the appropriate polymer for a given application requires an understanding of these qualities. The characteristics of polymers, such as their chemical makeup, molecular weight, and molecular structure, are studied using characterization techniques. Polymer characterization uses a variety of analytical techniques, including spectroscopy, chromatography, and thermal analysis.

8.1 INTRODUCTION

Pharmaceutical firms are constantly challenged to improve and develop new, effective manufacturing methods with the aim of offering pharmaceutical products that are both effective and economical due to the rapidly growing global competition [1]. The creation of new polymeric materials and modifying their characteristics for use in pharmaceutical and biological applications have been the focus of many different types of studies in recent years. The molecular weight, polymerization, configuration, percentage of crystallinity, and conformation of polymers are only a few of the numerous variables that influence their physiochemical characteristics. Drug formulation issues, particularly those involving controlled release delivery systems, can be addressed by tailoring these features [6].

Polymers are employed in the pharmaceutical business as protective agents, coating materials, binders, drug carriers, and taste-masking agents, and to enhance and regulate the administration of pharmaceuticals [3]. Due to the functionality, they offer, polymers have been responsible for notable developments in the biomedical area. In addition to being lightweight, cheap, electrically, and mechanically robust, and having high compatibility with both inorganic and organic materials, polymers are ideal for creating multifunctional hybrid systems. Polymers may be biodegradable or biocompatible [4]. Biodegradable polymers typically degrade in a natural environment, which can be viewed as a property that offers high potential for numerous medical applications, including coatings for implants, gene therapy, temporary implantable devices, drug delivery systems, tissue engineering, and temporary implantable devices [5]. This chapter covers the definition of polymers, their general properties, preparation techniques, major classifications based on various criteria, along with the properties and characteristics of each class, advantages of polymers over other materials, particularly in the medical field, and their various potential applications in the pharmaceutical and biomedical industries [2].

8.2 DEFINITION

Polymer is a word that originated from the two Greek words poly meaning "many" and Meros meaning "Parts" which means "many pieces." Polymers are generally defined as macromolecules. It consists of a long chain of several repeating units called monomers. Polymerization of monomers can be one type (homopolymer) or several types (i.e., copolymer). Each molecule is a polymer [7, 10].

8.3 DEGREE OF POLYMERIZATION

Polymers consist of many parts (monomers) which differ in the sequence of the polymer chain and its length. The size of the polymer can be expressed by the number of repeating units as well as their mass. The number of repeating units is called the degree of polymerization. The polymer ratio of molar mass is the relative molar mass of the monomer. However, the polymer composition and its average molecular weight alone are not sufficient to predict the final properties of a polymer. The structural polymer composition plays a large role in the behavioral differences. They have an exact composition and average molecular weight. The difference in behavior is represented by the topology of the entire polymer chains (linear, branched, cross-linked networks) [15]. The monomer sequence of the chain can be a block copolymer A-B, random A-B (copolymer), or the total number of monomers in a chain. Polymers are divided into two classes:

8.3.1 BIODEGRADABLE POLYMERS

Biodegradable polymers have minimum toxicity or lack of toxicity, they play an ideal role in various pharmaceutical and biomedical fields.

Biocompatibility with living tissue and cells. In addition, such a polymer is not required. It deteriorates over time and requires surgery before use. A few examples of biodegradable polymers include polyglycolic acid (PGA), polycaprolactone (PCL), polylactic acid (PLA), and polylactic glycolic acid (PLGA) [15].

8.3.2 NON-DEGRADABLE POLYMERS

Non-degradable polymers are the polymers that are mostly used because drug delivery systems can achieve zero-order kinetics for drug publication; Such polymers include polyvinyl A [15].

8.4 MONOMERS

A monomer is a single repeating unit that, after joining together in massive numbers, gives rise to polymers as the building blocks of polymers. For example: vinyl chloride, ethane, propylene ethers, etc. The number of repeating units (n) in the chain thus formed is called the degree of polymerization (DP = n). Polymers that have a high degree of polymerization are called high polymers, and those with a low degree of polymerization are called oligopolymers. Degree of Polymerization (DP) [8]. The degree of polymerization (DP) is a number that shows the range of repeating units (monomers) present in polymers, indicating the value of DP, which can be calculated by their molecular weight. The molecular weight of the polymer is calculated as [Molecular weight of polymer] = DP x molecular weight of each monomer. DP is denoted as 'n.'

Ex: $(CH_2–CH_2)$ n Polythene Here is the DP.

I) Calculate the molecular weight of polyethylene polymer given DP is 100.

Molecular weight of polythene = DP X Molecular weight of Polyethene = 100 X 28=2800 [8].

8.4.1 PRIMARY STANDARDS

8.4.1.1 CAPABILITY

The function of a monomer is the wide variety of sites for binding with other monomers under a given situation in polymerization reaction conditions. Due to a difunctional monomer, i.e., a monomer having two functional groups, it can combine with two other atoms under the right conditions. A polyfunctional monomer is one which can react with two or more molecules in a polymerization reaction. Definition: The number of binding and active sites with their functional groups present in a molecule. Example: The double bond of vinyl monomers ($CH_2 = CHX$) has a site for two vacancies [8, 9].

When a double bond breaks two single bonds become available for the combination.

$$H_2C{=}C{-}CH_2{-}CHX$$

a. When the monomer functionality is prepared by two bifunctional linear, straight-chain polymers.
Example:(a)vinylmonomers(b)adipicacid(c)hexamethylenediamine
Example of polymers-HDPE (high-density polyethylene)

-M-M-M-M-M-M-M-

b. When the ability of the monomer is three (trifunctional), the three-dimensional network polymer is Created.

Examples: phenol, glycerol Examples of polymers: Urea-formaldehyde, phenol-formaldehyde.

-M-M-M-M-M-M-M-

-M-M-M-M-M-M-M-

-M-M-M-M-M-M-M-

c. When the trifunctional monomer is mixed in small amounts with the bifunctional monomer, a branched chain of polymer is formed.
Example: LDPE (low-density polyethylene) [8, 9]

M-N-M-N-M-N-M-N-

M N

N M

M N

N M

M N

8.4.1.2 POLYMER NOMENCLATURE [8, 9]

Polymers are divided into four types

- **Homopolymers:** Polymers that consist of one of the types of monomers are called homopolymers. Polyethylene, PVC.
- **Heteropolymers:** Polymers that are composed of various types of monomers known as heteropolymers.
- **Linked Copolymer:** when the main chain consists of one monomer and the branched-chain consists of another monomer. It is called a graft copolymer.

-M-M-M-M-M-M-M-M

N N

N N

N N

N N

N N

- **Homo Chain Polymer:** The chain consists of only one of the type atoms, it is called Homo chain Polymer for example: polyethylene. -C-C-C-C-C-C-C-C-C-
- **Hetero Chain Polymer:** The chain contains different atoms, and is called Hetero chain polymer.

For example Polyester, Nylon-6.-C-O-C-O-C-O-C-O-

- **Copolymer:** Homopolymer of one monomer, copolymer of two (or more) monomers. Chain copolymerization (different monomers combined by growing polymer chains) can lead to,
- Substitutioncopolymer–ABABABABABABABABABABABA
- Block copolymer – AAAAAABBBBBBBBBBBBAAA
- Random copolymer – BAABBAABABBABABBBAABBAB
- Graft copolymer – another polymer branch is attached to the polymer chain

AAAAAAAAAAAAAAAAAAAAAABBBBBBBBBBBBBBBB (creating reactive sites on existing polymer chains, growing new chains) [8, 9].

8.4.1.3 TACTICITY

Having a change in configuration because of the direction of different functional groups. It is called the main chain respect tactic.

There are three types

1. **Isotactic:** When functional groups are placed at equal sides of the main chain, it is called isotactic.
2. **Syndiotactic:** when overall the functional groups arranged an alternative way of the structure called Syndiotactic.
3. **Atactic:** the functional groups having arrangement randomly in the main chain, it is called atactic [8, 9].

8.5 TYPES OF POLYMERS

There are many types of different polymers based on a number of special considerations. The common classifications of polymers are as follows: [8, 9]

8.5.1 CLASSIFICATION BASED ON SOURCE

8.5.1.1 NATURAL POLYMERS

These polymers are available in plants and animals.

For example, protein, cellulose, starch, rubber, and resin.

8.5.1.2 SEMI-SYNTHETIC POLYMERS

cellulose acetate (etrap) and cellulose nitrate, etc., cellulose derivatives have a typical example of this subcategory.

8.5.1.3 SYNTHETIC POLYMERS

Various synthetic polymers such as plastic (polyethylene), synthetic fibers (nylon 6,6), and synthetic rubber (Buna-S) are examples of man-made polymers.

1. It is hard and brittle at low temperatures and soft and malleable at high temperatures.
2. Soft and soluble in most hydrocarbon solvents and carbon disulfide (CS_2).
3. Water has high water absorption power.
4. This sticky substance is very difficult to handle Acid and base.
5. It has low tensile strength.
6. Low gastric resistance.
7. Susceptible to oxidative damage [8, 9]

8.5.1.4 VULCANIZATION

Vulcanization involves heating raw resin and sulfur at 100–140°C for 1–4 hours. Sulfur binds chemically to the double bond of various resin molecules and provides a connection between circuits. The degree of vulcanization depends on the amount of sulfur used.

Advantages of Vulcanization

1. The tensile strength of vulcanized rubber is very good. Around 10 times of raw rubber.
2. It has excellent durability, meaning that articles made from it convert into their original form when the deformation load is removed.
3. It has good resistance to moisture, oxidation, and digestion.
4. It has high resistance to wear and tear from raw rubber.
5. It is a widely used temperature range (–40–100°C) compared to raw resin's useful temperature range (10–60°C).
6. Better electrical insulator Ex: Ebonite
7. Resistant to organic solvents are

    i. Gasoline, benzene, CCl_4
    ii. Fats and oils, but Deep swelling.

8. There is only a little softness.
9. It has low elasticity. Their properties depend on the degree of vulcanization. Ebonite Almost no flexibility.
10. Vulcanized resins are very simple to manipulate to produce the desired shape [10].

Vulcanization was discovered in 1839 by Charles Goodyear. It consists of heating raw resin and sulfur at 100–140°C.

Chemically combined double bonds of various types of rubber form springs and connections between chains. During vulcanization, this cross-linking causes the resin to solidify and consequently prevents the individual movement of the rubber springs. The quantity of added sulfur determines the strength of vulcanized rubber.

For example, Normal Rubber (for batteries) can contain up to 30% sulfur [10].

8.5.1.5 RUBBER COMPOUNDS

Definition: "Mechanical mixing process (non-chemical) of various additives and polymer's reaction) that gives the final product some special properties called compounding plastic" [11]

1. Additives are added in the polymer to give a homogeneous mixture.
2. Additive function to change the polymer properties.
3. To introduce new properties for the created article.
4. Facilitate the processing of polymers
5. Making polymer products more attractive and colorful. They should be cheap is adequate in their function.
6. Must be stable in working conditions and services. It should not be divided into taste or smell.
7. The final compound should not be toxic for food-grade use processing.
8. It should not adversely affect the properties of the polymer. The main ingredients used in joining [11].

    i. Filler
    i. Plasticizer
    iii. Lubricant
    iv. Accelerator
    v. Dye

8.5.1.6 VULCANIZING ELASTOMERS

- **Definition:** Elastomers are very long polymers (500–1000%) in a state of stress, but start to fill up with re-emerging stress. One property of elastomers is called elasticity.

  This happens because it is covered by Elastomer structure [16, 17].

- **Properties of Elastomers:**

  Elastomer has a linear structure but is very porous.

  Chain link strength is irrelevant because this is a hydrocarbon.

  They have high flexibility.

  They can be pulled quickly.

These include:

8.5.1.7 BUNA S AND STYRENE RUBBER

Buna-S rubber is probably a form of synthetic rubber produced to copolymerization the butadiene (about 75% by weight) and the styrene (25%) [16, 17].

n(CH2=CH−CH=CH2)+n(CH2=CH−Ph)→(−H2C−CH=CH−CH−CH2−CH−Ph−)n

n−1,3-Butadiene (75%) Styrene (25%)

- **Properties**

  1. Styrene rubber is similar to natural rubber in processing characteristics and quality of the finished product.
  2. It has high abrasion resistance, high load-bearing capacity, and durability.
  3. However, it easily oxidizes, especially because of the presence of traces of ozone in the atmosphere.
  4. It also swells in oil and solvents.
  5. It can be vulcanized with sulfur or sulfur monochloride with natural resin (S_2Cl_2).
  6. However, vulcanization requires less sulfur but more accelerator [16, 17]

- **Uses**

  1. It is used for motorcycle tire manufacturers.
  2. Other applications of these elastomers are floor with tiles, shoe soles, newspapers, footwear Components, wire of cable insulation, carpet backing, adhesive, tank, etc.

8.5.1.8 NITRILE RESIN OR GR-A OR BUNA-N OR NBR

- **Preparation**

  Prepared by copolymerization of the butadiene and acrylonitrile emulsion system [12].

Copolymerization

$$mCH_2{=}CH{-}CH{=}CH_2 + nCH_2{=}CHCN(-CH_2CH{=}CHCH_2-)m{-}(CH_2{-}CH(CN)-)n{-}$$

1,3-Butadiene acrylonitrile is poly-butadiene conjugated to acrylonitrile the compounding and vulcanization method is similar to the natural resin method [12, 13].

- **Properties**

  In the presence of cyano groups, nitrile resin is less resistant to natural acids. It has excellent resistance to stains, chemicals, and aging (sunlight). As the percentage of acrylonitrile in nitrile rubber increases, resistance to acid, salt, oil, solvents, etc., also increases. However, tolerance to low temperatures decreases. Compared to natural rubber, nitrile resin (vulcanized) has a more heat-resistant effect at high temperatures. It also has good resistance to stomach acids even after immersion in gasoline or oil [12, 13].

- **Uses**

  1. To make conveyor belts, the bottom of tanks, containers
  2. Printing rollers, anti-oil foam
  3. Automotive parts and high-performance components of AC Houses [12, 13].

8.5.1.9 BUTYL RESIN

- **Preparation**

  Butyl rubber is prepared cationically. Copolymerization of aluminum chloride. Small amount of isobrene (1–5%) isobrene [13].

- **Properties**

  1. Amorphous in normal conditions, but crystallizes on stretching.
  2. Very resistant to oxidation due to unsaturation. Due to very low unsaturation, it can be vulcanized, but it cannot be made too hard.

3. Beans can be degraded to low molecular weight products by heat or light. Stabilization is imperative.
4. Being non-polar, it has good electrical insulating properties.
5. Compared to natural rubber, it has high permeability to air and other gases.
6. It does not crystallize on cooling, making it flexible down to –50°C.
7. Because of its hydrocarbon properties.

It dissolves in hydrocarbon solvents such as butyl resin Benzene but has excellent resistance to polar solvents which are alcohol, acetone, and aging chemicals (e.g., HCl, HF, HNO_3, H_2SO_4, etc.) [13].

- **Uses**

1. It is used to insulate high-voltage wires and cables in car tires
2. Conveyor belts for food and other materials.
3. Lining of tank hose [13].

8.5.1.10 THIOL

This is called polysulfide resin (or Gr-P). Can prepared according to Condensation Polymerization of sodium polysulfide (Na_2Sx) and ethylene dichloride [13].

- **Uses**

1. This applies to the factory of hoses, chemical-resistant pipes, and machine magazines;
2. Dies and Printing Rolls, diaphragms, and seals in Contact with Containers for transporting rockets and solid propellants for rockets, etc.

Poly Urethane Rubber: Polyurethane or isocyanate resin is produced by reacting poly alcohol and diocyanates.

$$N[OH–(CH_2)_2\text{-}OH+O=C=N–(CH_2)_2–N=C=O]–$$
$$[–O–(CH_2)_2–O–(CO)–NH–(CH_2)_2–NH–(CO)–]\ n\ [13]$$

- **Properties**

  - Polyurethanes are highly resistant to oxidation due to their saturated nature.

- It also has good resistance to many organic solvents, but acid and spices, especially concentrated and hot.
- Polyurethane foam is lightweight, can be used, and is resistant to heat, gas, chemicals, etc. [13].

- **Uses**

  For the production of surface coatings foams and spandex fibers.

8.5.1.11 SILICONE RESIN

Silicone resins have silicon-oxygen structures alternating with organic radicals' silicon atoms. Therefore, their structure is: R = alkyl or phenyl radical [12, 13].

- **Preparation**

  React silicon halide with alkyl halide or Grignard reagent. The reaction product is fractionally distilled to obtain various organosilicon chlorides. Then polymerized by carefully controlled hydrolysis.

  1. Dimethylsiloride is "bidional" and can produce very long-chain polymers.
  2. Ethsilchloride is "monofunctional" and is used for comparison to limit chain length.

  Monomethyl silicon chloroform joins the polymer. The ratio used in polymerization determines the amount of coupling that can occur [12, 13].

- **Properties of Silicones**
  - Depending on the application of different alkyl silicon halide preparations, silicones can be liquid, viscous liquid, semi-solid, rubber-like, or solid.
  - It exhibits excellent stability at high temperatures due to the silicon-oxygen bond, good water resistance, and good oxidation stability, but its chemical resistance is often inferior to other plastics.
  - Their specific gravity is limited from 1.03 to 2.1.
  - Their physical properties are less affected by temperature changes.
  - Non-toxic in nature [12, 13].

- **Types of Silicone and Their Uses**

  Liquid silicone or silicone oil is relatively low Molecular weight silicon, generally from dimethyl silicon. High humidity for metals, strength, low surface tension, and viscosity show very little variation in temperature [12, 13].

- **Uses**

  - Used as high-temperature petroleum, antifoaming agent, waterproofing agent
  - Heat transfer agents such as leather and textiles, moisturizers, and hydraulic fluids. Also
  - Used in cosmetics and steels.
  - Silicone oil is silicone oil obtained by adding fillers such as silica and carbon
  - Black, lithium soap, etc.
  - It is mainly used as a lubricant in very high and very low conditions [12, 13].

8.5.1.12 CONDUCTING POLYMER

Definition: A polymer that conducts electricity is called a conducting polymer Classified as follows [14].

- **Internal Polymer Conductor:** The polymer backbone has a large association responsible for conductivity.
  There are two types

  1. **Conducting of Conjugated Polymers:** Electrons in the backbone: There are such polymers conjugated – backbone electrons greatly increase their conductivity.

     Explanation: The overlapping conjugated electrons across the backbone lead to the development of the valence band. Additionally, the conduction bands extend across the polymer molecules. The valence band and conduction band are separated by a considerable distance. Therefore, electrical conduction takes place only after thermal or photolytic activation of electrons, providing enough energy to jump the gap and reach the lower level of the transmission band. All three types of polyanilines have conductivities on the order of 10^{-10} Scm^{-1}. The conductivity of this polymer is established in the presence of electrons in the backbone, which is sufficient for use in various applications.

2. **Applications of Transfer Polymers:** In rechargeable light-weight batteries based on the polyacetylene-lithium perchlorate-doped system.

    1. Approximately 10 times lighter than conventional lead acid batteries.
    2. It can produce a current density of up to 50 mA/cm^2.
    3. In electrochromic displays and optical filters: ICPs can absorb visible light to produce color.
    4. The product may be useful for displaying electrochromic and optical filters.

       Therefore, conductive polymer can used for an electro-chromic material, which is a switchable material color change in the electrochemical process of charge and discharge

    5. Wire in aircraft and aerospace.
    6. In the telecommunications network.
    7. In electromagnetic inspection materials.
    8. In electronic devices such as transistors and diodes.
    9. In solar cells, drug delivery systems on the human body, etc., in photovoltaics equipment.
    10. In molecular wire and molecular switch.
    11. Membranes formed by transfer polymers can exhibit boundary layer effects on ions, gases, etc., selective conductance useful for ion exchangers and drug addiction [14].

8.5.2 CLASSIFICATION BASED ON THE BACKBONE OF THE POLYMER CHAIN

8.5.2.1 ORGANIC AND INORGANIC POLYMERS

- **Organic Polymer:** A polymer is a chain. It is called an organic polymer composed mainly of carbon atoms. Depending on the side values of the backbone carbon atoms, hydrogen, oxygen, nitrogen, etc., they are synthetic organic polymers [18].

- **Inorganic Polymer:** Basically, chain backbone in general It does not contain a carbon atom, it is called an inorganic polymer. Glass and silicone rubber are examples of this [18].

8.5.3 CLASSIFICATION ACCORDING TO THE STRUCTURE OF POLYMERS

8.5.3.1 LINEAR POLYMERS

Polymers have long and straight chains. For example, high-density polyethylene and PVC can be mentioned.

8.5.3.2 BRANCHED POLYMERS

These Polymers consist of linear chains with multiple branches of polyethylene

8.5.3.3 CROSS-LINKING POLYMERS

Usually consists of difunctional and trifunctional monomers.

Strong covalent bond between linear polymer chains, e.g., vulcanized rubber, urea-formaldehyde Resins, etc. Cross-linked polymers are hard and in most cases do not melt, soften or dissolve.

Linear polymers Branched Polymers Cross Linked Polymers Alternating Random Block Graft [18].

8.5.4 CLASSIFICATION OF POLYMERS BASED ON COMPOSITION

8.5.4.1 HOMOPOLYMER

Monomer Polymerization: A polymer consists the number of repeating units.

8.5.4.2 COPOLYMER

When two different monomers are attached to the polymer chain is closed [10].

Now imagine that two monomers (A and B) are copolymerized in several different ways. A substitution copolymer is a copolymer in which two monomers are arranged alternately in a random copolymer; two monomers can follow each other in any order. In block copolymers, all types of monomers are grouped together, while the others are grouped together. Linked copolymers, block copolymers can be seen as two homopolymers connected together. End-to-end: one branched copolymer with one monomer in the backbone and another branched copolymer. Monomer copolymerization of other chains: They are heteropolymers or copolymers of two species of monomers, as opposed to homopolymers that use only one monomer. Copolymerization refers to the method used to chemically synthesize copolymers. Trade Suitable copolymers include ABS plastic, SBR, Nitrile resin, and styrene-acrylonitrile [10].

8.5.5 TYPE OF POLYMERIZATION

Polymerization reactions have four types:

8.5.5.1 ADDITION POLYMERIZATION

In this type of polymerization, molecules of the same monomer or various monomers combine on a large scale to form polymers. Monomers mostly used in this process (unsaturated compounds such as alkenes and salts) contain carbon-carbon double bonds in the form of polymerization. They can participate in the chain reaction. A chain reaction consists of three steps: initiation, propagation, and termination. In the initiation step, the initiator molecules are thermally decomposed or allowed to undergo a chemical reaction to create an "active form" or potential "active species." Free radicals, either cations or anions, then initiate polymerization by adding carbon to the carbon monomer double bond. Reactions occur when new free radicals cations or anions are formed. The initial monomer becomes the first repeating unit in the polymer chain. In the propagation step, a new "active species" is formed by adding another monomer as in the initial step. This process is repeated until termination. In the termination step, a growing chain ends by reacting with another growing chain, and then reacting with other species through polymerization or by the spontaneous cleavage of the active site. In certain circumstances, a final step is not needed to produce an anionic "living" polymer [19].

- Some Common Features of Addition Polymerization Are:

  1. After initiation, the polymer chain is formed very quickly.
  2. The concentration of activating species is very high. Bottom, Therefore, the polymerization mixture mainly consists of newly formed polymers and unreacted monomers

3. The carbon-carbon double bond in the monomer is two carbon atoms bonds in the polymer, so polymerization is often exothermic with cooling.

In addition polymerization mechanisms are broadly categories into two main classes, Free Radicals and Ionic Polymerization and Ion are classified into two types Cationic and Anionic Polymerizes [19].

8.5.5.1.1 Free Radical Polymerization

Various alkenes or dienes and their derivatives in the presence of free radical initiators (catalysts) such as benzoyl peroxide, acetyl peroxide, tert-butyl peroxide, etc. Free radicals can be defined as intermediate compounds that contain unpaired electrons, but do not carry an electric charge and are not free ions.

For example, the polymerization of hemp consists of heating or burning a mixture of polyethylene and a certain amount of benzoyl peroxide initiator.

1. The first step involved in the chain reaction is the initiation process. This process begins by adding phenyl free radicals formed by hydrogen peroxide to double bonds, thus creating new and larger free radicals.
2. The second step of the chain reaction is the propagation process, where one radical reacts with another radical to form ethene molecules and radicals of other sizes. These new and larger radicals promote the reaction, and this step is a step of chain propagation. The process stops in the last step of the chain reaction; the products produced react with other radicals to form polymerization products [19, 21].

8.5.5.1.2 Ionic Polymerization

Addition polymerization occurs due to the polymerization of ions is called ionic polymerization. It depends on the nature of the ion used to start the initiation process.

Ionic polymerization is divided into two types:

1. **Cationic Polymerization:** This depends on the use of cationic initiators containing reagents capable of cationic polymerization. They provide positive ions or H^+ ions. Common examples are water ($AlCl_3$ + H_2O) or boronated aluminum chloride trifluoride (BF_3 +

H_2O). It is effective with monomers like methyl ($-CH_3$) or phenyl ($-C_6H_5$), etc., electron-withdrawing groups such as Y including propylene lude ($CH_3CH = CH_2$) and styrene ($C_6H_5CH = CH_2$).

i. **Chain Initiation:** Cleavage of the initiator is defined as $BF_3 + H_2O \rightarrow H^+ + BF_3(OH)$. C–C alkene with form stable parking.

ii. **Chain Extension:** Carbocation Another monomer is added to the C–C double bond new carbocations.

iii. **Chain Termination:** The reaction ends by uniting with the carbocation (or) negative ion. With proton loss. This depends on the use of anionic initiators containing reagents which are capable [19, 21].

- **Anionic Polymerization:** Provide negative ions. Common catalysts include sodium in liquid ammonia, base metal alkyl, Grignard reagent, and triphenylmethyl sodium [$(C_6H_5)_3CNa$]. Effective with monomers containing electron groups such as nitrile (–CN) or chloride (–Cl), etc. Acrylonitrile [$CH_2 = C(CN)$], Vinyl Chloride [$CH_2 = C(Cl)$], and Methyl Methacrylate [$CH_2 = C(CH_3)\ COOCH_3$], etc.

i. **First Chain:** Potassium amide ($K + NH_2$) adds a double bond containing C–C to an alkene to form a stable carbanion. Where W is an electron-withdrawing group.

ii. **Chain Extension:** The carbanion adds a new form to the C–C double bond of another monomer molecule.

iii. **Carbanion:** There is no chain termination reaction of anionic polymerization. So, it is called living polymerization [19, 21].

8.5.5.2 COORDINATION POLYMERIZATION

This is a subgroup of addition polymerization. Usually included are Transition metal catalysts. Here, the "activities" are coordination chemistry. Including monomer carbon-carbon double bonds. It is a very important catalyst for controlled Polymerization called Ziegler-Natta catalyst, which is effective for alkene polymerization. Ziegler-Natta catalyst combines transition metal compounds such as titanium chloride and organometallic compounds [$TiCl_3$ and $Al(C_2H_5)_3$]. An important characteristic of these catalysts is their stereoregularity. When higher alkenes are polymerized, the polymer, for example, the polymerization of propylene, produces high polypropylene. Since there are no radicals involved, no branching occurs through this mechanism; the active chain The growing chain is directly bonded to the metal by the carbon atom. Ziegler-Natta catalyst: This separate co-coupling catalyst consists of two components, commonly called catalyst and cocatalyst. The catalyst component consists of chloride Titanium ($TiCl_3$ and $TiCl_4$) and the cocatalyst is an organometallic compound such as triethylaluminum ($Al(C_2H_5)_3$). Triethylaluminum [$Al(C_2H_5)_3$] acts as an electron acceptor, while titanium halide acts as an electron donor. Thus, the combination easily forms a coordination complex (Figure 8.1). The created complex is now an active center. The monomer is complex with the metal ion of the active center. The monomer is attached to the Ti-C and is formed; and complexed with Ti ions. The bond between R and Ti produces electron-deficient Ti. This transition state now leads to chain growth in the iron-carbon bond active center. By repeating the entire sequence, with the addition of a second monomer molecule, will do [24].

8.5.5.3 CONDENSATION POLYMERIZATION

This polymerization usually involves repetition. A condensation reaction (two molecules joining, with the loss of a small molecule) is a reaction between two bifunctional monomers. This polycondensation reaction may result in the loss of simple molecules like water, alcohol, etc., leading to the formation of high molecular weight condensation polymers such as Di. In this reaction, the product of each step has two types of functional groups, and the condensation sequence continues. Therefore, each step produces a separate and independent version of each other; this process is also called step-growth polymerization. The type of final polymer product depends on the number of end groups functionalized by condensation polymerization. Monomers that can react have one reactive group that stops the growing chain, resulting in the end product having low molecular weights and linear polymers. Polymers are formed using two reactive terminal monomers, while more than two cross-link groups lead to three-dimensional polymers. Polyester is made through ester bonds that contain monomer functionalities like carboxyl and hydroxyl groups (organic acids and alcohol monomers). Examples of polyester formation would be ethylene glycol, terephthalic acid, or Dacron [24].

8.5.5.3.1 Type of Polyamides [24, 25]

Polyamides are created by cross-linking functional monomers carboxyl and amine (organic acid and amine monomer). An example is nylon-6, which can be produced by condensation polymerization of hexamethylenediamine and adipic acid at high pressure and high temperature. This type of polymerization typically uses two difunctional monomers and undergoes normal organic reactions. For example, a diacid can be allowed to react with a diol in the presence of an acid catalyst to produce polyester. In this case, the development of the chain begins with the reaction of one of the carboxyl groups of the diacid with one of the hydroxyl groups of the diol. The free carboxyl or hydroxyl group of the resulting dimer can react with other functional groups of monomers or dimers. This process is repeated in the polymerization mixture until all the monomers are converted to lower molecular weight forms like dimers, trimers, tetramers, etc. These molecules, called oligomers, can then interact with each other to form free groups. Polymer chains can have an average molecular weight. Below are some common features of the polymerization:

1. Polymer chain formation is slow, sometimes it takes a couple of hours to a couple of days
2. All monomers are quickly converted into oligomers, so the concentration of the growing chain.
3. Most active chemical reactions have higher activation energy, resulting in higher temperatures for the polymerization mixture.
4. Step-by-step polymerization reactions are carried out to control the intermediate polymer molecular weight, i.e., <1,00,000.
5. There should be no branching or cross-linking; monomers with three or more functional groups are used.

8.5.5.3.2 Steps of Growth Polymerization

In stepwise polymerization, polymer formation occurs through functional reaction groups of monomers. The reaction is stepwise and takes place in the polymer formed generally slow work. General reactions involve organic functional groups. In-step polymerization, condensation, addition, ring opening, amidation, and others can be used. The most commonly used reaction is esterification. Stepwise polymerization reactions are always accompanied by the elimination of subgroups. Polycondensation is caused by two or more reactive monomers with functional groups such as hydroxyl, carboxyl, and amine groups that condense with each other. In the polymer phase, the mechanism is a simple extension of an organic condensation reaction in which a small molecule, such as H_2O or HCl, is excluded.

For example:

$$\text{R–COOH + H–OR} \qquad \text{R–C–O–R–}H_2O$$

In step-growth polymerization, the polycondensation reaction occurs. Oil Condensation reactions are caused by two or more functional monomer groups. If the monomer is the only functional group, the following reactions are possible.

$$\text{R–OH + HOOC–R} \qquad \text{R–OOC–R} + H_2O$$

Since there are no reactive functional groups left in the product, it cannot react with other reactive molecules. Consider two monomers, one mono-functional and the other bifunctional. However, it contains two reactive groups that can now react. The resulting molecule still has two active ends and can react with any monomer. Therefore, it is more reactive in forming large molecules.

The overall reaction can be depicted by stepwise polymerization or poly-condensation as follows:

i. Monomers must have two functional groups that are reactive for polymerization.
ii. Polymerization occurs through a stepwise condensation reaction between the functional groups of the reactants.
iii. A condensation reaction takes place between the two functional groups during polymerization.
iv. The resulting polymer consists of reactive functional groups at the end position of the chain (as an end group) and is therefore not considered "active" or "dead" as in chain polymerization. This type of polymer is called an "active polymer" [26]

8.5.5.3.3 Characteristics of Stepwise Polymerization

1. The assembly of polymer molecules proceeds only by stepwise inter-molecular reactions, and the reaction is involved in polymerization.
2. Monomer units can react with each other or with polymers of any size.
3. Functional groups at the ends of monomers are usually assumed to be identical, ensuring reactivity in polymers of any size.
4. High conversion of functional groups is necessary for the production of high molecular weight and mass production.

5. A reversible multi-step polymerization reaction.
6. Condensation polymerization is usually produced by stepwise polymerization, but not all stepwise polymerization is a condensation reaction [26].

This does not delete the product synthesized by polyurethane from diol and diisocyanate. Requirements for step-by-step polymerization: Polymerization development phase/polymers can be prepared from bi-functional monomers. This reaction can occur to form A-A-B-B polycondensation of di-carboxylic acid and diol. When a single bifunctional molecule self-condenses, the reaction can be described as Di. In both cases, partial condensation occurs when intermolecular condensation is preferred. Stepwise polymerization is a two-step reaction. Polysaccharides are monomers that form high molecular weight polymers through a number of steps. Most natural and synthetic polymers are formed through step-by-step polymerization, such as polyester, polyether, urethane, epoxy, and polyamide. Two notable examples of reactions involve dicarboxylic acids with diamines, forming polyamide (nylon), and their organic reactions of diacid and alcohol to form polyester, such as polyethylene terephthalate (PET). Due to the nature of the polymerization mechanism, the reaction must continue for a long time to achieve a high molecular weight polymer. This is the easiest way to think about the step-growth mechanism that most people struggle to understand. Each person has two hands (= two reactive sites), forming human circuits. The two most important polymerizations in construction are condensation and addition polymerization. In the condensation reaction, two monomers are combined and lost to form small molecules, usually alcohol, water molecules, or acids. The reaction involves binding other molecules with the formation of a single electron in a double bond [26].

Examples of EPDM-growth policies Examples of Step-Growth Polymerizations Functional Groups Linkage Polymer Type

–OH + –COOH –C(=O) –O–Polyester
–OH + –NCO –O–C(=O) –NH – Polyurethane
–NH_2 + –NCO –NH–C(=O) –NH – Polyurea
–NH_2 + –COOH –NH–C(=O) – Polyamide
–OH + –OH –O – Polyether.

8.5.5.3.4 Polyaddition Polymerization

Polyaddition polymerization reaction takes place. Atoms are transferred from one monomer molecule to another monomer molecule intermediate product.

Vinyl monomers are also pairs of monomers with reactive functional groups that can undergo additional polymerization. For example, styrene polymerizes in this way in the presence of perchloric acid. The mechanisms involved are an interesting way to form simple additions to polymer double bonds without the formation of free radicals, harmonics, or anions. However, as in chain polymerization, monomer units are joined one by one to form the polymer and remove all small molecules. Polymers are produced by the migration of hydrogen atoms from the monomer and their incorporation into a double bond, which is a high activation process compared to the process of free radicals or ionic polymerization. Energy for polymer growth is slower and carried out step by step, as opposed to the rapid growth of chain polymerization. Oligomers of vinyl monomers (e.g., 2.00 MW liquid polymers) are made by this method. The highly mobile hydroxyl group can easily contain monomers such as polyhydric alcohol and polyamine. Polyisocyanide is added to polyurethane to produce polyurethane [27, 28].

8.5.5.3.5 Polymerization Kinetics

With the Concept of Functional Group Reactivity. Molecular weight and step-wise polymerization kinetics are very simple. Examples include polyester and dibasic acid from glycerol. This reaction is catalyzed with acid and if no strong acid is added, the second molecule of acid-esterified Acts as a catalyst. It can be defined by reactions, reactions, and general chemical equations below [30]:

$$\text{A} + \text{B} + \text{Catalyst} \qquad \text{AB} + \text{Catalyst.}$$

$$\text{Rate} = [\text{A}]\,[\text{B}]\,[\text{catalyst}] \tag{3}$$

In an etherification reaction acid (H+) is the catalyst and if the stoichiometry of the functional group

A ‘and _B ‘is assumed.

$$[\text{A}] = [\text{B}]$$

Eqn. (3) can be written as:

$$\text{rate}= -d(A) = k[A]^2[H^+] \tag{4}$$

In self-catalysis polymerization poly etherification reaction

$$[H^+] = [A] = [B]$$

Eqn. (4) can be written as:

$$-[A] = k\,[A]^3 \qquad (5)$$

On integration of Eqn. (5):

$$\frac{\int}{3} -dc - \int kdt.$$

$$2kt = \frac{1}{-1} \qquad (6)$$

$$C2\ C2$$

It is convenient to introduce the extent of reaction (P), which is defined as the fraction of the functional groups that have reacted at time 't'

$$C=C_o\,(1\text{-}P) \qquad (7)$$

$$C_oK_t = 1/(1-P) - 1$$

$$1/(1-P) = C_oK_t + 1$$

$$(Xn)2 = CoK'\,t + 1.$$

$$(mn/mo)2 = CoK'\,t + 1 \qquad (3)$$

Kinetic polymerization and co-catalyst are examples of stepwise polymerization in the intermediate esterification reaction between alcohol and carboxylic acid. Development of the reaction to form polyester is easy following titration of unreacted acid in batch samples with respect to time. Simple esterification reactions are catalyzed by acids. If not a strong acid, the second acid molecule acts as a catalyst.

Therefore, the polymerization reaction rate can be written as:

$$d[COOH]/dt = k\cdot[COOH]_2.\ [OH]$$

Where; k is the stepwise growth reaction rate and [X] is the mol concentrations of the monomers [30].

The concentrations are usually defined as mole equivalents to the functional groups as per unit Volume. By this convention, we do not write a separate equation for each condensation Product (dimer, trimmer, etc.). However, this oversimplification is only true in hindsight K is the same rate regardless of the size (molecular weight) of the reacting species. If we choose the same concentration of hydroxyl and carboxyl groups, the equation above can be rewritten.

$$_{-}dc/dt = k\cdot c3 \text{ by integration}$$

We obtain an expression for the third-order reaction:

$$2kt = 1/c2 - 1/c02$$

Where; c0 is the initial concentration of functional groups.

The reaction is often prolonged Written as part of a group of time-reacted functions,

$$P = (c0 - c)/c0 \text{ Then}$$

C = C0(1 – p) and after replacing c in this term, the rate of polymerization. The equation reads:

$$2C02\cdot Kt = 1/(1 - p)2{-}1$$

In the esterification reaction, p can be calculated directly from the titer of the carboxyl group.

1/(1-p)2 Versus time, we find a linear relationship, which is the case for most esterification's Reaction of glycol and organic acid.

This is generally considered proof of all monomer units Similar speed series k.

$$Dc/dt\ 2\ W \text{ catalyst.}$$

The combination of these is K" t = 1/c – 1/c0

After substituting the concentration for the reaction expansion;

this expression reads C – p) – 1

In this case, the average of degree polymerization which determined

$$Xn = \text{Number of monomers}$$

$$\text{Number of monomer units} = 1/(1 - p) = c0/c$$

Increases linearly with reaction time, which is the most favorable condition to get a high average Molecular weight polymers from weak acid-catalyzed third-order reactions [30].

8.5.5.4 COPOLYMERIZATION

A polymerization reaction involves a mixture of one or more monomers allowing polymerization and copolymer formation. Copolymerization is not limited to chain growth polymerization but proceeds step by step. Each monomer is used multiple times in the same polymer chain. For example, a mixture of styrene and 1,3-butadiene forms a copolymer called styrene-butadiene resin (SBR). When a mixture of two (or more) monomers undergoes polymerization, it is referred to as copolymerization. Polymers formed this way are known as copolymers. Plastics are high polymers that can be shaped by applying heat and pressure. During the molding process, there is no chemical change. Plastics have a linear structure. Examples include polyethylene, polyvinyl chloride, polystyrene, nylons, and polyesters [30].

8.5.5.4.1 Thermoplastics

Thermoplastic polymers soften when heated and harden when cooled, and the process is repeated several times. They undergo reversible changes in heating. For example: Polyethylene (PE), polypropylene (PP), polyvinyl chloride (PVC), Polytetrafluoroethylene (PTFE or Teflon), polystyrene (PS), plexiglass, nylon, polyester [25].

1. It is formed by addition polymerization.
2. It is soft and less brittle, consisting of tightly interlocked linear polymer chains.
3. It is soluble in the organic solvents.
4. Softening heating and therefore can be made into any shape. In the heat of cooling, the resin becomes hard and stiff Retain in the shape of mold.

It can cause reheating and reshaping in any way, however many times the chemical nature changes. Thermosetting polymers or thermosets: Thermosets are polymers that undergo chemical changes and cross-linking, resulting in permanent hard, solid, and non-burning properties. They will not soften upon heating after installation. Examples include phenol-formaldehyde resin, urea-formaldehyde resin, epoxy resin (araldite), melamine, and Bakelite [25].

8.5.5.4.2 Thermosetting Resin

1. It is mainly produced by condensation polymerization.
2. Mainly branched polymer chains with the potential to form 3-dimensional structures.
3. Be firm and firm in heating during molding. Not Soluble in common organic solvents [26].

- **Glass Transition Time (Tg):** A normal rubber ball, when cold below –70°C, becomes hard and brittle, so it breaks like a glass ball hitting the hard ground. Does this happen because there is a temperature range for amorphous and polycrystalline polymers? The material is initially soft, flexible, and rubbery, but at low temperatures, the polymer becomes hard, brittle, and glass-like. This is known as the low-temperature state where the polymer softens. Rubber in an elastic state is called viscoelasticity. At higher temperatures, if not interconnected, the polymer becomes a highly concentrated fluid and starts to flow, known as a viscous fluid state [27].
- **Definition:** The temperature at which the internal energy of polymer molecules increases, allowing the chain segments of the polymer to transition from a frozen state to a mobile state where they can leave the lattice site, is known as the glass transition temperature. A property associated with this transition is a low volumetric expansion coefficient, which results from the change in the slope of the volume vs. temperature curve. This phenomenon is called the glass transition temperature (Tg) [27]. This behavior is naturally displayed in rubber. Another common phenomenon is observed in the amorphous region of partially crystalline polymers, which also exhibit glass-like behavior. The Tg is independent of the crystal level, and the magnitude of the Tg-related phenomenon decreases with decreasing amorphousness content. Therefore, it is sometimes challenging to induce defects in crystalline polymers with high Tg. There is a correlation between glass transition rates and polymer molecular weight.

The glass transition of temperature polymers affects molecular weight and subsequent reactivity 20,000, molecular weight effect is not pronounced. Two mathematical relationships are established.

Tg = Tg 1/Mn (1)

1=1/+??/Mn (2)

The glass transition temperature (Tg) and the melting temperature (Tm) are influenced by molecular weight and the specific structure of the polymer. Numerous studies have been conducted to explore the relationship between Tg and Tm as below:

Tg = 1/2Tm (1) (symmetrical polymers)
Tg =2/3 Tm (2) (foymmymers) A combined form of these two equations can occur regardless of molecular symmetry.

½ </< 2/3 (3)

Characteristics of Tg:
Tg depends upon the following.

i. The chain of polymer length.
ii. Cross-section of the polymer
iii. Physical obstruction to internal circulation.
iv. The rate of heating and cooling.
v. Linear polymers have a higher Tg than cross-linked polymers below Tg,
vi. The polymer Is hard and brittle.
vii. Cross-linked polymers have almost no Tg, but have a decomposition temperature.
viii. Tg decreases with cross-section at a given temperature [27]

Examples are (i) polyethylene which consists of one repeating unit of ethylene.

In this case, Tg is very low because there are no strong intermolecular forces. Substituents on carbon atoms are only hydrogen and not bulky groups. Nylon-6, a polyamide, has a high Tg (50°C) due to the presence of many polar groups. Molecules leading to strong intermolecular hydrogen bonds between them include the C=O and N-H groups. Side groups affect chain mobility and hence the glass transition temperature. Polystyrene and PMMA are examples where the presence of CH_3 groups impacts chain motion. In

PMMA, the backbone of the chain is free for rotation around the C-C bond, inhibiting chain motion and resulting in increased Tg values around 70°C and 100°C for polystyrene and PMMA, respectively. Polyvinyl carbazole has a high glass transition temperature (150°C) due to large groups impeding freedom of movement. Aromatic rings present in the chain of polyethylene terephthalates increase chain flexibility and result in high Tg values at 69°C. Higher Tg values are mainly for cellulose derivatives, e.g., cellulose nitrate, due to the rigid ring structure of the chain [27].

- **Glass Transition Temperature and Plasticizers:** Plasticizers are not low molecular weight. The substance is volatile and mostly in liquid form, which cures when the polymer is added. Flexibility, ability, and therefore utility. A plasticizer will significantly reduce swelling in a very amorphous polymer because even if the amount is very small, lowering the Tg of the polymer. This effect is due to the reduction of the pulling force between polymer chains. Its size is very small compared to the plasticizer molecules. Polymer molecules can therefore intercalate in the polymer matrix to form polar and the force of attraction between chain segments. These attractive forces are responsible for reducing unity, increasing the force and mobility between the polymeric chains, and lowering the Tg value. Examples of plasticizers are NS (naphthyl-salicylate), tri-cresyl-phosphate, Ms (methyl-salicylate), and MA (methyl-acetate). It can be seen from the graph that all plasticizers significantly reduce the value of Tg. Degradation (plasticization effect) depends on other factors such as resolution parameters, density, and polarity. Many plasticizers include dibutyl phthalate, dioctyl phthalate, and tri-cresyl phosphate. They are used to lower the Tg value of PVC [27, 28].
- **Glass Transition Temperature and Copolymers:** Copolymers are mixtures of two homos Polymers. The Tg value of individual homopolymer sands, two monomeric ratios Components.

  Tg (A), Tg (B)–Tg (AB)–Copolymer glass transition temperature.

  WA, WB – Weight fraction of each monomer component of the copolymer. From this equation, it can be seen that the glass transition temperature of the copolymer will be between that of the corresponding homopolymer. Glass transition temperature [28]. Can decompose many homopolymers in this way. Copolymerization with a small amount of other monomers. This process is usually called internal plastic.

8.6 PROPERTIES OF POLYMER

Polymers possess properties such as lighter weight, excellent mechanical properties, and satisfactory optical qualities. Desirable characteristics include resistance to compounds, suitability for rapid and large-scale production, and the ability to be produced in a variety of colors. Recently, the thermal, structural, optical, electrical, and other qualities of polymers have attracted significant attention for their application in optical and electronic devices [31]. Polymer properties are largely categorized into various modules based on the level at which their property is well-defined and their physical origin. The most fundamental quality of a polymer is the recognition of its basic monomers. Secondly, a set of properties known as microstructure. Materials such as ceramics, polymers, glass, and bio-composites are non-conductive resources. When these non-conductive materials are placed in an electric field, they alter the electric field and undergo significant changes, acting as a storage of electrical charges. If the core function is to store charge, then these materials can be referred to as dielectrics [32–38].

8.6.1 OPTICAL PROPERTIES

Polymers such as HEMA and PMMA-MMA are utilized as mediums that are also known as polymer lasers. Such polymers have a greater surface quality and are also very translucent so the laser properties are acquired by the laser dye used to fix the polymer matrix [38].

8.6.2 THERMAL PROPERTIES

A true pillar for polymer description is its thermal analysis, using Differential Scanning Calorimetry. Changes in the compositional and structural parameters of the material usually affect its melting transitions or glass transitions, which can be linked to many performance parameters. For semi-crystalline polymers, it is an important method to measure crystallinity. Thermogravimetric analysis can also give a signal of a polymer's thermal stability and the effects of excipients such as flame retardants. Other thermal analysis techniques comprise the combinational usage of basic thermal techniques that include differential thermal analysis, thermomechanical analysis, dynamic mechanical thermal analysis, and dielectric thermal analysis [38]. Mechanical Properties: The representation of mechanical properties of any polymer is used to imply its physical strength. Factors like molecular weight, structure, and intermolecular forces determine how polymers resist stress [42]. With increasing molecular weight and hence the level of intermolecular forces, polymers display superior properties under applied stress. Longer chains with high molecular weights exhibit unique properties of polymers as materials. The mechanical behavior may involve stress-strain curves where the stress (force per unit area) needed to stretch the material to a certain elongation is plotted. In order to experimentally generate these curves, tension stress is applied to a polymer sample of known dimensions, which is elongated until it breaks [40, 53].

8.6.3 *MECHANICAL PROPERTIES OF SOLID POLYMERS*

- **Stress-Strain Behavior:** The effect of strain rate on mechanical behavior is vital. In general, reducing the rate of deformation has an equal influence on the stress-strain characteristics as growing the temperature; that is, the material converts softer and more ductile [35].
- **Viscoelastic Deformation:** An amorphous polymer may perform like a glass at lower temperatures, an elastic solid at intermediary temperatures above the glass transition temperature, and a viscous liquid as the temperature is added. For intermediary temperatures, it is rubbery which shows the combined mechanical properties of these two extremes; the phenomenon is called viscoelasticity [35].
- **Fracture of Polymers:** The fracture strengths of polymers are low compared to the strength of ceramics and metals. Basically, during the fracture process, cracks are formed at the regions where a localized stress concentration is applied (i.e., scratches, notches, and sharp flaws). In the case of metals, the stress is intensified at the edges of these cracks causing the crack propagation and fracture. Covalent bonds in the network or cross-linked structure are severed during fracture [35].

8.6.4 *MISCELLANEOUS MECHANICAL CHARACTERISTICS*

- **Impact Strength:** The degree of resistance of a polymeric compound to impact loading may be of concern in some applications. Both polymers and metals may exhibit ductility or fragile fracture under impact loading circumstances, depending on the specimen, temperature, strain rate, size, and type of loading [35].

- **Fatigue:** Polymers may experience fatigue failure due to conditions like cyclic loading. Similar to metals, fatigue occurs at stress levels that are low relative to the yield strength. Fatigue data are plotted in the same manner for both types of material, and the resulting curves have the same general shape. As expected, fatigue strengths and fatigue limits for polymeric materials are much lower than for metals [35].
- **Tear Strength and Hardness:** The capability to resist tearing is a vital property of some plastics, mainly those used for thin films in packaging. The mechanical parameter that measures tear strength is the energy required to tear apart a cut specimen with a standard geometry. The magnitudes of tensile and tear strengths are related. The influence of water intrusion and the related mobilization of polymers on mechanical properties is due to the interconnected pore network. Water intrusion during wet storage is a 3D problem. The more hydrophobic the re-dispersible powder is and the higher its quantity, the lower the intrusion rate. As the polymers seal the pores, they reduce the degree of connectivity of the pores and also the intrusion rate.

8.6.5 MOLECULAR WEIGHT

Complexes made up of small molecules show a distinctive, well-defined molecular weight; whereas polymers display a distribution of molecular weights since not all the polymer chains of a given sample will have the same molecular weight or chain length. Therefore, in order to characterize a given polymer sample, it is crucial to either describe the full molecular weight distribution (MWD) or some average quantities related to the distribution [40].

8.6.6 CRYSTALLINE AND AMORPHOUS POLYMERS

Polymers display different thermal, physical, and mechanical properties depending on their structure, molecular weight, linearity, and intra- and intermolecular interactions. If the structure is linear, polymer chains can pack together in regular arrays. With increased temperature, the crystal cells (crystallites) begin to melt, and the whole polymer mass suddenly melts at a specific temperature. Above the melting temperature, polymer molecules are in continuous motion, and the molecules can slip past one another. In many cases, the structure of a polymer is so irregular that crystal formation is thermodynamically infeasible. Such polymers form glass instead of crystal domains. A glass is a solid material existing in a non-crystalline (i.e., amorphous) state. The amorphous arrangement is formed due to either the quick cooling of a polymer melt in which crystallization is prohibited by quenching or due to the lack of structural regularity in the polymer structure [42].

8.6.7 GLASS TRANSITION TEMPERATURE

Tg is an expression of molecular motion, which is reliant on many factors. At temperatures below the Tg, amorphous polymers are hard, stiff, and glassy, although they may not necessarily be brittle. On the other hand, at temperatures well above the Tg, polymers are rubbery and might flow. From a pharmaceutical standpoint, Tg is an important factor for solid dosage forms. Pharmaceutically acceptable polymers with their Tg values close to the service temperature of 37°C would be the best candidates [42].

8.6.8 MOLECULAR WEIGHT

A very narrow molecular weight distribution is very much desired for a polymer that is intended to be mechanically strong. On the other hand, a polymeric adhesive may have a wide distribution of molecular sizes. In general, a given polymer cannot be identified as a molecule with a specific molecular weight. There are different ways that molecular weights of a polymer can be expressed: by the number of chains, by the weight of the chains (the chain size), or by viscosity. However, the two most common ways are number (Mn) and weight (Mw) average calculations [42].

8.6.9 VISCOELASTIC PROPERTIES

Polymers are neither a pure elastic nor a pure fluid material. They have the ability to store energy (display elastic behavior) and dissipate it (display viscous behavior). For this reason, most polymers are viscoelastic materials. Over time, the polymer intermolecular forces will essentially become weaker and hence, the polymer becomes softer. For many applications, there is a working range of molecular weights that a given polymer in a solid or solution state can successfully be processed [33, 34, 42, 43].

8.7 WHAT IS CHARACTERIZATION

The analytical area of the polymer field is known as polymer characterization. Characterization of polymeric materials is a key topic in this area on several levels. The material's performance is often improved due to the characterization. As a result, several characterization methods have to be connected, if possible, to the material's desired attributes, such as its strength, impermeability, thermal stability, and optical qualities [56].

Polymer characterization focuses on the molecular structure and mass characteristics of polymer materials, both natural and artificial. Decision-makers can benefit from polymer characterization since these characteristics have an impact on the behavior of polymers. Other specialties, like polymer morphology, fall under the umbrella of polymer characterization. The primary research areas in this extremely specialized field are polymer structures and interactions between monomers and molecules. The molecular mass, molecular structure, shape, thermal characteristics, and mechanical properties are often determined using characterization procedures. Characterization refers to the compositional and structural characteristics of a material (including faults) that are important for specific preparation, the investigation of attributes, or application and are sufficient for the material's replication [57].

The characterization of polymers can be done using a number of experimental techniques. Physical chemistry techniques are frequently used in molecular characterization, which often include polymer solutions. Spectroscopic techniques can sometimes be applied. Mass spectrometry (MS), infrared (IR), Raman, electron spin resonance (ESR), and UV-visible absorption spectroscopy are a few examples of typical spectroscopic methods. The measurement of macroscopically important characteristics is sometimes referred to as classic polymer characterization. It entails using a macroscopic polymer specimen, often in its finished solid state, and carrying out tests that reveal the characteristics of that polymer.

Understanding the characteristics and functionality of polymers requires polymer characterization, which is a crucial step. A multitude of methods are used to analyze the physical, chemical, and mechanical characteristics of polymers, each of which offers crucial details on various facets of the polymer.

8.7.1 WHY POLYMER CHARACTERIZATION IS IMPORTANT?

Through polymer characterization, researchers can discover more about the properties of polymer materials. Since polymer materials have so many applications, polymer characterization is crucial to a variety range of industries and businesses. We look at a few of the many applications for polymer characterization in the section that follows and talk about their importance.

8.7.1.1 QUALITY CONTROL IN THE PLASTICS MANUFACTURING INDUSTRY

Polymers are widely employed in the food and beverage business, from flexible films used to wrap microwave meals to plastic water bottles. The pharmaceutical and healthcare industries, which are subject to stringent packaging and safety rules, also use polymer polymers. Manufacturers may closely monitor quality control and make sure the plastics are up to par by using polymer characterization.

8.7.1.2 IMPROVING PRODUCT PERFORMANCE

Polymer polymers are highly sought after because of their robustness and longevity. For instance, yachts frequently employ high-density polyethylene (HDPE), which is made from ethylene monomers and has good weatherproof qualities. Even though HDPE is widely used, manufacturers are always looking for ways to enhance product performance and create even more weather-resistant polymers. Manufacturers of HDPE can have a better knowledge of how a material will function thanks to polymer characterization. One of the most well-known polymers in the world is Teflon, a synthetic substance used to produce non-stick pots and pans. The fluorocarbon tetrafluoroethylene is polymerized to create Teflon, also known as polytetrafluoroethylene (PTFE). Like HDPE, producers are always seeking new methods to improve the product and boost performance.

8.7.1.3 DEVELOPING NEW POLYMER MATERIAL

The polymer industry's research and development is fueled by polymer analysis methods like mixed matrix membranes and composites based on graphene. For instance, polymer characterization is assisting engineers in creating novel eco-friendly materials called "green" polymers manufactured from renewable resources including biomass, thermoplastic starch (TPS), and cellulose derivatives. The environmentally friendly polymers produced by the British chemical company INEOS are used to line the interiors of dairy and beverage cartons. The oil-based, renewable naphtha is obtained from the Finnish business UPM Biofuels and used to create mass-balanced polymers. The naphtha produced as a byproduct of forestry industry pulping procedures leaves little of an environmental trace. The director of sales and marketing of UPM Biofuels, Juha Rainio, stated "We arc committed to replacing fossil-based feedstocks with renewable alternatives." The sustainable byproduct of the processing of wood pulp, tall oil, is used to make our UPM BioVerno naphtha, which is produced at Finland's distinctive Lappeenranta Biorefinery.

8.8 POLYMER CHARACTERIZATION TECHNIQUES

The study of techniques for disclosing the structure or properties of polymers is known as polymer characterization. As part of polymer characterization, the properties of a polymer material may be measured. It includes both molecular characteristics, such as degree of crystallinity, microstructural information, molecular weight, etc., and macroscopic property measurement, such as time dependency of features, mechanical properties, thermal properties, microstructural information, etc.

8.8.1 POLYMER MOLECULAR (MASS) WEIGHT CHARACTERIZATION

The characterization of polymer molecular weight is significant because it controls a variety of physical characteristics. Because polymerization processes produce a variety of molecular weights and morphologies, a polymer's molecular weight differs from that of typical molecules. The average molecular weight by number, average molecular weight by weight, and polydispersity can be used to summarize the distribution of molecular weights. Collaborative property measurements, viscosimetry, light scattering techniques, and size exclusion chromatography are the popular ways to find these properties. It is possible to directly compute the molecular weight distribution features based on the hydrodynamic volume of the polymer using gel permeation chromatography, a kind of size exclusion chromatography. The monitoring of diffusive processes in soft matter, such as macromolecular solutions, is well-served by dynamic light scattering (DLS). By dispersing a laser beam from the sample at a specific angle, this technique, also referred to as photon correlation spectroscopy or quasielastic light scattering, measures the correlation in time of the laser beam. Macromolecules exhibit a diffusive Brownian motion when they are in solution.

8.8.2 SPECTROSCOPY METHODS CHARACTERIZATION

The main methods for figuring out chemical composition and molecular morphology rely on polymers' ability to absorb electromagnetic radiation. The three main analytical methods covered in this course are IR, Raman, and NMR spectroscopy, making up the majority of the content. An inelastic phenomenon known as absorption involves the transfer of energy from electromagnetic radiation to a substance.

8.8.2.1 VIBRATIONAL SPECTROSCOPY METHODS CHARACTERIZATION

Infrared spectroscopy, sometimes referred to as IR spectroscopy, is the branch of spectroscopy that examines the region of the electromagnetic spectrum (EMR) with a longer wavelength and lower frequency than visible light. There are several different techniques included, many of which are based on absorption spectroscopy. It is used to identify and investigate compounds, much like any spectroscopic method. The infrared spectroscopy method or technique generates an infrared spectrum for a particular sample, which may be solid, liquid, or gaseous, using a device known as an infrared spectrometer (or spectrophotometer). The near, mid, and far infrared portions of the electromagnetic spectrum (EMR) are typically separated into these three groups based on how closely they resemble the visible spectrum. The fact that molecules absorb particular frequencies that are distinctive to their structure is taken advantage of in infrared spectroscopy. These absorptions have resonance frequencies, which means that the frequency of the radiation absorbed corresponds to the transition energy of the vibrating bond or group. The energies depend on the atom masses, the associated vibronic coupling, and the shape of molecular potential energy surfaces. Raman spectroscopy is the spectroscopy method for observing low-frequency modes in a system, including rotational, vibrational, and other modes. Raman spectroscopy is frequently used in chemistry to give molecules a fingerprint that can be recognized. Especially sensitive to vibrations with little or no change in dipole moments are symmetric vibration modes, which require a change in polarizability [3].

8.8.2.2 SPIN SPECTROSCOPY RESONANCE METHODS CHARACTERIZATION

NMR spectroscopy, sometimes referred to as nuclear magnetic resonance spectroscopy (NMR), is a research method that makes use of the magnetic characteristics of certain atomic nuclei. It establishes the atomic or molecular characteristics of atoms or the molecules containing them. It is based on the phenomena of nuclear magnetic resonance (NMR) and may give precise details on the kinetics, reaction state, and chemical environment of molecules and their structure and dynamics. Resonance frequency fluctuates as a result of the intramolecular magnetic field surrounding an atom in a molecule, providing access to information about a molecule's electronic structure.

8.8.2.3 ELECTRONIC SPECTROSCOPY METHODS CHARACTERIZATION

Absorption or reflectance spectroscopy in the ultraviolet-visible spectral band is referred to as ultraviolet-visible spectroscopy or ultraviolet-visible spectrophotometry. This indicates that it makes use of light in the visible and nearby light spectrums. The theory behind ultraviolet-visible absorption states that molecules with n or Pi electrons, which are not involved in bonds, can absorb energy from ultraviolet light for the excitation of electrons into higher anti-bonding molecular orbitals [58]. The light having a longer wavelength can absorb, the more easily excited the electrons are (i.e., the lower energy gap between the HOMO and the LUMO). Electron paramagnetic resonance (EPR), commonly known as electron spin resonance (ESR) spectroscopy, is a technique for analyzing materials containing unpaired electrons. EPR's core concepts are analogous to those of nuclear magnetic resonance (NMR), with the exception that electron spins are energized rather than atomic nuclei's spins. Free radicals must be present in the sample for electron spin resonance, or ESR, spectroscopy to be effective. Synthesis and degradation are often shown in derivative mode.

8.8.2.4 SCATTERING SPECTROSCOPY METHODS CHARACTERIZATION

In addition to inelastic absorption processes, electromagnetic radiation, and materials may interact elastically, resulting in diffraction and scattering effects. The study of X-ray diffraction data in processed plastics uses various distinctive analytic methodologies due to the tiny crystallite size and predominance of crystalline orientation. The tools employed in diffraction that are particular to polymeric materials will be the emphasis of this course [59]. Using the X-ray crystallography technique, which involves the crystalline atoms causing an incoming X-ray beam to diffract into a range of various orientations, the atomic and molecular structure of a crystal is identified. These diffracted beams' intensities and angles may be utilized to create a three-dimensional image of the density of electrons inside the crystal. This electron density can be used to get information about the crystal's atoms' typical locations, chemical bonds, disorder, and other characteristics. Small-angle neutron Scattering (SANS) explores these length ranges from the near-atomic (nanometer) to the near-optical (micrometer size). In study fields including polymers, complicated fluids, biology, and materials science, it has been a key characterization technique.

8.8.2.5 MICROSCOPY METHODS CHARACTERIZATION

The common type of microscopy is optical microscopy. It can observe objects as small as 0.2 micrometers and works with visible light. Confocal microscopy, a modern advancement of this method, focuses on certain layers inside the specimen to achieve depth resolution. One nanometer is the smallest size scale that electron microscopy can see. When imaging surfaces or thin samples, electron microscopy is used in either the transmission mode (TEM) or the scanning mode (SEM). Stains are applied to samples to increase contrast.

8.8.2.6 TRANSMISSION ELECTRON MICROSCOPY METHOD CHARACTERIZATION

A beam of electrons is carried through an incredibly thin specimen using the microscopy method known as transmission electron microscopy (TEM), interacting with the material as it does so. When electrons passing through a material interact, an image is created. This image is then amplified, focused, and applied to a fluorescent screen, among other imaging devices.

8.8.2.7 SCANNING ELECTRON MICROSCOPY METHOD CHARACTERIZATION

An electron microscope called a scanning electron microscope (SEM) uses a concentrated electron beam to scan a sample to create pictures of it. The sample's atoms and electron interactions lead to a variety of detectable signals that reveal details about the sample's surface topography and composition. SEM has resolution capabilities greater than 1 nm. In environmental

SEM, specimens can be viewed under wet conditions, in high vacuum, low vacuum, and a variety of increased or cryogenic temperatures. Detection of secondary electrons generated by atoms stimulated by the electron beam is the most used SEM technique. The angle at which the beam strikes the specimen's surface, or the topography of the specimen, is one factor that affects the number of secondary electrons that may be detected. A picture showing the topography of the surface is produced by scanning the sample and collecting the secondary electrons with a unique detector.

8.8.2.8 THERMO-ANALYTICAL METHODS CHARACTERIZATION

Calorimetry involves measuring the heat released and/or absorbed during chemical processes or thermodynamic changes. Differential scanning calorimetry (DSC) is used to calculate the amount of heat absorbed or released by a sample and reference during heating and/or cooling cycles. The DSC signal is made up of the differential in heat flow. In DSC, the reference and measured samples must stay at the same temperature. The heat transfer to the two samples is constant in the Differential Thermal Analysis (DTA) method.

8.8.2.9 DIFFERENTIAL SCANNING CALORIMETRY

The difference between the quantity of heat needed to raise a sample's temperature and the temperature of a reference as a function of temperature is determined using the thermo-analytical technique known as differential scanning calorimetry. Throughout the experiment, the sample and the reference are both kept at temperatures that are nearly identical. According to the temperature program generally used for a DSC investigation, the temperature in the sample holder should increase linearly as a function of time. Over the temperature range that will be scanned, the reference sample should have a clearly defined heat capacity.

8.8.2.10 DIFFERENTIAL THERMAL ANALYSIS

Differential thermal analysis (or DTA) is a thermoanalytical method. DSC has mostly supplanted the previous approach known as DTA. While subjecting the substance under investigation and inert reference to the same thermal cycles, DTA records any temperature differences between the sample and reference [60]. The DTA curve is then used to depict this differential temperature against time or temperature. Relative to the inert reference, changes in the sample, whether endothermic or exothermic, can be found. A DTA curve, therefore, offers information on the changes that have taken place, including glass transitions, crystallization, melting, and sublimation. The heat capacity of the sample has no impact on the area beneath a DTA peak, which represents the enthalpy change. The reference and sample are warmed by the same source.

8.8.2.11 THERMO-MECHANICAL ANALYSIS

Thermal analysis, a field of materials science that examines how a material's characteristics vary with temperature, uses a technique called thermo-mechanical analysis (TMA). Thermo-mechanometry (TM) is a subfield of thermo-mechanical analysis [60]. In essence, dynamic mechanical spectroscopy and dielectric spectroscopy are expansions of thermal analysis that show additional phase transitions by altering the volume with temperature.

8.8.2.12 THERMO-GRAVIMETRIC ANALYSIS

Thermal analysis is also known as thermogravimetric analysis (TGA) and evaluates changes in a material's physical and chemical properties as a function of temperature (with a constant heating rate) or as a function of time (with a constant temperature and/or continuous mass loss) [61]. Physical processes like second-order phase transitions, such as absorption, vaporization, adsorption, sublimation, and desorption, may be studied using TGA. The same is true for other chemical processes that TGA can shed light on, such as chemisorption, and desolvation (particularly dehydration) [62]. TGA is frequently used to analyze certain properties of materials that show mass gain or loss owing to oxidation, breakdown, or loss of volatiles (such as moisture). Applications of TGA are:

1. Determination of a sample's inorganic content, such as ash.
2. Determination of organic content in a sample.
3. Investigations of reaction kinetics and degradation processes.
4. Characterization of materials by examination of distinctive degradation processes.

8.8.2.13 MECHANICAL PROPERTIES METHODS CHARACTERIZATION

Characterizing a polymer's mechanical characteristics often involves calculating the thickness of a polymer layer. The stress-strain characteristics of polymer films can be best described by the tensile strength and Young's modulus of elasticity. The most common approach used to describe this viscoelastic behavior is dynamic mechanical analysis. Rheology, hardness, and viscometry are additional methods.

8.8.2.14 DYNAMIC MECHANICAL ANALYSIS

Material analysis and characterization methods include dynamic mechanical analysis (DMA, sometimes called dynamic mechanical spectroscopy). The viscoelastic behavior of polymers can be best studied using this method. The complex modulus can be determined by applying a sinusoidal load to the material and measuring its strain. This method can be used to determine the material's glass transition temperature and to identify transitions corresponding to different molecular movements. The temperature of the sample is often varied, leading to fluctuations in the complex modulus.

8.8.2.15 YOUNG'S MODULUS

Linear elastic solid materials have a mechanical characteristic known as Young's modulus, also referred to as the elastic modulus. It describes how strain (proportional deformation) and stress (force per unit area) interact in a material.

8.8.2.16 OTHER CHARACTERIZATION METHODS

- **Viscosity Method Characterization:** A viscometer, sometimes known as a viscosimeter, is a device used to gauge a fluid's viscosity. For liquids whose viscosities vary based on the flow conditions, an instrument called a rheometer is utilized. Viscometers are limited to measuring a single kind of flow. Typically, an item either moves through a fluid while it is stationary or moves through a fluid while it is stationary.
- **Rheology Method Characterization:** The experimental methods used to ascertain the rheological qualities are collectively referred to as rheometry. The establishment of connections between deformation or flow and applied stress is one of the objectives of rheology. Rheometers are used in continuous stress or constant shear modes of operation. The rheological characteristic that has to be assessed

determines the best experimental approach to use. This might be the linear viscoelastic characteristics (complex viscosity corresponding to elastic modulus), the constant shear viscosity, the elongational properties, etc. The measured property will rely on the flow conditions during which it is being tested for all genuine materials, even though for certain materials this dependency is vanishingly minimal under specific circumstances. Rheometry is very important for smart fluids since it is the main way to measure the relevant qualities of materials like a magnet or rheological fluids and electro-rheological fluids.

- **Hardness:** The hardness of a solid determines how resistant it is to various kinds of permanent shape changes when a compressive force is applied. Stronger than other materials, metal is one of them. Strong intermolecular interactions frequently characterize macroscopic hardness; however, there are other measures of hardness, such as scratch hardness, indentation hardness, and rebound hardness, due to the complexity of the behavior of solid material with the applied stress. A substance's hardness can be influenced by its plasticity, ductility, strain strength, viscosity, elastic stiffness, toughness, and viscoelasticity. Hard matter is distinguished from soft matter by typical examples such as concrete, certain metals, ceramics, and very hard materials.

8.9 POLYMERS' ROLE IN PHARMACEUTICAL APPLICATIONS

Numerous polymers can be created with desired properties for particular applications, especially in the pharmaceutical and biomedical industries, thanks to the availability of these materials and the ability to modify their various chemical, physical, or biological properties. In this part, the most typical applications will be briefly covered.

8.9.1 POLYMERS USE IN TABLET DOSAGE FORM

To achieve desired dosage form parameters, such as uniformity of content and weight, disintegration time, friability, hardness, dissolution time, crushing strength, tensile strength, and polymeric materials are frequently used in tablet dosage forms for various applications [63]. For instance, when the API was wet granulated by certain polymeric blends, such as maize starch, sodium starch glycolate, and pregelatinized starch, powdered cellulose, repeatability of some tablet manufacturing lines by fluid bed granulation and drying was attained [64]. Polymeric substances are employed as diluents because they are cheap, tasteless, and compatible with the API and other ingredients in the formulation. The majority of the polymeric diluents used in tablet manufacturing are natural substances including pregelatinized starch, alginates, maltodextrin, microcrystalline cellulose, and alginates. Methacrylic acid copolymers (polymethacrylates) are among the synthetic polymers used as fillers [63]. As binders, some polymers that support the formulation's compatibility and free-flowing characteristics are employed, including carbomer, alginates, microcrystalline and powdered cellulose, chitosan, gelatin, polyethylene glycol (PEG), maltodextrin, polycarbophil, polyvinyl pyrrolidone, polydextrose, starch, pregelatinized starch, and cellulose derivatives like ethylcellulose (EC), Alginates, pregelatinized starch, and starch are further used as disintegrants [65]. To limit drug release, several forms of starch derivatives have been created by chemical, physical, or enzymatic changes. Some of these types are included in Table 8.1 [66]. In addition to their earlier uses, smart polymers and stimuli-sensitive polymers are crucial for creating more sophisticated drug delivery systems that will enhance the therapeutic effects of specific medications.

Physical stimuli including temperature, light, pH, ultrasounds, electric fields, and mechanical stress cause these kinds of polymers to vary in their physiochemical characteristics [67].

TABLE 8.1 Different Starch Derivatives are Used as Excipients in Extended-Release Tablets

Modified Starch	Tablets Structural Features	Drug Release Characterizations
Retrograded Starch	Denser Gel; less swollen and more enzyme-resistant structure	A constant drug release for over 6 h
Cross-linked Starch	A gel layer rapidly formed on the surface on the surface of the tablet	A quasi zero-order release
Carboxymethyl Starch	A pH-sensitive hydrophilic matrix; lower pH inducing a fast swelling; stable gel layer formed in the presence of Na+ ion	A sustained release property (degree of substitution [DS] 0.1–0.2); a delayed release property (DS 0.9–1.2)
Starch Acetate	Less hydrophilic matrix; cracks formed on the central horizontal plane of the tablet	Fickian diffusion
Grafted Starch	A grafting chain-derived acrylic layer led to enzymatic resistance and hydrophobic behavior	Fickian diffusion is the principal mechanism in most formulations. To some drugs, the release is an anomalous transport
Debranched Starch	Slow penetration of a solvent front into the tablet; good compatibility	An almost constant (zero-order) drug release; a swelling-controlled solvent

8.9.2 POLYMERS USED IN CAPSULE MANUFACTURING

The majority of capsule shells (hard shells) are manufactured from gelatin obtained from animals. They are harmless, reasonably priced, and have a high safety profile. Nevertheless, to fulfill the needs of certain patients, a non-animal-derived alternative is required for a number of factors. Hard capsule shells were initially created using hydroxypropyl methylcellulose (HPMC) or hypromellose approximately 20 years ago [68]. A supplementary gelling ingredient like kappa-carrageenan was present in those capsule shells. Since HPMC does not gel on its own at low temperatures, the use of a secondary gelling agent is required. There are more recent HPMC capsules on the market that have not been produced with a supplementary gelling ingredient or gelling promoter [69]. In contrast to firm gelatin, once the temperature rises beyond 30°C, the dissolving and disintegration of HPMC capsules will slow down. As a result, drinking cold water is recommended when taking HPMC pills. Due to the fact that some of these capsules' contents are from vegetable sources, they are more widely accepted by consumers than hard gelatin capsules [70]. Polymers like cellulose acetate phthalate (CAP) and other substances (such as methacrylic acid copolymer and formaldehyde) have been employed as coating materials in delayed-release capsules (like diclofenac capsules). With less than 10% drug release in acid and more than 75% in buffered conditions, the resulting capsules demonstrated gastro-resistant qualities [71].

8.9.3 POLYMERS USED IN DISPERSE SYSTEMS

By making the fluid more viscous, certain polymers are utilized as suspending agents in suspensions. Alginates, acacia gum, tragacanth, guar gum, bean gum, carrageenan, xanthan gum, microcrystalline cellulose, and powdered cellulose are examples of natural polymers. Semi-synthetic polymers include methylcellulose, sodium carboxymethylcellulose, HPMC, hydroxyethyl cellulose, and carboxymethylcellulose. Synthetic polymers include carbomers [72]. Cross-linked polymers called carbomers are employed to increase viscosity because, when neutralized, they may retain and absorb water molecules. Numerous different types of carbomers are employed in numerous dosage forms for pharmaceuticals. Carbopol 934, Carbopol 971P, Carbopol 974P, and PEMULEN TR-1 are the polymers that are most often utilized in dispersed systems [73]. Methylcellulose (MC) comes in a variety of MWs and polymerization levels. Additionally, in syrups that use sugar as a base, MCs are employed instead of sugar. Due to its qualities that enhance viscosity, MC in suspensions can lengthen the period that therapeutic drugs are in touch with the stomach and increase the amount of time required for settling [71].

8.9.4 POLYMERS USED IN TRANSDERMAL DRUG DELIVERY SYSTEMS

The API is mostly integrated into a polymeric matrix and is found in relatively high concentrations in transdermal drug delivery systems (TDDS). This sort of drug delivery system's polymers should enable optimal drug diffusion and release. In addition to being stable, safe, cheap, and simple to make, polymers should be able to include a significant amount of the active ingredient [74]. In contrast to the matrix system where the polymeric particles interact with the drug and govern the releasing mechanism, the reservoir system produces a controlled release mechanism by separating the polymer matrix from the adhesive layer and introducing a rate-controlling membrane [75]. A range of polymers may be utilized in this kind of drug delivery system for a variety of objectives. A biodegradable polymer called rosin is made from pine trees. Despite having cutaneous toxicity, rosin has a film-forming capacity that makes it useful for creating drug delivery systems based on films. Polymerized rosin doesn't exhibit dermal toxicity, though. In the study, rosin, polyvinyl pyrrolidone, and dibutyl phthalate were polymerized to generate the polymeric matrix of a transdermal drug delivery system for diltiazem hydrochloride [74]. The resulting films were smooth flexible films with a greater percentage of elongation that had increased tensile strength. These patches showed promising pharmacodynamics and pharmacokinetic capabilities after being tested in animal models [76].

8.9.5 POLYMERIC OCUSERTS

Ocuserts are ocular inserts that are available as solid devices. They are made to administer medications and offer a longer period of drug retention in the eye than liquid versions. You insert ocuserts into your conjunctival sac. They frequently consist of a polymeric vehicle that has been drugged [77]. Based on their potential biodegradability and solubility behavior, the polymeric materials utilized in this drug delivery system may be divided into 3 kinds

(Figure 8.1). The first kind of ocusert is completely soluble and doesn't require removal after being deposited into the eye. However, because of the glassy nature of the insert, there is a higher risk of expulsion with this kind of ocusert. Due to the solubilization process that occurs when the components of the polymeric ocusert are exposed to lacrimal fluid, vision is blurred [23]. Soluble polymers used in this drug delivery system include natural polymers (e.g., collagen) and synthetic polymers (e.g., cellulose derivatives such as hydroxypropyl cellulose, hydroxyethyl cellulose, methylcellulose, and hydroxypropyl cellulose, divers such as ethylene vinyl acetate copolymer and polyvinyl alcohol, cellulose acetate phthalate, polyvinyl-pyrrolidone as a complexing agent, and bioadhesives such as polyacrylic acids [77]. The second category is the insoluble ocuserts, which may be divided into 3 categories: hydrophilic contact lenses, osmotic systems, and diffusion systems. Systems including a liquid, colloid, gel, carrier holding the medicine, or a solid matrix are called diffusion and osmotic systems. Organic, inorganic, hydrophilic, hydrophobic, synthetic, or natural polymers can all be used to make drug carriers. Although contact lenses are typically used to correct vision, they can also be used as a drug delivery system and for vision correction when presoaked in a drug solution [79, 80]. The biodegradable ocular insert, which is the third form of ocusert, is made from bioerodible polymers that go through hydrolysis, such as polyester derivatives and cross-linked gelatin derivatives. The system's capacity to adjust its degradation rate is its key positive feature [81].

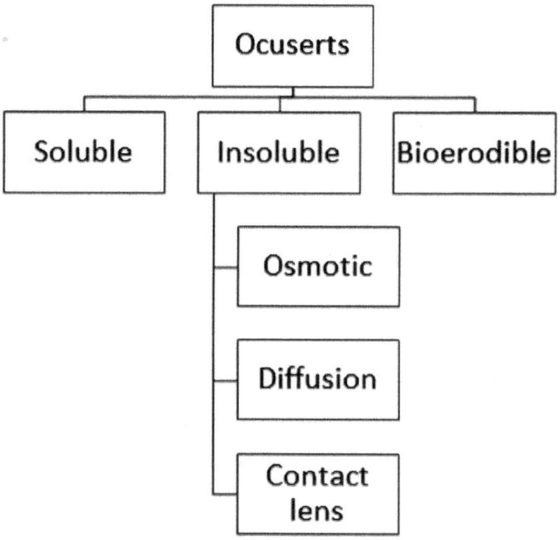

FIGURE 8.1 Classification of ocuserts.

8.9.6 PROGESTASERT SYSTEM

Systems for Progestasert were created about 40 years ago. Daily progesterone release from these has effective contraceptive effects without impairing ovulation [82]. The first hormone-releasing intrauterine device (IUD), Progestasert, has 38 mg of progesterone microcrystallized. Progestasert was pulled off the market about 30 years ago due to a multitude of problems, including the possibility of an ectopic pregnancy [83]. In an intravaginal therapeutic system, the controlled release of medications is possible using specific non-absorbable thermoplastic polymers. IUDs frequently utilize polymers with enhanced stiffness, such as ethylene-vinyl acetate copolymers [84]. PE, poly (methyl acrylate), plasticized PVC, poly(butyl methacrylate), plasticized nylon, plasticized PE terephthalate, polyacrylonitrile, poly-4,40-isopropylenediphenylene carbonate, poly(trifluorochloroethylene), and PE vinyl esters are further thermoplastic polymers that are utilized [85].

8.9.7 POLYMERS USED IN SWELLING CONTROLLED DRUG DELIVERY SYSTEM

Devices that exclusively use swelling to control the medication's release rate are referred to as swelling-controlled drug delivery systems. By continuously introducing liquid into a polymer, swelling, a characteristic of polymer chain relaxation occurs, which breaks the hydrogen bonds created during tablet compression and triggers the creation of new hydrogen bonds [86]. This kind of regulated drug delivery system is frequently built on somewhat hydrophilic polymers, primarily cellulose ethers, including hypromellose (e.g., HPMC and MC) [86, 87]. Another illustration is cross-linked gelatin, which may be used as a swelling-controlled drug release system. An example is the gelatin B emulsion cross-linked by genipin, which results in tiny nanoparticles that exhibit this sort of release mechanism [88].

8.9.8 POLYMER-DRUG CONJUGATES

In polymer therapies, the therapeutic substance is chemically attached to a polymer rather than being contained within it. This results in polymer-drug conjugates. Ringsdorf proposed a paradigm for macromolecular prodrugs in the mid-1970s. His model included a bioresponsive linker, a targeting moiety, and a therapeutic medication with a low molecular weight that is covalently bonded to the polymer [89, 90]. Polymer-drug conjugates offer significant advantages over pristine drugs, such as increasing the water solubility of lipophilic drugs, protecting drugs from degradation, extending plasma half-life, changing biodistribution, increasing drug targeting, and controlling drug release under specific conditions like a specific pH or the presence of certain substances. The conjugation of drugs results in significant changes in their behaviors [91]. Polyglutamic acid (PGA), PEG, and HPMA copolymers make up the majority of polymer carriers in polymer-drug conjugates. Unfortunately, biopersistent carriers like PEG and HPMA are responsible for several drawbacks associated with the prolonged administration of parenteral medications or with the use of large dosages, including the potential to develop lysosomal storage disease syndrome [92]. PEGylation is a common technique that involves attaching a medicinal substance, usually a protein, to PEG. PEG-protein conjugates are significant due to a number of characteristics, such as their capacity to shield the protein from enzymatic degradation and lessen its absorption by the reticuloendothelial system; these characteristics are provided via straightforward steric hindrance [89]. Many PEGylated Protein Conjugates, including PEG-L-Asparaginase for the treatment of Leukemia, have been approved for clinical use. The three most important polymer-drug conjugates of the first generation are polyglutamic acid with paclitaxel, polystyrenemaleic anhydride-neocarzinostatin conjugates, and camptothecin (CT-2106) conjugates [94]. The clinical development stage has been reached by a number of drug-polymer conjugates, including linear cyclodextrin-bound camptothecin (IT-101) and HPMA copolymer-doxorubicin galactosamine (PK2) [95]. Antineoplastic medication doxorubicin has a limited therapeutic index and cardiotoxic side effects. To decrease systemic toxicity and improve medication targeting, many nanodrug carrier systems were created. Doxorubicin has been conjugated with the HPMA copolymer, and the drug conjugation demonstrated enhanced cardiovascular tolerance as neither blood pressure nor heart rate was impacted in rats treated with the drug conjugate as opposed to other rats given free-form doxorubicin. When employing HPMA carriers for sustained drug administration, care must be exercised since HPMA copolymers were discovered to induce microscopic damage to the heart tissue [96]. In order to create a medication delivery system with decreased toxicity, the antitubercular medicine isoniazid was conjugated with a PEG derivative. When compared to pure isoniazid, the cytotoxicity of the PEG-bis(isoniazid) polymer-drug conjugate was reduced. The capacity to reduce dosage frequency was made possible by increased medication accumulation and bioavailability at the injection site [97]. In addition to systemically delivered medications, the creation of inhalable polymer-drug conjugates has been studied using PEG conjugates. PEG-based drug conjugates improve the drug's penetration through the mucosal layer, which can lower the exposure of the mucosa to medications that harm the mucosa. Alendronate, for example, has the ability to harm mucus because of its structural resemblance to phosphatidylcholine, which results in the competitive displacement of mucosal phosphatidylcholine and results in mucosal injury. Rats receiving pulmonary injections of alendronate combination and low molecular weight PEG experienced less lung mucosal damage than those receiving the medication in its free form [98].

8.9.9 POLYMERIC NANOPARTICLES

Due to their tiny size, polymeric nanoparticles (PNPs) have several special characteristics. The potential to integrate therapy with imaging (theranostics), regulated release, an enhancement in the therapeutic index, and preservation of drug molecules are all benefits of using PNPs as carriers for therapeutic agents. PNPs have been created using both synthetic and natural polymers, such as proteins and sugars [99]. PNPs come in both nanosphere and nanocapsule forms. The therapeutic chemicals can either be enclosed within the nanosphere or adsorbed on its surface since nanospheres are matrix particles, which means their whole mass is solid. Nanocapsules are vascular systems that function as reservoir systems. They are made of a solid material shell that encloses a liquid core that is either filled with water or oil [100]. Gelatin, sodium alginate, albumin, and chitosan are a few of the natural polymers typically used in PNP formulation. In addition, commonly used synthetic polymers include polylactide (PLA), PLGA, polyanhydrides, polyglycolide (PGA), polyorthoesters (POE), PCL, polycyanoacrylates, poly(malic acid) (PMLA), polyglutamic acid (PGA), PEG, poly(N-vinyl pyrrolidone) (PVP), poly(methacrylic acid) (PMAA) [101], polyglutamic acid, and polyorthoesters [102]. Abraxane, which is used to treat breast cancer and is made up of paclitaxel loaded in an albumin protein-based nanoparticle, is an example of a polymeric nanoparticle drug delivery system that is currently on the market. Abdoscan, a dextran and iron oxide-based nanoparticle utilized in diagnostic imaging of the liver and spleen, is yet another illustration [103]. It has been examined whether oral chemotherapy treatments may be encapsulated in nanoparticles to increase their absorption and provide a different route of administration for medications that are now exclusively given intravenously. In a recent study, docetaxel exhibited enhanced absorption in guinea pigs when it was enclosed in nanoparticles. This study demonstrated that these nanoparticles were transported through the lymphatic system, avoiding the first-pass effect. This elevated drug level in the blood was greater than that of intravenously administered Taxotere, leading to greater tumor growth inhibition and greater survival rates in mouse models with metastatic lung cancer [104]. Another study used sodium alginate nanoparticles, a biodegradable natural polymer, to encapsulate the anticancer medication carboplatin. It has been demonstrated that encapsulating various formulations increases effectiveness by 52.24%–68.70%. When compared to nonencapsulated formulations, which only had a 3-hour duration, the release pattern *in vitro* demonstrated a prolongation of the drug's duration of action to almost 12 hours. In mice, greater concentrations of parenterally given encapsulated formulations have been seen in the liver, spleen, and lungs [105]. When compared to pure rifampicin, gelatin nanoparticles containing the antibiotic showed better biodistribution and a higher concentration of the particles was localized in different organs. These nanoparticles had sustained release, improved the drug's area under the curve (AUC), and significantly decreased the number of germs in mice with TB infections in the lungs and spleen [106]. In a different study, doxorubicin-loaded PEG-PLGA nanopolymersomes demonstrated lower toxicity and increased accumulation at the tumor site in mice with murine breast tumors. According to *in vivo* research [107], these nanopolymersomes were stable in blood circulation and prevented doxorubicin from being released. Additionally, polymeric nanoparticles have been researched as vaccine delivery systems. In a recent study, Newcastle disease viruses (NDV) were added to chitosan-based nanoparticles, and the resulting particles demonstrated a high level of safety, minimal toxicity, and favorable biological properties. When compared to chickens immunized with the commercial Newcastle disease vaccine, chickens immunized with the chitosan-based nanoparticles elicited a stronger immune response [108]. Polymer-based nanomaterials have their advantages and disadvantages in versatile areas of applications; some of these are summarized in Table 8.2 [109].

TABLE 8.2 Polymeric Nanomaterials, Their Applications, Advantages, and Disadvantages

Classification	Polymer	Advantages	Disadvantages	Application Area
Natural polymeric material	Cellulose	Extensive sources.	Rare adverse reactions.	Pharmaceutical adjuvant.
	Starch	Extensive sources low prices, on-toxic.	Poor mechanical properties, and poor blocking performance.	Hemostasis material, tissue-engineered scaffold.
	Chitosan	Biocompatibility, antimicrobial.	Poor spin-nability, poor strength.	Hemostasis material, medical dressing'
	Alginate	Hypo toxicity, biocompatibility.	Bad biodegradability.	Pharmaceutical excipient.
Chemosynthesis Material [Copolymer]	Polymethyl methacrylate resin	good biocompatibility.	The monomer has cytotoxicity.	dental materials, artificial crystal.
	Poly(lactic-glycolic acid)	Controllable biodegradability.	Higher cost.	Absorbable suture.
	polyurethane	Low cost, rich resource.	Degradation speed slow	Excipient medical bandages
Biosynthesis Material	Polylactic acid	safe, non-toxic.	Poor toughness, degradation speed slow.	Antiadhesion materials, patches.
	Poly B-hydroxybutyrate	Biodegradable, safe.	bad thermal stability.	Drug delivery carrier.

Peptides, proteins, and nucleic acids are examples of chemical substances that either undergo fast enzymatic degradation or metabolization upon delivery. The notion that nanotechnologies may be used to alter or even regulate the medication distribution at the tissue, cellular, or subcellular levels has arisen as a result of this. Polymer-based nanoparticles, which have been created since the early 1980s when advancements in polymer chemistry permitted the fabrication of biodegradable and biocompatible materials, are one of the methods used for drug targeting. Nanoparticles are submicron (1 μm) colloidal systems that are often made of polymers. Nanoparticles are thus colloidal structures that are 7 to 70 times smaller than red blood cells in size. They can be injected intravenously without embolization risk. Nanoparticles can be made into nanospheres or nanocapsules, depending on the technique used to make them. Nanospheres are matrix systems where the medicine is evenly distributed throughout the polymer of the particle. Contrarily, nanocapsules are vesicular systems made of an aqueous or lipophilic liquid core that contains the medicine and is encircled by a single polymeric membrane.

8.9.10 POLYMERS USED IN OCULAR DELIVERY SYSTEMS

By absorbing water to create viscoelastic gels, both synthetic polymers and natural polymers are employed for their viscosity-enhancing qualities, which improve the residence duration in ocular medication formulations. Polyvinyl alcohol, cellulose derivatives (like HPC and HPMC), poly (vinyl pyrrolidone), natural mucopolysaccharide, and carbomers, which are made of weakly cross-linked polyacrylic acids, are polymers used in ocular delivery systems to increase the viscosity of the preparation. In contrast to systems that increase viscosity, *in situ* gelling systems have the benefit of undergoing a phase change from liquid to gel. Thermogelling systems such as tetronics, pluronics, and poloxamers are some of these [110].

The ability of mucoadhesive polymers to attach to conjunctival mucin via noncovalent bonds—that is, the fact that they are macromolecular hydrocolloids with several hydrophilic functional groups, including hydroxyl (OH), carboxyl (COOH), sulfate, and amide—can also lengthen residence time [111]. Polysaccharides (xanthan gum, carrageenan, gellan gum, pullulan, scleroglucan, and guar gum), chitosan, pectin, xyloglucan, polygalacturonic acid, hyaluronan (HA), and cellulose derivatives are a few of the polymers that have mucoadhesive qualities [112].

8.9.11 POLYMERIC DRUG DELIVERY

Due to their biodegradability, a variety of natural polymers and synthetic polymers can be employed as drug delivery methods. The medicine may be confined within a polymer membrane or enclosed in a polymeric matrix, from which it may diffuse into neighboring tissues upon implant. In rare circumstances, the polymer's erosion and dissolution might also have an impact on the release mechanism. Chitosan is one of the polymers that has been successfully used to deliver proteins, peptides, anti-inflammatory drugs, antibiotics, growth factors, gene therapy, and bio-imaging applications. It is also used as a drug conjugate and a biodegradable release system. In addition to polyanhydrides, poly (ortho esters), polyacetals, polyphosphoesters, polylactide, polycarbonates, and PAs are also employed as drug delivery systems [117]. Due to their adjustable and adaptable sensitivity, smart polymers provide the opportunity to improve accuracy and create more programmed medication administration. According to Table 8.3, each class of these polymers has distinct benefits and drawbacks when used as drug delivery systems. Temperature-sensitive polymers are classified into two classes—UCST systems and LCST systems—based on how they react to various temperatures, as was previously noted in Section 8.9.10. Because UCST systems need higher temperatures than are suitable for thermolabile drugs, the LCST system is more appropriate as a drug delivery system [118]. For instance, block copolymers of PPO and PEO are biodegradable thermoresponsive polymers that go through the sol-gel transition at physiological and ambient temperatures. In a recent experiment, the MR contrast agent [Gd(HPDO$_3$A)(H$_2$O)] and the anticancer medication doxorubicin were combined in temperature-sensitive liposomes and injected into tumor-bearing rats (rhabdomyosarcoma). In comparison to the free form of the drug, the study showed a significant increase in the drug concentration inside the tumor and increased cellular absorption of doxorubicin into the interstitial space. It used a non-intrusive heating system with accurate temperature control. Due to its LCST of 32°C, which is reasonably near to the physiological body temperature, poly(N-isopropyl acrylamide), or PNIPAM, is a temperature-sensitive polymer that has undergone significant study and has been used in several drug delivery systems. In reaction to temperature variations, the PNIPAM solution experiences reversible and quick phase separation/phase transition events. A number of methods, such as the addition of salts or surfactants, or the copolymerization of hydrophobic or hydrophilic monomers, can change the LCST of PNIPAM. When PNIPAM is crosslinked into hydrogels, the volume of the gel will quickly decrease due to the phase change from the coil to a globule, which releases the entrapped solvent and drug in a fast manner, allowing for more linear diffusion and a controlled drug release [119]. Additionally, PNIPAM has been created as microgels, which offer the beneficial qualities of both hydrogels and nanoparticles. In PNIPAM microgels that were manufactured by many researchers and loaded with insulin, the microgel carrier demonstrated its capacity to prevent denaturation of the insulin. The hydrogel can increase a drug's bioactivity while shielding it from low pH and enzymatic activity. The physical or chemical cross-linking of polymers can be used to modify or alter the characteristics of hydrogels. Hydrophobicity, chemical makeup, and cross-linking density all affect a hydrogel's capacity to gel. Numerous applications involving the delivery of genes, proteins, and drugs have made use of hydrogel [120].

TABLE 8.3 Smart Polymeric Drug Delivery Systems: Major Advantages and Limitations

Types of Smart Polymers Based on Stimulus	Advantages	Disadvantages
Ultrasound	Controllable protein release	Surgical implantation is required for a nonbiodegradable delivery system, Specialized equipment for controlling the release
Mechanical stress	The possibility of achieving drug release	Controlling the release profile is difficult.
pH	Suitable thermolabile drugs	Lack of toxicity data
Light	The trigger mechanism is easily controlled accuracy of stimulus control	The low mechanical strength of the gel.
Electric field	Drug release is pulsative with the electric current change	Surgical implantation is required.
Temperature	Active moieties can be easily incorporated	Injectability issues under application conditions

8.9.12 SHAPE MEMORY POLYMERS

Due to their mechanical utility, shape memory polymers have attracted a lot of research.

These polymers can transform from a preliminary shape to a final shape and vice versa. A magnetic field, a pH shift, light, or heat can all trigger a form change in an object. These polymers' capacity to alter form, particularly by expanding, has made them ideal for minimizing invasion during procedures that include placing a certain type of implant within the body. The replacement of heart valves, sutures, medication administration, orthodontic uses, and catheters are more examples of biomedical applications. Polyesters such as poly-p-dioxanone (PPDO), polyhydroxyalkanoates (PHAs), PCL, and PLA are examples of biodegradable shape memory polymers (BSMP) [121]. Numerous types of shape memory polymeric stents have been researched, and some of them have shown promising results, such as the ability to fully expand at 37C in 25 s for stents containing 25 wt% poly[(R)-3-hydroxybutyrate-co-(R)-3-hydroxyvalerate] (PHBV) [122]. A self-expanding, biodegradable stent has been created using blends of PLGA and PLA in a specific ratio. The stent demonstrated strong hemocompatibility and self-expansion, making it a superior choice over metallic stents. A crosslinked PEG-PCL copolymer network with a melting point close to body temperature was used in another study to create a shape memory drug-eluting stent. The stent discharged the loaded pharmaceuticals (curcumin and mitomycin C) after its shape memory performance. Results from the investigation of a crosslinked polycaprolactone and poly (sebacic anhydride) (PSA) composite as a drug-eluting system demonstrated that this system has significant promise in biomedical applications, particularly as drug-eluting stents. Segmental bone defects were studied as a potential therapy for shape memory polymers (SMPs). Degradable shape memory polymers have the major benefit of eliminating the need for the postoperative removal of external fixators or metal plates, which are utilized to stabilize the body throughout the healing process. After being implanted in male rats for a year, biodegradable nanoparticles with a polyhedral oligomeric silsesquioxane (POSS) core and eight poly(D, L-lactide) (PDLA) arms showed no pathologic abnormalities, making them a potential candidate for tissue repair applications [123]. Shape memory polymer (SMP) foam formulations demonstrated excellent results in *in vitro* studies of peripheral embolization devices to evaluate their efficacy and safety, including the safety of the mechanical properties of these devices that reduce the risk of rupturing or perforating blood vessels, and lower device stiffness compared with conventional embolic devices, which provide better access to tortuous vessels. However, *in vivo* research is necessary for biocompatibility testing as well as additional safety and efficacy analysis [124].

8.9.13 STIMULI-SENSITIVE POLYMERS FOR CANCER THERAPY

The main issue with cancer therapy is that it selectively targets cancer cells. One of the targeted options being investigated is the use of a tumor stimuli-sensitive delivery system that depends on the typical physiological imbalances of tumors, including hypoxia, increased production of certain enzymes, tumor acidosis, and redox imbalance. In order for an appropriate quantity to reach the tumor site, the polymers utilized in this system need to be stable in physiological fluids. They need to release the therapeutic agent quickly and precisely where it is needed. N-(2-hydroxypropyl) methacrylamide (HPMA), a copolymer that when conjugated with pirarubicin by hydrazone link generates a drug conjugation that releases pirarubicin in the acidic surrounding media of the tumor, is an example of a polymeric material that is employed for this purpose [125]. PEG-polyaspartate block copolymer, which is conjugated to epirubicin by a hydrazone linker, is another polymeric conjugation that is triggered by acidity. In recent work, prodrug conjugates were created by establishing azobenzene linkages between combretastatin (CA4) and polyethylene glycol hexane thiol (PEG-C6). These prodrug conjugates are accelerated under hypoxic circumstances. To enclose the anti-cancer agent, these drug conjugates have the capacity to self-assemble into micelles. There have been studies on a number of nanocarriers, and while the results are encouraging, more work has to be done to address biocompatibility and biodegradability issues. In cancer treatment, thermos-responsive devices have undergone extensive research.

For the purpose of targeting and imaging prostate cancer, iron oxide nanoparticles were coated with a thermosensitive polymer and coupled with R11 peptides. The iron oxide nanoparticles are guided to the desired location by a magnetic field. When compared to nonconjugated nanoparticles, the nanoparticle conjugate demonstrated higher accumulation in the tumor site. In magnetite nanohydrogels made of PEG, poly(N-isopropylacrylamide-co-itaconic anhydride) (P[NIPAAm-co-IA]), and Fe_3O_4 nanoparticles, doxorubicin hydrochloride was loaded. The resulting system released very little medication under physiological settings, but it released the most medication at a higher temperature of around 41°C and a pH level of about 5.3. The delayed drug release profile and great drug loading capacity of this system were also present. This smart drug delivery system's *in vitro* cytotoxicity studies and cellular uptake results were higher, making it a promising drug delivery system. In cancer treatment, several different thermoresponsive systems have been researched [126]. Over the past 10 years, researchers have looked into ultrasound-responsive drug delivery systems. They discovered that ultrasound caused therapeutic agents to be released from liposomes, multilayer capsules, microemulsions, and polymeric micelles. Deep tissue penetration by ultrasound is noninvasive. Following the application of ultrasound, a number of physical factors, including localized hyperthermia, pressure changes, cavitations, and acoustic fluid streaming, can cause the release of drugs. To improve the targeting of cancer cells, drug delivery methods, and stimulus-responsive polymers that react to one or more of these stimuli have been created. Poly[bis(p-carboxyphenoxy)] alkane anhydrides, polylactides, and polyglycolides are examples of polymers that degrade more rapidly under ultrasound. To circumvent the efflux pump-mediated mechanism of drug resistance in hepatocellular cancer, ultrasound-responsive nanobubbles filled with paclitaxel and siRNAs targeting antiapoptosis genes were created. It was shown that applying low-frequency ultrasound to the tumor location enhanced medication penetration into the tumor, which reduced tumor volume in mice. The capacity to regulate drug release by endogenous or external stimuli is the major benefit of stimuli-responsive drug delivery systems. Unfortunately, these systems are easily influenced by a variety of outside factors, which can lead to issues like poor drug release accuracy and undesirable side effects. Delivery devices with multi-stimuli responsiveness have been developed to address these issues and improve the precision of medication release in the challenging diseased setting [127].

8.9.14 POLYMERS USED IN IMPLANTS

The fundamental goal of an implant is to prolong the functional range of biological systems beyond their natural lifespans in order to protect human lives. The most important prerequisite for implants and devices is compatibility with the biological environment. The substrate used to create implants and the packaging sheet used to enclose devices both need to be biocompatible. Specific conditions must be met for implanted item survival and patient comfort, and these needs vary depending on the implant's functioning and the implantation site [134]. Polyesters like Dacron, silicone rubber, and high-density polyethylene are examples of synthetic polymers that have been investigated for use in medical applications. In a number of clinical applications, inert synthetic polymers have recently been produced and are being utilized effectively. For instance, heart valves, orthopedic fixation, hip and knee implants, coatings on implants to improve blood compatibility, and cardiovascular devices like pacemakers [135]. Teflon-coated hip implants minimize friction, which is likely to increase persistence throughout the device's anticipated lifespan. To encase implanted devices, polymer coatings including perylene, silicone rubber, epoxy, and PTFE have been employed as coating materials. These materials have, however, only partially shown to be effective in preventing water from entering the implanted device. Polyurethane elastomers based on polyether are more processable, stable in the biological environment, and biocompatible. These characteristics have made them suitable for usage in a range of tissue- and blood-contacting biomedical devices. Due to its great mechanical strength, PLA, a polyester polymer, is mostly utilized in load-bearing biomedical applications such as internal fixation devices (including pins, screws, plates, and rods) to keep shattered bones together. However, several studies revealed that PLA disrupts the process by causing an imbalance in the number of osteoclasts and osteoblasts during the *in vivo* bone remodeling phase. Therefore, the vast majority of commercially available polyester-based products are for non-load-bearing applications. Another polyester known as PCL has been utilized to create tissue repair patches, including those made by Ethicon Inc. (Edinburgh, United Kingdom), and it is also employed as a filler to cover non-load-bearing bone voids [136].

8.9.15 FILM COATINGS AND ORAL FILMS

Tablet, pellet, and capsule film coatings have often employed polymer mixes. A secondary polymer may be used with a main film former to enhance substrate adherence, alter the coating's mechanical characteristics, increase moisture resistance, or regulate medication delivery. Another justification for mixing polymers might be cost-saving measures. Many film coating systems that are sold commercially are based on polymer blends. These include both functional (controlled release) and non-functional (immediate release) coating products. Non-functional coatings are used to disguise flavors, identify products, improve stability (against moisture and light), and improve aesthetic appeal. Low-viscosity HPMCs are the polymers that are utilized most frequently for non-functional coating. Plasticizers must be added to HPMC when it is used to coat tablets in order to increase film flexibility. For HPMC films, polyethylene glycol (PEG) is frequently employed as a plasticizer. Film-substrate adhesion is another mechanical characteristic that may be changed by combining other polymers with HPMC. An example of a polymer that has been combined with HPMC to enhance film adherence is hydroxypropyl cellulose (HPC). When combined with HPMC, HPC has also been shown to lessen logo bridging and edge cracking in films [128]. Polymeric blends are also used in nonfunctional film coatings to speed up the coating process, produce dry, completely forming coating systems, and enhance medication stability by lowering water vapor permeability. Polymer mixtures have been utilized a lot as functional film coatings for applications requiring controlled release. Functional film coatings can be applied using organic solvent-based solutions; however, latex dispersions in aqueous systems are preferred. These systems are colloidal dispersions of polymers like ethylcellulose and polymethacrylates that are insoluble in water. In addition to altering drug release rate and mechanical properties, polymeric additives may be used to improve the controlled-release film coatings' long-term dissolution profile stability [129]. The tableting of enteric-coated particles is an illustration of the usage of a polymer mix to provide optimal mechanical characteristics. Films of the enteric polymer Eudragit® L 30D-55 have a tendency to be brittle and shatter under compression when tableting coated particles. While lacking enteric functionality, the polymer Eudragit® NE 30D is incredibly flexible and its films, when coated onto particles, maintain integrity during compression. The markets for nutraceuticals and nutritional supplements are drawn to film coatings made from natural or generally regarded as safe (GRAS) components. This is due to daily consumption limitations on the common enteric polymers used in medicinal goods. Utilizing GRAS polysaccharides with anionic functional groups has been one method for creating enteric film coatings for nutritional products. Based on this idea, Nutrateric®, a coating technology, was created. It uses sodium alginate and methylcellulose together. Ethyl cellulose is the main film former in this mixture. Due to the carboxylic acid groups of the alginate polymer being ionized in a pH-dependent manner, it offers intestinal protection. It is insoluble at low pH levels, but dissolves at high pH levels, creating holes in the ethyl cellulose covering that aid in drug release and disintegration. A mixture of alginate and pectin-based enteric coating has also been documented [130]. Lingual (for enteral), sublingual, and buccal medication formulations all employ oral thin films. ODFs, also known as orodispersible films, are a practical dose form for children and people who have dysphagia. As matrix systems, oral films' matrix properties—including drug loading, handling, mucoadhesion, and disintegration times—as well as their mechanical characteristics and handling, are all crucial. In a patented method, the polymer ratio is adjusted to control the rate of drug release, mucoadhesion, and tearing resistance of the film made of PEO and cellulosic polymers. Additionally, polymers like PEO and HPC can be used in oral films created by wet casting in place of low molecular weight plasticizers due to their inherent flexibility [131].

8.9.16 BIOAVAILABILITY ENHANCEMENT AND MELT EXTRUSION

Polymer blends have been utilized to improve drug solubility, especially when using processes like hot melt extrusion (HME), which may be used to create solid dispersions. The HME technique uses high mixing and heat energy to melt or soften materials, which helps medicinal compounds dissolve or disperse finely in a polymeric matrix. It is feasible to create amorphous or microcrystalline dispersions with enhanced drug solubility under specific circumstances. To avoid crystallization, manage medication dissolution, and avoid supersaturation, solid dispersions may need to have a balance between hydrophobicity and hydrophilicity. Multiple polymers may be required to meet these requirements. For binary polymer blends, synergistic effects on ritonavir crystal growth inhibition have been documented. Intriguingly, spray-dried isradipine solid dispersion combinations comprising a hydrophilic polymer (HPMC) and hydrophobic polymer (zein) were found to lessen drug crystallinity and improve drug dissolving more efficiently than the individual polymers [132]. Melt extrusion is used to make the Noxafil® and Nucynta® ER products. HME is offered as an alternate way of producing sublingual films, which are typically produced using wet casting methods. In solid dispersions, blends of low molecular weight polymers like poloxamers and polyethylene glycols are utilized. In the patented Meltdose® technology, tacrolimus is dissolved in a solution of PEG 6000 and poloxamer 188 to produce a solid dispersion that is sprayed onto a lactose carrier. Implants and inserts are two examples of delivery devices made using HME. The first pharmaceutical melt-extruded product was Lacrisert®, an ocular insert intended to treat dry eye syndrome and manufactured entirely of hydroxypropyl cellulose. DurystaTM (bimatoprost), a recently authorized (2020) ocular sustained-release implant that is delivered via intracameral injection, is an example of a device that mixes many polymers. Patients with open-angle glaucoma or ocular hypertension can use the product to lower intraocular pressure. The biodegradable polymers poly(D, L-lactide), poly(D, L-lactide-co-glycolide), and poly(D, L-lactide) with an acid end group are used to create the implant. When compared to using eye drops every day, the DurystaTM implant offers a significant advantage in that bimatoprost is intended to be released in 3–4 months. Extruding a mixture of polymers and active ingredients into filaments, which are subsequently cut to the implant's dimensions, produces the implant. These polymers hydrolyze *in vivo*, producing natural carboxylic acids, as opposed to the HPMC-type polymers discussed above, which are physiologically inactive *in vivo*. The length of drug release and mechanical characteristics of the implant may be customized using various biodegradable polymer chemistries. PLA degrades more slowly than PLGA because it is less hydrophobic, whereas polymers with (carboxylic) acid endcaps are more hydrophilic than those with ester endcaps. By stimulating the hydrolysis of ester connections within the polymer backbone, the acid end group additionally accelerates the rate of breakdown [133].

8.9.17 *MODIFIED-RELEASE DOSAGE FORMS*

It is now widely acknowledged that employing quick-release dosage forms for many therapeutic drugs leads to poor treatment and/or systemic adverse effects. Modified-release dosage forms have been created by pharmaceutical experts in an effort to alleviate the drawbacks of traditional oral dosage forms.

8.9.18 *EXTENDED-RELEASE DOSAGE FORMS*

Drugs with a short biological half-life may have a greater therapeutic impact if they are formulated in prolonged or sustained-release dosage forms. Extended and sustained-release dosage forms decrease the number of doses the patient needs to take to maintain a therapeutic effect, enhancing compliance while extending the duration that systemic drug levels are within the therapeutic range. Cellulose derivatives such as cellulose acetate and ethyl cellulose, and polyvinyl derivatives like polyvinyl acetate, are the most often utilized water-insoluble polymers for extended-release applications. In contrast to ethyl cellulose, which is available in a variety of grades with varying viscosities and forms stronger and more lasting films, quaternary ammonium groups are different between Eudragit RS and RL, making Eudragit RS less permeable to water.

8.9.19 *GASTRORETENTIVE DOSAGE FORMS*

An alternate method for producing an extended-release profile is to use gastroretentive dosage forms. In this method, the formulation will stay in the stomach for an extended amount of time, releasing the medication *in situ*. The drug will then dissolve in the liquid contents and slowly flow into the small intestine. Such a delivery system would solve the issues associated with medications that are absorbed preferentially from particular sites in the gastrointestinal tract (GIT).

8.9.20 POLYMERIC MICELLES AS PHARMACEUTICAL CARRIERS

Polymeric micelles have a variety of desirable qualities as drug delivery systems. They may be loaded with a wide range of pharmacological drugs that are poorly soluble, efficiently accumulate in diseased body regions with weakened vasculature (infarcts, tumors), and can be targeted by attaching different particular ligands to their surface. They are stable both *in vitro* and *in vivo*. Micelles for medicinal and diagnostic purposes may be easily made in large quantities. Micellar carriers look to have a bright future.

8.9.21 POLYMERIC VESICLES

A range of macromolecular amphiphile topologies, including random graft copolymers, and polymers with hydrophobic low-molecular-weight pendants or terminal groups, block copolymers can be used to create polymeric vesicles. These particles, with sizes ranging from nanometers to micrometers, can be employed for medication targeting and responsive release.

8.9.22 POLYMERS USED FOR THE DELIVERY OF GENES IN GENE THERAPY

It has been discovered that a variety of polymers are good candidates for the transfer of genes across the numerous biological barriers mentioned in the preceding paragraph because they contain a cationic charge at physiological pH. The goal of the optimal gene delivery system is to safely transport the gene to the target tissue's nucleus with as little exposure to potentially damaging factors as possible.

8.10 INTRODUCTION OF POLYMERS IN BIOMEDICAL APPLICATION

A biomaterial is known as any natural or synthetic substance prepared to interact with the biological system of a human being and treat medical incompatibilities. It must be compatible with the system, meaning that it should elicit an appropriate host response. Polymers have significant potential to be applied in the biomedical sector due to their diverse physical and mechanical properties. To meet the needs of the biomedical community, polymers play an important role, such as being biodegradable or non-biodegradable [137]. Biodegradable polymers degrade and are absorbed without the need for surgical revisions, making them highly relevant in the biomedical sector today. Non-biodegradable polymers improve the lives of patients through applications like dental fillings, contact lenses, vascular prostheses, orthopedic implants, and artificial lips. These polymers, known as biostable or non-integrable polymers, work over the lifetime of the patient. Non-biodegradable polymers receive more attention than metals and ceramics due to their easy processing and wide range of physical, mechanical, chemical, and thermal properties [138]. Biomedical polymers have been intensively developed for promising use in the biomedical sector such as drug administration, diagnosis, biosensing, regenerative medicine, and disease treatment. For example, polymer-based carriers have made great advances in enhancing bioavailability and therapeutic index through spatiotemporal drug delivery, leading to improved treatment of diseases such as cancer, organ grafting, and infection. The random rotating design of intelligent polymeric support, which can provide positive feedback to external or intrinsic signals at specific lesion sites and trigger the release of payloads inside diseased cells, greatly enhances therapeutic efficacy.

8.11 CLASSIFICATION OF POLYMER (FIGURE 8.2)

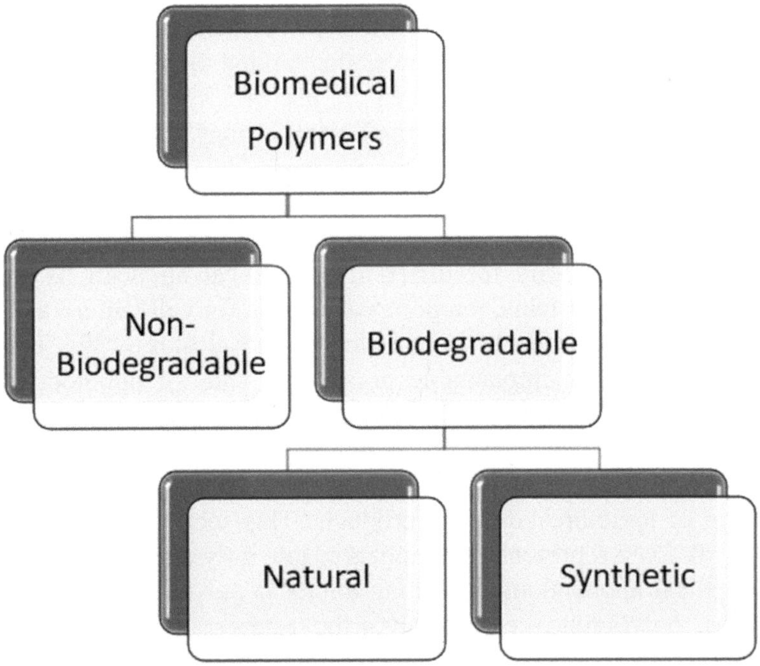

FIGURE 8.2 Classification of polymer.

8.12 SYNTHESIS OF BIOMEDICAL POLYMER

Both manufactured and naturally generated biomedical polymers have been categorized. Proteins and polysaccharides are examples of naturally occurring polymers that have biodegradability and connections with cells and tissues but also have restricted mechanical qualities, unpredictable degradation, and potential immunogenicity. Synthetic polymers, in contrast, are typically deficient in intrinsic bioactivities but exhibit good controllability in composition, structure, mechanical properties, and degradation behaviors. Additionally, synthetic biomedical polymers can be further divided into biodegradable synthetic polymers and nonbiodegradable synthetic polymers based on their biodegradability *in vivo*. A range of nonbiodegradable synthetic polymers has been developed for different purposes in biomedical applications, including poly(ethylene glycol) (PEG), ultrahigh molecular weight polyethylene (UHMWPE), polymethacrylate derivatives, poly(tetrafluoroethylene), poly(dimethylsiloxane), poly(urethanes) (PUs)/poly(urethane-urea), polyether ether ketone, polyethyleneimine (PEI), and poly(N-isopropylacrylamide). Many of these polymers are being researched or used as permanent implantable prostheses for partial or full replacement of the functions of damaged organs due to their nonbiodegradability and bioinert nature *in vivo*. For instance, the United States Food and Drug Administration (FDA) has authorized the use of a synthetic vascular graft made of poly(ether-urethane-urea) in clinical settings. A meniscal substitute made of PU-based implants and a composite of UHMWPE and a CoCrMo alloy were both approved by the FDA for use in orthopedic applications. A wide variety of biodegradable polymers have also been developed and evaluated in recent years for different biomedical applications, such as controlled delivery systems, temporary scaffolds for cell culture and tissue regeneration, and absorbable devices for wound closure [139]. This is in contrast to permanent implantable prostheses made of nonbiodegradable polymers. After performing their respective roles *in vivo*, these polymers may be broken down, metabolized, and/or eliminated from the body without noticeable hazardous side effects, a sustained load on the body, or the production of toxic breakdown byproducts. This section will focus on the synthesis of some representative biodegradable polymers that are designed for biomedical applications, including aliphatic polyesters, aliphatic polycarbonates, poly(amino acid)s, poly(ortho esters), poly(propylene fumarate) (PPF), poly(β-amino esters), poly(ester amide)s, poly(amidoamine)s, polyanhydrides, poly(ester urethanes), poly(organo)phosphazenes, and polyphosphoesters. In addition, the (possible) biological applications of these polymers are briefly discussed, with their biodegradation tendencies (Table 8.4).

TABLE 8.4 Classification of Polymers

	Biodegradable Polymer			Non-Biodegradable Polymer
	Natural	Synthetic		
Animal	Gelatin, Hyaluronan, chitin	Polyesters	Polylactic acid, polyglycolic acid, poly(c-caprolactonc)	Poly(ethylene), Poly(propylene), Poly(tetrafluroethylene),
Plant	Cellulose, starch, hemicellulose, agar, pectin, guar Gum, psyllium gum acacia, lignin	Cross-linked polyesters	Polyethylene glycol	Poly(methylmethacrylate), Poly(dimethylsiloxane), Poly(urethanes),
Microbes	Xanthan, gellan, hyluronian, luzdharn	Polyanhydrides	Poly(alkylcynoacrylate)	Poly (ethylene tetraphthalate),
Algae	Cazzageenan, alginate, agar	other	Poly(orthoesters), poly(phosphazenes), poly (D-3-hydroxybutyric Acid), poly(dioxanones), poly(oxalates).	Poly(sulfone), Poly(ethylene oxide).
Fungus	Schizophyllan, cardian, scleroglucan, pullulan, Chitin			

8.12.1 *ALIPHATIC POLYESTERS*

Predominantly researched biodegradable polymers for biomedical applications are synthetic aliphatic polyesters, which commonly include poly(lactic acid) (PLA), poly(glycolic acid) (PGA), poly(lactic acid-co-glycolic acid) (PLGA), and poly(-caprolactone) (PCL). There are four different forms of PLA based on residue chirality: poly(L-lactic acid), poly(D-lactic acid), poly(rac-D,L-lactic acid), and poly(meso-lactide). PLA, PGA, and PLGA have been accredited by the FDA for various clinical uses, such as absorbable sutures (DEXON® (PGA-based), Vicryl Rapide® (PLGA(10:90)-based)), bone fixation devices (Biofix® (PGA- or PLLA-based), Fixsorb® (PLLA-based)), facial fillers (Sculptra® (PLLA-based)), and delivery of drug systems (Eligard® (Leuprolide acetate/PLGA)). In comparison to the polyesters described above, PCL has a longer hydrolytic breakdown time of more than 2 years. The copolymer of GA and -caprolactone (-CL) is the foundation of a monofilament suture that is readily available on the market. The aforementioned aliphatic polyesters are commonly made by employing tin(II) 2-ethylhexanoate (Sn(Oct)2) as a catalyst within ring-opening polymerization (ROP) of a variety of cyclic lactones, such as lactide (LA), GA, and -CL and their mixed monomers. In addition, polyhydroxyalkanoates (PHAs), a group of aliphatic polyesters frequently produced by different bacteria, have drawn a lot of interest for biomedical applications. The most extensively researched PHA is poly(3-hydroxybutyrate; P3HB), which has been developed commercially as biodegradable plastics. P3HB degrades more slowly than polyesters based on LA and/or GA due to its high crystallinity and hydrophobicity, which results in the production of harmless 3-hydroxybutyric acid. Other hydroxyalkanoate units can be added to modify the crystallinity and rate of degradation. In addition to the microbiological source, a chemical pathway through the ROP of -butyrolactone monomer for the synthesis of P3HB has also been reported. The FDA's clearance of an absorbable suture made of poly(4-hydroxybutyrate) (P4HB) for clinical purposes encourages research into the biomedical uses of PHAs. It is interesting that PHAs generated from bacteria include endotoxin [140].

8.23.2 ALIPHATIC POLYCARBONATES

A carbonyl bond joining two ether bonds is seen in polycarbonates (PCs). PCs are often synthesized using the ROP of cyclic carbonates. Due to the simple processes to produce functionalized cyclic carbonates among the reactions between diol-containing compounds and triphosgene or ethyl chloroformate, aliphatic PCs exhibit engrossing advantages in the incorporation of functional moieties when compared to aliphatic polyesters. Although aliphatic PCs are generally hydrolytically stable, the presence of enzymes can hasten their deterioration. Because of its excellent flexibility, poly(trimethylene carbonate) (PTMC), a representative aliphatic PC, has been studied for soft tissue regeneration and drug delivery systems, and non-acidic microenvironment brought on by the consequences of degradation. Teflon is another name for PTFE, a fluorocarbon-based polymer. The ideal features that can make the material more suited for artificial vascular grafts, catheters, and sutures are chemical and biological inertness, nondegradability, hydrophobicity, high crystallinity, hemocompatibility, and biocompatibility. Additionally, the material should also have a low elastic modulus and tensile strength. The surface of ePTFE grafts has been modified to increase endothelial adherence, giving them a strong thromboresistance property to prevent blockage in smaller-diameter arteries. Expanded PTFE (ePTFE) allows tissue to grow inside it [141].

8.12.3 POLYPEPTIDE

Polypeptides, another name for synthetic polypeptides, are made up of amino acid residues connected by peptide bonds. By utilizing solid-phase peptide synthesis or ROP of α-amino acid N-carboxyanhydrides (NCAs) with amine-containing initiators, polypeptides can be chemically synthesized. It is simple to produce polypeptides with reasonably high molecular weights based on the ROP of α-amino acid NCAs. It is interesting to note that the ROP of α-amino acid NCAs frequently follows the amine mechanism and the activated monomer mechanism in the presence of an amino-based initiator. The ROP of NCAs is typically initiated by the electron withdrawal attack of primary amines on the CO-5 of NCAs in the presence of primary amines or highly nucleophilic secondary amines. Polypeptides and their copolymers with regulated molecular weights and shapes can be produced via the primary amine-initiated ROP. When tertiary amines or some secondary amines with bulky substituent groups are used as initiators, however, only a small portion of NCAs are first deprotonated at NH-3 by the initiators, and the ROP of the remaining NCAs is then triggered by the activated NCA monomers. High molecular-weight polypeptides can be created using this method.

Polypeptides have received a great deal of attention in the biomedical sector due to their distinctive secondary structures that are comparable to natural peptides, high biocompatibility, and enzymatic degradability. Polypeptides break down into α-amino acids, which are nontoxic, nonimmunogenic, and metabolizable, in front of proteinases. Phase I, II, and III clinical trials have assessed a number of ROP-derived polypeptides. As an illustration, the phase III clinical trial for CT-2103, a polymeric prodrug based on poly(L-glutamic acid) (PLG) conjugated with paclitaxel (PTX) through an ester link, has begun in the United States for the cure of advanced non-small cell lung cancer in conjunction with carboplatin. Additionally, cisplatin-loaded poly(ethylene glycol)-block-poly(L-glutamic acid) micelles (NC-6004) have entered phase I/II clinical trials in the United States and Europe for the cure of various cancers in conjunction with antibodies and/or chemotherapeutics [142].

8.12.4 POLY(ORTHO ESTERS)

Three hydrolyzable and acid-sensitive geminal ether linkages are present in the backbones of poly(ortho esters) (POE). There are four POE families that have been investigated. For example, POE II is created by reacting a diol with a diketene acetal while using a small amount of an acidic catalyst. By controlling this reaction, the molecular weight of the polymerization products can be effectively managed. In a preparation procedure similar to that of POE II, oligo(lactide)s or oligo(glycolide)s are typically inserted into the polymer backbone to alter the degradation rate in neutral environments. As a result, POE IV was created, which has garnered more attention for potential medicinal uses than other types of POE in recent decades. Polyethylene, a versatile biomedical polymer, is classified as low-density polyethylene, high-density polyethylene (HDPE), and UHMWPE based on molecular weight. Due to their chemical inertness, mechanical strength, low tissue reaction, and biostability, HDPE and UHMWPE have attracted the most attention for use as biomaterials. Because its mechanical properties resemble those of natural bones, HDPE has a long history of use in clinical settings as a bone and cartilage substitute [143]. To correct craniofacial abnormalities, reconstruct external ears, and implant augmentations for the chin, malar, and nasal regions, MedPor®, a sintered porous scaffold made with HDPE microbeads, was created. Early 1980s researchers Bonfield et al. created HAPEXTM (hydroxyapatite/HDPE composite), which combines inorganic fillers with an organic matrix to approximate the composition of bone. Their HAPEXTM approach, which aimed to increase mechanical stiffness through chemical couplings such as silane and acrylic acid grafting, was successful in replacing minor load-bearing bones, such as the middle eardrum and the cheekbone. For load-bearing calcified tissue applications like bone, teeth, and joints, Nath et al. combined alumina particles with hydroxyapatite present in the HDPE matrix to improve mechanical qualities like rigidity and hardness. Due to its increased wear resistance, cross-linked UHMWPE or the inclusion of additives like vitamin E to the UHMWPE matrix have been employed for complete hip and knee replacements rather than standard UHMWPE. This protects the implant from wear-mediated osteolysis [144].

8.12.5 POLY (PROPYLENE FUMARATE)

One biocompatible, biostable polymer, poly(propylene), is utilized frequently in clinical settings for everything from sutures to load-bearing implants. Ventral incisional hernia has been treated with poly(propylene), which has fiber-forming properties. When compared to polyethylene, poly(propylene) has excellent rigidity and strength. Superior mechanical performance in temperature resistance and fatigue delivers enough mechanical properties even at normal body temperature, and it is able to withstand millions of loading-unloading cycles in orthopedic implants. By supplying tetracycline, an antibacterial agent, this can also be employed in root canal therapy. Teflon, also known as PTFE, is a fluorocarbon-based polymer. Its key features for medical use include chemical and biological inertness, nondegradability, hydrophobicity, high crystallinity, hemocompatibility, and biocompatibility. For applications like artificial vascular grafts, catheters, and sutures, the material should also possess low elastic modulus and tensile strength. ePTFE grafts have modified surfaces to enhance endothelial adherence, providing strong thromboresistance to prevent blockages in smaller arteries. Expanded PTFE (ePTFE) permits tissue to grow inside of it [144].

8.12.6 POLY (METHYL METHACRYLATE)

Despite being a hard, brittle substance, PMMA has a long history in medicine, having been used as bone cement, intraocular lenses, and dental fillings since the 1940s. This is due to its versatility in handling, processing, and polishing, as well as its proven biocompatibility in the oral environment. PMMA has a long history in dentistry. Bioinertness, inertness to degradation, resilience to UV light, and transparency make PMMA a common implant material for intraocular lenses worldwide due to its refractive index of 1.5 and ability to maintain surface smoothness when in contact with vascular tissues. Heparin surface modification has been demonstrated to improve biocompatibility compared to unmodified PMMA lenses by lowering postoperative inflammation and corneal endothelial cell damage. Due to its instantaneous strong fixing capabilities, PMMA can be employed as bone cement for osteoporotic bone as well as for implant fixation. The principal drawbacks include the toxic nature of methyl methacrylate, brittleness, exothermic reaction of polymerization, poor adhesion to bone tissue, lower fracture toughness, and poor wear resistance. It can release wear particles during compression due to its low wear resistance, which causes a long-lasting inflammatory response.

8.12.7 POLY(DIMETHOXYSILANE)

Next biostable synthetic polymer applied in biomedical applications, like drug delivery systems and blood-contacting biomaterials, is poly(dimethylsiloxane, or PDMS). The optimal characteristics of this elastomeric polymer include nontoxicity, transparency, blood compatibility, elasticity, biocompatibility, and durability. Additionally, the PDMS polymer backbone's flexibility exposes its methyl group at a number of low-interaction interfaces, resulting in the minimum amount of interaction at their surface. This bioinertness prevents microbial growth, which increases its appeal for use in biomedical applications. Due to its high gas permeability, PDMS is more suitable for medical applications like contact lenses and wound dressing because it permits the diffusion of gases like oxygen and carbon dioxide. Although cell grip is a crucial component of the angiogenesis and wound healing processes, the hydrophobic nature of PDMS discourages it. The silicone surface has been altered using a variety of techniques, including plasma treatment and laser treatment techniques to enhance its wettability and cell-adhesive properties. Due to surface reorganization, these approaches do not produce a stable cell-adhesive property. In order to achieve long-lasting cell adhesion, cell-adhesive proteins such as fibronectin, laminin, and collagen have been coated onto the PDMS surface. Due to proteolytic action, the use of such proteins causes immunological reactions, disease transmission, and poor stability, which limits their use in medicinal applications. Utilizing peptide conjugation, modified PDMS surfaces have been created and used as stable cell-adhesive surfaces.

8.12.8 POLY (AMINO ESTERS)

A class of biodegradable polymers known as Poly (-amino esters) (PBAEs) have tertiary amino groups and ester linkages in their backbones. The Michael additions between primary monoamine or secondary diamines and diacrylate-conjugated monomers are often the basis for the manufacturing of linear PBAEs. A hyperbranched or cross-linked product may be produced by the addition of a multifunctional monomer. In recent years, PBAEs have been investigated as carriers for the controlled delivery of genes or other therapeutic agents, showing electrostatic interactions with various negatively charged genes/therapeutic additives and responsiveness to physiologically relevant pH changes of the tertiary amino groups. This kind of polymer typically degrades hydrolytically within 3 days, producing noncytotoxic diols and bis(-amino acids) [145].

8.12.9 POLY (ESTER AMIDES)

The backbones contain both ester and amide linkages. Because of the combination of polyester's biodegradability and polyamide's thermomechanical capabilities, this kind of polymer has also gained a lot of interest in biomedical applications. PEAs can be made using a variety of processes, including Passerini reactions, polycondensation reactions based on monomers containing diamide (or bis (amino acids)) and diester moieties, and ROP of morpholino-2,5-dione derivative intermediates [146].

8.12.10 POLY (AMIDOAMINES)

Synthetic polymers known as poly(amidoamines) have amide bonds and tertiary amino groups in their backbones. Due to their electrostatic interactions with genes and other biomacromolecules, along with their biocompatibility and biodegradability, poly(amidoamine)s has been researched for a number of biomedical applications. Similar to PBAEs, stepwise Michael addition reactions between primary and secondary amines and bisacrylamides are generally used to make linear poly(amidoamine). By substituting bisacrylamides for diacrylate-containing monomers, poly(amidoamine) s are more hydrolytically stable than PBAEs. You can create linear or hyperbranched polymers based on the stoichiometric ratios of the di- or trifunctional amines and bisacrylamides. For instance, the reaction between a primary monoamine or secondary diamine monomer and a bisacrylamide monomer produces linear poly(amidoamine)s. It is important to note that poly(amidoamine) dendrimers have also gained a lot of interest as vehicles for the transport of drugs and genes because of their benefits, which include precise control over their size and structure and the abundance of functional groups [147]. Divergent and convergent techniques are common strategies for the manufacturing of poly(amidoamine) dendrimers. For instance, a divergent synthesis method can produce poly(amidoamine) dendrimers through an iterative two-step process that involves exhaustive Michael addition of methyl acrylate to the amino groups of primary diamines and subsequent amidation of the surface esters with the primary diamines [148].

8.12.11 POLYANHYADRIDES

A type of biodegradable polymers known as polyanhydrides have anhydride connections in their backbones. Because anhydride linkages are water-sensitive, polyanhydrides show faster hydrolytic degradation than polyesters and PCs. The synthesis of polyanhydrides is typically based on condensation polymerization, which may result in rather broad polydispersity indices, as opposed to the exact processing of the polymers using ROP. For instance, the melt condensation of the mixed anhydride prepolymers of 1,3-bis(p-carboxyphenoxy) propane and sebacic acid (Scheme 2i) yields the representative biomedical polyanhydride, poly[bis(p-carboxylphenoxy)propane sebacic acid anhydride] (PCPP-SA). FDA approval and commercialization of Gloadel®, an absorbable implant based on PCPP-SA for regulated release of carmustine, are used to treat brain tumors [149].

8.12.12 POLY (ESTER URETHANE)

Due to their high biocompatibility, hydrolytic stability, and superior mechanical qualities, polyurethanes (PUs) with microphase-separated soft and hard segments have received extensive research as long-term implants. PUs are frequently made by polycondensing monomers comprising diisocyanates with diols or polyols. Poly (ester urethanes) have been created by including biodegradable polyester diols or triols as the soft segments in the PU backbones in order to modify the biodegradability of PUs. By adjusting the polyester segment's makeup, the degradation rate can be adjusted. Another class of synthetic polymers called polyurethane is widely employed in heart valves, vascular grafts, and other blood-contacting devices. This polymer exhibits an elastic quality akin to rubber because its glass transition temperature (Tg) is lower than that of ambient air. Additionally, the phase-separated microstructures offer superb hemocompatibility.

Such a polymer's malleability and elastic properties resemble natural blood arteries. Therefore, polyurethane's attributes, including its blood compliance, biocompatibility, fatigue resistance, durability, compactability, and elasticity, may be suitable for temporary vascular applications. However, prolonged use of oxidation- and hydrolysis-prone polyurethane leads to cleaning and calcification, which is not advantageous for cardiovascular applications [150, 151]. The ability of polyurethane to calcify provides it with a better substrate for bone regeneration. The creation of wound dressings is more appealing due to polyurethane's excellent barrier property and oxygen permeability. Due to its poor degradation behavior, this polymer has been deemed to be non-degradable. The creation of numerous medical devices, including catheters and cardiovascular devices, heavily relies on the diversity of polyurethane features due to the material's simplicity in bulk and surface modification. Dacron® is a brand name for the thermoplastic polyester known as poly(ethylene terephthalate). Sutures and vascular grafts are just a couple of the medicinal uses for Dacron [152]. The material is more promising for use in biomedical applications because of its qualities of hardness, stiffness, biochemical and structural stability, as well as biocompatibility.

8.12.13 POLY (ETHELENETETRA PHTHALATE)

Dacron's biostability is primarily a result of the hydrophobic aromatic groups' existence, which also limits hydrolytic breakdown due to their high crystallinity. However, the lack of bioactivity restricts its application in tissue engineering [153]. Improved polymer-cell contact has been achieved through surface functionalization. Another issue with using knitted Dacron was that blood must clot before its implantation; however, this issue has been solved by impregnating the knitted Dacron with albumin, collagen, and gelatin to make it more impermeable [154].

8.12.14 POLYSULFONES

Excellent strength, stiffness, strong resistance to radiation, dimensional stability, temperature, and biocompatibility are some of the distinguishing characteristics of polysulfone. These characteristics are similar to those of lightweight metals. Additionally, polysulfone characteristics do not alter during surgery for sterilization. Polysulfone's hydrophobic character, however, restricts its potential utility in applications involving blood contact and tissue engineering. Polysulfone membrane has been utilized extensively in hemodialysis applications because it permits toxin diffusion from blood without endangering the blood proteins. Additionally, it has been applied to the creation of bioartificial livers (BAL). The prolonged drug release capabilities of polysulfone capillary fiber have been demonstrated in intraocular applications [155].

8.12.15 POLYETHYLENE OXIDES

Poly(ethylene oxide), another artificial water-soluble polymer, has received FDA approval for a number of medical uses. PEO has the ability to produce hydrogels, making it appropriate for creating drug administration systems and scaffolds that can regenerate [156]. This polymer also limits protein adhesion, which lowers the body's levels of immunogenicity, antigenicity, and complement activation. This makes it a great substrate for applications involving cell-based tissue engineering. This polymer is more appealing for biomedical use due to its nontoxicity and solubility in both water and organic solvents. PEO is a key component of the micellization process and has also been employed in the creation of amphiphilic copolymers (pluronics, PEOPPO-PEO). More focus has been placed on the further addition of PEO to the surface of biomaterials in order to enhance surface compatibility by preventing platelets, bacteria, and plasma proteins from adhering to the surface through steric repulsion [156, 157].

8.12.16 POLY(ORGANO) PHOSPHAZENES

An example of an inorganic-organic hybrid polymer is called a poly(organo) phosphazene, which has two organic pendant groups attached to an alternating phosphorus-nitrogen backbone. In most cases, the ROP of hexachlorocyclotri-phosphazene or the live cationic polymerization of a phosphoranimine utilizing phosphorus pentachloride as a catalyst is used to synthesize the polymer precursor, poly(dichlorophosphazene). After that, poly(dichlorophosphazene) is substituted with monofunctional, occasionally protected nucleophiles like amino acid ester derivatives, sodium salts of glycolate or lactate ester, and sodium salts of protected D-glucose to produce biodegradable poly(organo) phosphazenes. Despite the hydrolytic stability of the phosphorus-nitrogen bonds, the correct organic side groups must be added in order for poly(organo) phosphazenes to hydrolyze [158]. Additionally, there is a strong bonding between the physicochemical characteristics of poly(organo)phosphazenes and their hydrolytic degradation rate. Additionally, the organic side groups have a significant impact on the hydrolytic degradation rate and physico-chemical characteristics of poly(organo)phosphazenes. Certain benefits are the reason why certain biodegradable polymers have been looked into for potential biomedical uses. As stent coatings and embolizing microspheres, an FDA-approved product based on poly[bis(trifluoroethoxy)phosphazene] (polyzene-F®) has been tested [150, 151].

8.12.17 POLYPHOSPHOESTERS

Repeating phosphoester linkages and two variable groups—one in the backbone and the other in the pendant group—make up polyphosphoesters

(PPEs). The phosphoester linkages can be broken down by enzymes and hydrolysis. The highly diverse physicochemical characteristics and biodegradation behaviors of these polymers, which have been studied for a variety of biomedical applications, are caused by the varying moieties in both the backbone and side group. Various methods have been explored for the manufacturing of PPEs, including ROP, polycondensation, and polyaddition. For example, Sn (Oct)2 and a hydroxyl-containing molecule can be used as the starting material and catalyst, respectively, in the ROP of cyclic phosphoester monomers to produce PPE [152, 162]. To correct craniofacial abnormalities, reconstruct external ears, and implant augmentations for thc chin, malar, and nasal regions, MedPor®, a sintered porous scaffold made with HDPE microbeads, was created. In the early 1980s, researchers Bonfield et al. created HAPEXTM (hydroxyapatite/HDPE composite), which combines inorganic fillers with an organic matrix to approximate the composition of bone.

8.13 BIOMEDICAL POLYMER APPLICATIONS

8.13.1 BIOMEDICAL POLYMERS FOR THE TRANSPORT OF SMALL MOLECULES

As a result of significant advancements in therapeutic index and bioavailability, spatiotemporal drug delivery made possible by vesicles, micelles, hydrogels, and other polymer-based carriers has enabled effective treatment of diseases like cancer, infections, and cardiovascular disease. Over the past few decades, this field has made enormous progress. One notable example is the development of anticancer drug delivery systems, which will be the main focus of this section.

8.13.1.1 THE RELATIONSHIP BETWEEN THE IN VIVO FATE OF NANOMEDICINES AND THEIR PHYSIO-CHEMICAL QUALITIES

Although current nanomedicines for chemotherapy can reduce side effects, they frequently fall short of increasing therapeutic efficacy. The relationships between the physicochemical properties and *in vivo* behavior of nanoparticles were developed by researchers to improve therapeutic outcomes and to guide the design of nano-delivery systems. An intravenously injected nanomedicine's drug delivery mechanism includes circulation, accumulation, penetration, internalization, and release. The *in vivo* success of nanomedicines depends on their nano properties, including size, shape, and surface properties. Basically, nanoparticles between 100 and 150 nm have the best circulation and tumor accumulation, but those under 30 nm are better at penetrating tumors [163]. Undoubtedly, PEGylation is a well-known advancement in the creation of functionalized nanoparticles that, by extending blood circulation and enhancing tumor accumulation, make it easier to use nanoparticles successfully in clinical settings. For instance, Wang et al. [164] used polymeric nanoparticles with controllable PEG density or length but similar other nano-properties by incorporating hydrophobic homopolymer and regulating the PEG block molecular weight to establish a reliable relationship between surface PEG density or length and nanoparticle's *in vivo* behaviors [165].

8.13.1.2 NANOMEDICINES THAT RESPOND TO STIMULI

It has been seen from the numerous basic research that the necessary characteristics of nanomedicines frequently conflict in different steps, with the majority of these conflicts falling under the categories of PEG, charge, and size issues. Through the recent development of stimuli-responsive nanomedicines, we are now able to modify the characteristics of nanomedicines to increase the effectiveness of drug delivery. These nanomedicines have the ability to self-adapt to the needs during the delivery process and solve the problems [166]. Smart nanomedicines can react to external stimuli like temperature, light, gamma/X-ray irradiation [90], and ultrasound as well as endogenous stimuli including pH, redox gradients, enzymes, and light. When the pharmacokinetics of drug activation are well-controlled, such a design greatly lowers the toxicity of pharmaceuticals to normal tissues and enhances therapeutic efficacy. The neutral/PEGylated/shielding functional groups in the healthy tissues to the cationic/dePEGylated/exposing functional groups in the lesion sites is a stealthy-to-sticky transition that nanomedicines can achieve [167].

8.13.1.3 DRUG DELIVERY TECHNIQUES WITH ACTIVE TARGETING

Another key approach for site-specific medication delivery is active targeting. Identifying ligands, such as proteins (like transferrin and antibodies), peptides, aptamers, and tiny molecular ligands (like folate and monosaccharides), have all been described with positive outcomes both *in vitro* or in animal investigations. Preclinical and clinical settings have been promoted for a variety of targeting tactics. Because of their inherent stealth qualities, natural homing ability, and inherent capacity to penetrate biological barriers, circulating cell-targeting techniques, in addition to molecular targeting ligands, offer the possibility of active accumulation in the target tissues. For instance, lesion sites can efficiently amass nanoparticles that have hitched a ride with immune cells and erythrocytes. Shuai and colleagues [168] reported another instance of the use of an anti-PD-1 antibody-linked nanocarrier that could bind to circulating PD-1+ T cells and then follow their infiltration into the tumor, significantly enhancing the antitumor therapeutic impact. An illustration is the self-amplifying tumor-homing nanotherapeutic system described, which contained the coagulation-targeting peptide-decorated poly(L-glutamic acid)-graft-maleimide poly(ethylene glycol)/combretastatin A4 conjugate (A15-PLG-CA4). A15-PLG-CA4 then released CA4 in the tumor, a vascular disruptor that has the ability to damage ingrained tumor blood vessels and cause hemorrhage. Following the hemorrhage, blood coagulation factor XIII (FXIII) was activated to FXIIIa, which improved A15-PLG-CA4 homing to tumors and released more CA4 in the tumors. The cycle then restarted to begin the next. As a result, A15-PLG-CA4 started a chain reaction that allowed more medications to be delivered to the tumor, and it effectively treated large C26 tumors (>500 mm^3). Gu and others [169] reported a relay drug delivery method that boosted anticancer therapy and the targeted signal. By starting the inflammation cascade signal, a signal transmission nanocarrier containing tumor necrosis factor (TNF-) could cause tumor vascular damage. A platelet membrane-coated execution biomimetic nanocarrier then caught the amplified inflammation signal and accumulated at the tumor site. Overall, compelling improvements have been made in the fundamental comprehension and design of delivery systems. The number of papers and patents on the topic is rising quickly. Functional nanomedicines that can get past biological barriers are frequently too complex to be translated, though. The development of nanomedicinal therapeutic products that can finally fulfill their promise of enhancing patient benefit requires a great deal of attention.

8.13.2 GENE DELIVERY USING BIOMEDICAL POLYMERS

That can correct pathogenic abnormalities at the genetic level. The ability to treat diseases at the genetic level is quite promising. Plasmid DNA (pDNA) and messenger RNA (mRNA), among all therapeutic nucleic acids, up-regulate particular genes or proteins; small interfering RNA (siRNA), short hairpin RNA (shRNA), and microRNA (miRNA) mimic down-regulate gene expression at the mRNA level via RNA interference (RNAi); synthetic miRNA inhibitors specifically inhibit miRNA functions. Gene treatments have received FDA approval for the cure of acute lymphoblastic leukemia, large B-cell lymphoma, and RPE65 mutation-associated retinal degeneration. Globally, more than 2,000 gene therapy field trials are either underway or finished. Because nucleic acids are hydrophilic biomacromolecules that are permeable to cell membranes but susceptible to hydrolytic degradation, effective cytosolic delivery into target cells *in vivo* is essential for the success of gene therapy. Due to their intended biocompatibility, low immunogenicity, and structural variety, polymers comprise a significant category of nucleic acid delivery materials.

8.13.2.1 SYSTEMS FOR DELIVERING NUCLEIC ACIDS BASED ON POLYMERS

Because nucleic acids are negatively charged, cationic polymers, also known as polycations, are frequently used to electrostatically condense nucleic acids transferred to nano-sized complexes (polyplexes) that can be ingested by cells. Polycations that are frequently employed include chitosan derivatives, PEI, poly(L-lysine), poly(amidoamine), poly (-amino esters), poly(L-histidine), poly(2-dimethylaminoethyl methacrylate), poly[(2-acryloyl)ethyl(p-boronic acid benzyl)dimethylammonium bromide], and poly(glycidyl). The binding affinities with nucleic acids can be improved by adding functional moieties to polycations. For example, altered PLL with p-toluyl sulfonyl arginine can significantly increase the DNA binding ability through a variety of interactions, including electrostatic contact, salt bridging, hydrogen bonding, and hydrophobic interaction. The polyplexes' instability in salt, caused by the ionic screening effect, which reduces the electrostatic interaction, is a significant drawback. Salt stability could be significantly improved by polyplex micelles made of copolymers with hydrophobic and hydrophilic blocks that are positively charged. Polymersomes, which are likewise made of amphiphilic block copolymers, have an aqueous cavity that can contain hydrophilic nucleic acids, a hydrophilic shell, and a durable membrane. A short, positively charged block could be added to the polymersomal structure, which is preferentially positioned towards the inner aqueous cavity to boost the gene encapsulation force. This structure gives them the requisite salt stability. In addition to these systems, templated nanoparticles—which consist of an inner template (such as an inorganic or PLGA nanoparticle) and a surface-positioned polymer—are also frequently applied for gene delivery. This is because they have a controllable size, morphology, and aspect ratio, as well as inherited properties from the template, such as photo, thermal, and magnetic properties [170].

8.13.2.2 POLYMER CONCEPTS TO COMBAT PHYSICAL LIMITATIONS

Polymer designs for overcoming physiological obstacles to the transport of nucleic acids. The numerous intracellular barriers must be surmounted by polymeric carriers in order to accomplish effective gene transfection. The initial intracellular obstacle to the introduction of nanovehicles into the cytosol is the cell membrane. Positively charged nanovehicles are preferentially attacked by cells through endocytosis because of their electrostatic affinity with cell membranes. Cationic polymers can be predominantly designed with increased molecular weight, higher charge density, and multivalent structure to reinforce the transmembrane efficiency. Additionally, altering the cationic polymers' hydrophobicity or aromaticity or adding targeting ligands can significantly improve membrane affinity. Endo/lysosomes frequently trap nanovehicles after endocytosis, hence an effective endo/lysosomal escape mechanism is essential to allow the cytosol with the nanovehicles. The so-called "proton sponge effect" is caused by polymers containing buffering domains, such as protonatable amines with varying pKa, which can cause an osmotically induced endosome burst. Alternatively, endo/lysosomal escape can be mediated by polymers conjugated with lipids or cell-penetrating peptides (CPPs) through membrane fusion or membrane disruption. When exposed to light, photosensitizer-embedded nanovehicles can also release reactive oxygen species (ROS), which can weaken endo/lysosomal membranes through the process of photochemical internalization (PCI). Smaller and more positively charged nanovehicles can get away from endo/lysosomes more readily. Endocytosis could not be a reliable method for delivering intracellular genes; hence, polymeric nanovehicles with non-endocytic internalization capabilities are preferred. In order to achieve this, artificial cationic polypeptides with stable α-helical structures and guanidine side groups have been designed to resemble natural CPPs and cause the direct translocation of gene payloads into cells [171]. The effective release of gene payloads into the cytosol is a significant hurdle to intracellular gene delivery. While tight encapsulation of nucleic acids promotes cytosolic delivery and systemic circulation/distribution, the high binding affinities between the delivery vectors and the gene cargoes will impede the liberation of cytosolic genes. Degradable or transformable polymeric nanovehicles have been widely developed to solve this problem. This lowers the binding ability with positively charged, hydrophobically modified polyplexes smaller than 50 nm are advantageous for nuclei entry in DNA delivery because DNA must be carried into the nuclei to begin transcription. Nuclear localization signal peptides that bind importin to the target nuclear pore complex can be used to facilitate the intranuclear transport of big nanoparticles. The nanovehicles must also get past a number of extracellular barriers in the *in vivo* environment to reach the target cells. The nanovehicles need to withstand serum and circulate in the blood for a long time to be injected systemically. Because gene delivery frequently uses polycations, it is frequently necessary to decorate surfaces in a way that prevents negatively charged serum proteins from adhering. PEGylation and polyanion coating of the surface to increase serum stability and extend blood circulation are typical techniques [172] However, PEG's immunogenicity will result in accelerated blood pressure. Such a problem could be effectively solved by the application of zwitterionic polymers like polysulfobetaine (PSB) and polycarboxybetaine (PCB), which adopt an overall neutral charge and have high hydration capability to repel serum proteins. Clearance of PEGylated nanovehicles after repeated injections. The compromise of the contact with target cells is a different problem associated with PEGylation or polyanion coating. In response to triggers like an acidic pH or an upregulation of matrix metalloproteinases in tumor tissues, cleavable PEG or charge-reversible polyanions (from negative to positive) are developed, which can shed off in the extracellular environment to expose the positively charged inner core. Alteration of polycations like dendri by fluorocarbons. Because fluorocarbons have a low surface energy and a hydrophobic yet lipophobic nature, they could be used to modify polycations like dendrimers and PEI to resist serum while enhancing membrane activities [173]. Mucosal delivery is another significant administration method with good patient compliance, in addition to systemic injection. Positively charged nanovehicles are captured by the mucus layer that lines the epithelia because it contains a lot of negatively charged mucin glycoproteins.

8.13.2.3 GENE-FUNCTIONALIZING POLYMERS

pDNA is frequently used to compensate for target gene deficiencies, which are translated into mRNA and protein through mRNA transcription. Because pDNA is so big, it frequently forms stronger molecular bonds with polymers. Therefore, a reliable cytosolic release mechanism is absolutely necessary. Additionally, the delivery system must be able to transport DNA into the nuclei, where it starts transcription. You et al. [124] described bioreducible cationic nanomicelles with a fluorocarbon core and cationic shell connected by disulfide connections. The nano micelles can release DNA in reaction to intracellular glutathione (GSH) and can effectively condense DNA even at an N/P ratio of 1. This is due to fluorocarbon-assisted assembly. Shen and others go into cells via macropinocytosis to avoid endo/lysosomes and considerably improve the stability of polyplexes at reduced concentrations by optimizing the hydrophobicity of polycations. The transfection efficiency was decreased by the unmodified polyplexes being dissociated or stuck in the lysosomes at low concentrations. Highly branched poly (-amino esters) (HPAEs) with multivalency-reinforced DNA delivery capabilities were reported. When they screened a library of HPAEs with guanidines and backbone-embedded disulfides in multipotent adipose-derived stem cells and astrocytes, respectively, the optimized HPAEs resulted in transfection efficiencies of 77% and 52% for minicircle DNA. mRNA is frequently used in protein-replacement therapy and vaccine development. Unlike pDNA, which must first be transported to the nucleus and then translated into the target protein, mRNA can be loaded directly into the ribosome. As a result, mRNA distribution is frequently more effective than pDNA and carries a lower risk of insertional mutagenesis; however, it must be carefully managed due to its susceptibility to enzymatic hydrolysis. Effective mRNA transfection of APCs by means of intranasal delivery led to efficient local and systemic immune responses. Charge-altering releasable transporters for anti-cancer mRNA vaccines were created by Levy et al. [174]. In order to effectively deliver mRNA into APCs, oligo(-amino ester) cations formed polyplexes with mRNA; nevertheless, to release mRNA, they lost the cationic charges through charge-neutralizing intramolecular rearrangement. As a result, the CART vaccines outperformed standard synthetic viral peptide mixes in their ability to induce an antigen-specific immune response against mRNA-encoded viral epitopes *in vitro* and *in vivo*. The difficulty in polymer synthesis and quality control may be the reason why polymer-based mRNA vaccines are frequently less effective than the reported lipid-based mRNA vaccines. By utilizing synthesized siRNA or miRNA mimics to degrade mRNA, the RNAi process can be applied to reduce the overexpression of genes associated with illness. Small and linear siRNA or miRNA mimics are extremely susceptible to enzymatic degradation. Thus, siRNA/miRNA mimics must be tightly encapsulated by nanovehicles [175]. They don't need to be lifted into the nuclei because they work in the cytoplasm by recognizing the target mRNA. To deliver siRNA (targeting Bcl-2 and P-glycoprotein) and DOX simultaneously, a DNA origami-based nanodevice was described. This device is co-assembled from functional moieties such as trans-activator of transcription (TAT) and disulfide-containing DNA locks. The DNA locks secure the nanodevice during circulation but open the nanodevice in reaction, and TAT enhances cellular uptake and tumor retention [175].

8.13.3 BIOMEDICAL POLYMER FOR PROTEIN DRUG DELIVERY

Restoring protein function is the easiest treatment approach because it is related to many disorders [176]. Protein therapy acts specifically on biological processes and targets, helping to avoid off-target effects and reduce systemic cytotoxicity compared to small molecule medications and gene therapy. Clinical uses of therapeutic proteins are still in their infancy, partly due to their fragile tertiary structure, short plasma half-life, high immunogenicity, unwanted biodistribution, and cell membrane impermeability. To address these problems, new types of biomedical polymers have been employed for the application of therapeutic proteins [177].

8.13.3.1 USING BIOMEDICAL POLYMERS TO TRANSFER EXTRACELLULAR PROTEINS

The pharmaceutical sector has been completely transformed by protein therapy. Monoclonal antibodies are said to be 7 of the top 10 most popular medicines in 2018. However, after intravenous injection, the host defense system quickly recognizes the majority of exogenous proteins, causing the protein to be cleared and degraded before the protein can exert its therapeutic effects [178, 179]. Therefore, for the successful application of therapeutic proteins, decreasing immunogenicity, extending plasma half-life, and optimizing biodistribution is crucial. These polymer-based protein delivery strategies are discussed in this section. Recent developments in this field primarily center on conjugating hydrophilic polymers onto proteins and encapsulating proteins within cross-linked polymer networks [180]. Poly(ethylene glycol), the most used polymer for extracellular protein delivery, is PEG. The most popular method for delivering proteins at the moment is PEGylation, which involves directly conjugating PEG to protein molecules. PEG chains engage hydrophobically with proteins as they circulate in the bloodstream, causing them to collapse on their surfaces. They then bond with water molecules to produce a layer of hydration around the proteins. Such a hydration layer can effectively shield the protein from the mononuclear phagocyte system's (MPS) recognition and elimination, giving the protein a longer circulation duration, better bioavailability, and decreased immunogenicity. The molecular weight, chain topology, and density of PEG are all strongly correlated with its shielding effect [181]. Recently reported a PEG-based polymeric nanogel for the simultaneous administration of insulin and glucose oxidase (GOx) for hyperglycemia control. Insulin and GOx have enhanced bioavailability and plasma half-lives as a result of the coating of the PEG layer, which effectively shields them from immune clearance during blood circulation. A PEG-based multienzyme nanocluster for the simultaneous administration of catalase (CAT) and uricase (UOx) was also presented. Due to the confinement of UOx and CAT in nanoscale space, uric acid might be split down by UOx without CAT. Due to the confinement of the UOx and CAT in nanoscale space, UOx was able to break down uric acid without producing the harmful byproduct H_2O_2. PEG-based polymers have also been utilized to deliver antibodies and cytokines extracellularly in addition to enzymes. For instance, proposed a PEG-based polymeric micelle for the simultaneous administration of paclitaxel (PTX) and the immune checkpoint inhibitor anti-PD-1. They used PEG with a molecular weight of 20 kDa to protect anti-PD-1 during blood circulation in their research. Anti-PD-1 and PTX were released from the micelle as they entered the tumor microenvironment (TME), resulting in efficient immune activation. Despite PEG, approximately 25% of patients will develop anti-PEG antibodies after the first or ongoing administration of PEGylated proteins, despite the fact that PEG has been extremely successful in improving protein stability and bioavailability. The as-produced anti-PEG antibodies cause opsonization to be activated, which speeds up the removal of PEG-based protein delivery systems from the circulation. Therefore, protein treatment calls for the creation of new polymers that can efficiently protect proteins in blood circulation while preventing opsonization [182]. Zwitterionic polymers, such as carboxybetaine-, sulfobetaine-, and phosphobetaine-based polymers, are defined as polymers that include both negatively charged and positively charged groups in their structure. Zwitterionic polymers hydrate more strongly by electrostatically driven hydration than PEG and other hydrophilic polymers that do so through hydrogen bonding, creating a special surface that is not amenable to protein adsorption. Zwitterionic polymers are a good antifouling coating material for extracellular protein distribution because of their resistance to nonspecific protein adsorption and high biocompatibility. For instance, introduced a protein nanocapsule based on poly(2-methacryloyloxyethyl phosphorylcholine) (PMPC) and discovered that this PMPC-based nanocapsule could greatly extend the circulation duration of UOx *in vivo* without inducing an immunological response. In addition to offering an antifouling surface, Lu et al. found that endothelial cells' nicotinic acetylcholine receptors (nAChRs) and choline transporters (ChT) could also bind to MPC. After making this finding, they created another PMPC-nanocapsule and successfully delivered trastuzumab (Tras) and nimotuzumab (Nimo) to the central nervous system (CNS). These investigations show how zwitterionic polymers have enormous potential for the creation of fresh protein treatments. The proximity architecture of multienzymes ensures that the harmful intermediate formed in enzyme therapy can be efficiently removed by another enzyme. The polyacrylamide layer can significantly improve the circulatory stability of multienzymes in blood. Together, these enzymes catalyze the conversion of alcohol to acetate without the production of H_2O_2, successfully preventing the development of alcoholism caused by acetaldehyde and liver damage caused by H_2O_2. Additional natural polymers that can be used include chitosan, dextran, hyaluronic acid, etc., demonstrated a dextran-based microsphere for vascular endothelial growth factor systemic administration.

8.13.3.2 USING BIOMEDICAL POLYMERS TO TRANSPORT PROTEINS TO THE CYTOSOL

Even though protein therapy has had significant success, it is important to keep in mind that practically all FDA-approved protein medicines work on extracellular targets such as membrane proteins and secretory proteins. This is mostly due to proteins' inherent hydrophilic and macromolecular properties, which prevent them from passing through cell membranes. However, less than 30% of the proteins encoded by the genome are found outside of cells, and the remaining 70% of intracellular proteins are no longer viable targets for protein therapy. Furthermore, the cytosolic delivery of proteins like cyclization recombinase (CRE) and clustered regularly interspaced short palindromic repeats (CRISPR) associated protein 9 (Cas9) is subject to high demands from contemporary biotechnologies involved in molecular biology and genome editing. This section discusses the most current developments in the creation of biomedical polymers for the administration of cytosolic proteins.

- **Polymers Based on Polyethyleneimine:** Gene delivery has made extensive use of PEI, a type of synthetic cationic polymer with a high charge density and proton buffering capability. However, due to its low protein loading capacity, PEI's delivery efficiency for protein delivery is significantly reduced. As a result, several PEI compounds have been created to enhance protein loading. For instance, Zuber et al. described a pyridylthiourea-modified PEI (PEI) that can carry proteins into the cytoplasm. Hydrogen bonding, hydrophobic, and ionic interactions were all improved by this alteration between the PEI and proteins. As a result, monoclonal antibodies were successfully transported into cells via PEI to control biological processes. Another often utilized modification group is guanidyl (GUA), which helps polymers better load proteins by establishing salt bridges and hydrogen bonds with the protein.
- **Polymers Based on Dendrimers:** Dendrimer is a polymer with numerous amino groups that resembles a tree and has a high proton buffering capability, similar to PEI. Recent research has quickly created dendrimer-based polymers and demonstrated their promise for cytosolic protein delivery. As an illustration, Cheng et al. Dendrimers altered with guanidinobenzoic acid (GBA) (DGBA) have recently been created to transport proteins [163]. A dendrimer scaffold, a membrane-disrupting region, and a protein-binding surface make up DGBA. With the cooperation of these three components, DGBA successfully loads and delivers cytosolic proteins with a number of attractive characteristics: (i) proteins don't need to be altered; (ii) cytosolic administration is highly effective; and (iii) several proteins can be used.
- **Additional Polymers for Delivery of Cytosolic Proteins:** Other cationic polymers besides PEIs and dendrimers have also been researched for cytosolic protein delivery. They described a technique for protein encapsulation that may create a thin coating of poly(N-(3-aminopropyl)-methacrylamine) around each protein molecule. They introduced a chitosan-based nanoparticle for the delivery of protein, and they discovered that the ratio of chitosan to TPP, the molecular weight, the concentration, and the protein feeding ratio were the main factors affecting the delivery effectiveness. Shi et al.

[182] introduced mixed-shell polymeric micelles (MSPMs) based on mannose-modified poly(ethylene glycol)-b-poly(-caprolactone) (Man-PEG-b-PCL) for the cytosolic delivery of OVA. On the upper face of MSPMs, hydrophobic domains created by the interaction of the hydrophilic PEG and the hydrophobic PAE were employed to load OVA. After delivering OVA to APCs, MSPMs encouraged OVA uptake by APCs and accomplished lysosome escape following internalization. Adding cell-penetrating poly(disulfides) to proteins can increase membrane permeability while maintaining protein functionality.

8.14 BIOMEDICAL POLYMERS FOR BIOIMAGING BIOSENSING AND THERMOSTATICS

To link signals with the molecular state in living subjects, probes are necessary. Exogenous probes, including tiny molecular probes, have been extensively developed for bioimaging in recent years. Small molecular probes frequently have issues with their water solubility, photobleaching speed, aggregation-induced quenching, and observation window, which restrict their use *in vivo*. Polymer probes can accumulate in disease areas for instant and accurate diagnosis because of their enhanced water solubility, good photostability, prolonged *in vivo* circulation time, and ease of functionalization when combined with hydrophilic or amphiphilic polymers [183]. Depending on the polymer configuration, there are two ways to create these biomedical polymer-based probes. The first kind of polymer-based probes is created by covalently bonding probes to hydrophilic polymers. Hydrophilic polymers provide probes with better water solubility. Probes of the second kind are physically incorporated into micelles of amphiphilic polymer. This section provides a summary of current developments in the creation of biomedical polymer-based probes with imaging capabilities for *in vivo* bioimaging, including fluorescence, photoacoustic, and other imaging modalities. In biological research and clinical practice, real-time spectrofluorimetric imaging in living patients demonstrates the benefits of nonradiative, high sensitivity, and high spatial and temporal resolution. The FDA recently approved the fluorescent probe pafolacianine (Cytalux) as an adjuvant for intraoperative imaging-guided ovarian cancer surgery. To enable fluorescence imaging of physiological and pathological processes at the cellular or molecular level in biological tissues, fluorescence probe development is crucial. Activatable fluorescent probes have been developed for disease diagnosis in accordance with the number of disease biomarkers' effects on fluorescence signal changes (such as pH variations, hypoxia, ROS, and disease-associated enzymes). An extracellular pH-activatable nanoprobe was created by Gao and colleagues [184] using the poly(ethylene glycol)-b-poly(2-(hexamethyleneimino) ethyl methacrylate copolymer with the near-infrared dye Cy5.5. The fluorescent signal of Cy5.5 dye inside the nanoprobe was quenched at blood pH (7.4) as a result of homofluorescence resonance energy transfer-induced signal quenching. The nanoprobe might become dissociated in an acidic pH environment, which would significantly boost the Cy5.5 dye's fluorescence signals. With a 102-fold fluorescence amplification between blood pH (7.4) and the tumor microenvironment pH (6.7), the nanoprobe consequently displayed a strong pH transition at pH 6.9. Because of this, the nanoprobe showed an extreme pH response following conjugation with cRGD for precise tumor pH imaging in different tumor models with various cancer characteristics, and it is currently being tested in clinical studies. By conjugating the macromolecular probe (Ir-PVP) with another distinctive characteristic of most solid tumors is hypoxia [185]. A number of poly(N-vinylpyrrolidone) (PVP)-based iridium complex (Ir III) activatable probes for hypoxia imaging have been created by Jiang and colleagues [185]. By fusing the biological polymer PVP with an Ir complex (Ir-PVP), they first reported a macromolecular probe (Ir-PVP). Ir-PVP was quenched in healthy tissues but displayed potent near-infrared signals in anoxic environments, enabling highly sensitive detection of a variety of hypoxic tumor models and even 1,000 cancer cells at the start of tumor development. Additionally, the synthesized nanoprobe made possible by coprecipitating IrPVP with poly(-caprolactone)-b-poly(N-vinylpyrrolidone) block copolymer (PCL-PVP) allowed for the accurate detection of tumor metastasis in the lung or lymph nodes. Additionally, they added more modifications to the PEG-iridium complex using an acidity-sensitive [185].

8.14.1 BIOMEDICAL POLYMER USED FOR BIOIMAGING

The biomedical polymer-based probes for bioimaging offer high specificity and sensitivity, long circulation times, and strong water solubility. Biomedical polymer-based probes could be further functionalized to increase their bioimaging applications thanks to the simple preparation procedure. Even though considerable advancements have been investigated, there is still much room for development. For instance, mononuclear phagocyte system (MPS) uptake of biomedical polymer-based probes may result in significant buildup in the liver and spleen and sluggish clearance due to their size, which could have *in vivo* cytotoxic consequences. Further research is needed to address the *in vivo* safety concerns of biomedical polymer-based probes, including long-term physiological effects, pharmacokinetics, and pharmacology. Generally speaking, with these encouraging developments, biomedical polymer-based probes have the ability to move from the bench to the bedside. Biomarkers can be found *in vitro*, on the surface of the body, and *in vivo* using electrical, optical, and other methods.

8.14.2 BIOMEDICAL POLYMERS FOR BIOSENSING

Due to their real-time and precise detection at the desired position, implanted biosensors have drawn more and more attention. Long-term stable monitoring typically depends on a strong contact between implanted biosensors and surrounding tissues, presuming the biosensor has good biocompatibility, is miniature, and has a mechanical fit with the tissues in flexibility. Traditional biosensors, on the other hand, are built on stiff silicon substrates and metal electrodes, and they can cause significant inflammation when used for prolonged *in vivo* monitoring. As a result, flexible thin-film and fiber biosensors are created using other polymers that have moduli that are many orders of magnitude lower, enabling long-term stable monitoring *in vivo*. Polymers, including substrates, insulating layers, electrodes, active materials, surface modifications, and aided implantations, with various characteristics have been included in biosensors [186–188].

8.14.2.1 SUBSTRATES MADE OF POLYMERS AND INSULATING LAYERS

Typically, flexible biosensors require thin insulation layers to entirely cover non-detection areas, and polymer substrates to support electrodes and active components, and flexible biosensors. Due to their high biocompatibility, stability, and compatibility with biosensor fabrication processes, polyimide (PI), polycarbonate (PC), SU-8, and polydimethylsiloxane (PDMS) have all been extensively investigated for film biosensors. Organic field-effect transistor (OECT) biosensors have been made using flexible fiber substrates made of nylon fibers, resulting in little inflammation and retaining the fixing of the interface. Parylene has been widely employed as flexible substrates for film biosensors and as insulating layers for both film and fiber biosensors. Parylene can be vapor-deposited to create thin self-standing films.

8.14.2.2 ELECTRODES MADE OF POLYMERS

For a mechanical fit with tissues and quick transit of electron impulses, the electrodes must be flexible and conductive. Therefore, by coating them on polymer substrates, conducting polymers like polypyrrole (PPy) and poly(3,4-ethylenedioxythiophene) (PEDOT) have been studied as electrodes. Their flexibility and conductivity typically need to be balanced by changing the process techniques. Another successful method for polymer electrodes that offers more flexibility in material selection and manufacturing process design is the combination of flexible nonconductive polymers with conductive nanomaterials, like carbon nanotubes and graphene. Assembling aligned carbon nanotube films into macroscopic hierarchical fibers or winding them on polymer fibers represents another innovative and successful technique developed in recent years that offers high electrical conductivity while maintaining flexibility.

8.14.2.3 ACTIVE MATERIALS BASED ON POLYMERS

The basic components of biosensors are active materials, which can recognize the target analytes and define the performance, such as sensitivity, selectivity, limit, and range of detection. Since the electrode/neural interface's low impedance will result in a high signal-to-noise ratio when neural signals are being detected, PEDOT: poly(styrene sulfonate) (PEDOT:PSS) has been extensively researched to increase the interface's capacitance and decrease the interface's impedance. The potential change of the sensor is attributed to the detection of particular ions, such as calcium ions. Ion-electron transduction from an ion-selective membrane is thought to be responsible for the potential shift in the sensor for the detection of particular ions, such as calcium ions. As a result, the middle ion-electron transduction layer for ion sensors was also made of PEDOT:PSS. Additionally, PEDOT:PSS is ideal for the OECTs channel due to its ion-doping capability and straightforward processing [187, 188]. In enzyme-based biosensors like the glucose sensor, glucose is oxidized by O_2 catalyzed by glucose oxidase. The resulting H_2O_2 can then react with platinum nanoparticles to cause variations in current. When an enzyme needs to be immobilized for stability, some polymers, including chitosan, are utilized, and conducting polymers like polyaniline (PANI) with redox characteristics are typically used as an intermediary to increase the activities. They primarily act as three-dimensional conductive scaffolds to immobilize enzymes and platinum nanoparticles, increase sensitivity, and extend the linear range [188, 189]. *In vivo,* stability issues for enzyme-based biosensors are another concern, and some non-enzyme sensors have also been investigated to achieve targeted detections. Specifically, molecularly stamped polymers have been created by polymerizing analytes with polymers, such as PPy, and then removing the analytes. Although molecularly stamped polymers are very promising, they still need to be tested *in vivo.* In general, conducting polymers are preferred as the active components in biosensors.

8.14.2.4 MATERIALS FOR SURFACE MODIFICATION BASED ON POLYMERS

To enhance the biosensor's anti-interference property, interface stability, and anti-biofouling, the surface is frequently treated using a polymer. Hydrophilic zwitterionic polymer and microporous (0.2 μm) polysulfone membranes are frequently investigated for their ability to withstand biofouling *in vivo*, including that of blood cells and proteins. For instance, polydopamine inspired by mussels was used to bind zwitterionic polymers to the surface of biosensors, boosting sensing performance and stability while reducing inflammatory reactions [190]. As the top layer of fiber sensors, a hydrogel of calcium ion cross-linked sodium alginate with good biocompatibility has also been reported. It could be directly implanted because it has an *in vitro* rigidity modulus of greater than 1 GPa. After coming into contact with body fluid for a given amount of time inside the body, the hydrogel layer softened and its modulus was decreased to the kPa level for mechanical compatibility with the tissues. These flexible polymer layers enabled the biosensors to be implanted successfully and to form stable contacts with tissues [191].

8.14.2.5 POLYMERS ASSISTED IMPLANT MATERIAL

While all-polymer biosensors are uncommon, many implanted biosensors with polymers as different components have been reported to monitor a variety of physical and chemical information *in vivo*. Future research should focus on developing and synthesizing biological biomedical polymer design and synthesis should take into account flexibility, performance, and processability while maintaining high sensitivity and low limit of detection in order to meet the demand for implantable all-polymer biosensors. Additionally, it will be possible to create multifunctional integrated sensing systems for the simultaneous detection of numerous biomarkers, such as ions, neurotransmitters, and blood glucose, by taking advantage of how simple it is to modify polymers. To achieve *in-situ* amplification of signals in the body, high-performing semiconductor polymers should be created and introduced because some biomarkers have incredibly low concentrations with minute variations. In order to achieve *in-situ* amplification of signals in the body, high-performing semiconductor polymers need to be created and introduced. This is because some biomarkers have extremely low concentrations with minute variations. Finally, combining implantable biosensors with other imaging techniques may enable the discovery of physiological mechanisms that are more complex. For instance, some work has been done to create polymer-based biosensors that are compatible with fMRI and to research the brain using several models with complementing temporal and spatial resolutions.

8.14.3 BIOMEDICAL POLYMERS USED FOR THERANOSTIC

Theranostic systems based on polymers can target sick regions of the body without harming healthy organs or tissues. Theranostic systems may then provide details on the severity of the disease and, if applicable, provide disease response to therapy once the location of interest has been identified. The uses of biology are covered in this section. Optical, ultrasonic photoacoustic, nuclear, and MRI imaging modalities are among the imaging modalities that medical polymers used in theranostics are described in detail.

8.14.3.1 BASED ON OPTICAL IMAGING, THERANOSTICS

In order to give a real-time, non-invasive, and non-ionizing readout of illness information at the cellular or even molecular level, optical imaging, in particular fluorescent imaging, is preferred. However, because the majority of fluorescent theranostic systems rely on emitted photons for the signal readout, they invariably have low SBR and limited tissue penetration due to the tissues' light absorption, scattering, and autofluorescence. For *in vivo* applications, these problems may be successfully solved by the employment of near-infrared (NIR) fluorophores. Numerous theranostic polymer-based systems have been created recently with NIR fluorophores for imaging [191]. Real-time light excitation is necessary for fluorescence imaging, but doing so also stimulates tissue autofluorescence, which reduces imaging sensitivity and specificity in living patients. Afterglow luminescent probes have the ability to store excitation energy in flaws and gradually produce photons after the light is no longer excited. As a result, afterglow imaging significantly increases the signal-to-background ratio *in vivo* with no autofluorescence interference from biological specimens. PPV-based SPNs with a T/2 of around 6 minutes were reported by Pu et al. [192] to be able to emit long-NIR afterglow luminescence. SPNs produced an afterglow that was more intense than that of inorganic afterglow substances by a factor of over 100. With an SBR up to 127 times greater than that produced by traditional NIR fluorescence imaging, high-contrast lymph node and tumor imaging in living mice could be made possible. These PPV-based SPNs' afterglow luminescence also exhibited temperature dependence, which was used for in-the-moment temperature monitoring during photothermal therapy.

8.14.3.2 THERANOSTICS WITH MRI

MRIs are another effective diagnostic technique due to their high resolution, lack of radiation exposure, and noninvasive imaging. MRI frequently needs contrast agents, which are divided into positive agents (gadolinium complexes for T1-weighted imaging) and negative agents (SPIO particles for T2-weighted imaging), to boost the sensitivity. The most widely used Gd(III) chelates for T1 MRI in clinical use are diethylenetriaminepentaacetic acid-Gd (DTPA-Gd, Magnevist) and Gd-diethylenetriaminepentaacetic acid bismethylamide. Macromolecular Gd chelates, especially those based on polymers, have been shown to have better imaging quality, longer blood circulation durations, and the ability to preferentially aggregate in sick areas. Real-time monitoring of the effectiveness of cancer therapy can be gained by integrating polymer-based MRI. An enzyme-responsive polymeric prodrug-based theranostic system with cyanine 5.5 and Gd-chelates. The nanoparticles significantly increased MRI contrast while demonstrating substantial cytotoxicity to cancer cells, comparable to free PTX. In 4T1 mouse tumor models, the imaging-guided treatment showed strong anticancer activity. Additionally, this system made it possible to accurately track the chemotherapeutic response in real time. The creation of T2-type theranostic agents was motivated by worries about the nephrotoxicity and long-term brain deposition of Gd-based T1 contrast agents. Polymeric nanoparticles loaded with SPIO nanoparticles (SPIONs) and DOX were applied to both tumors targeting with magnets and noninvasive MRI imaging. This theranostic device was anticipated to map the tumor's real-time properties in each patient and enable earlier illness detection, more precise prognostic data, and improved treatment efficacy monitoring. After being given to animals with tumors, a sizable accumulation of the nanoparticles was noticed [193].

8.14.3.3 ULTRASOUND IMAGING-BASED THERANOSTIC

Another clinically common tool for illness diagnosis and monitoring is ultrasound imaging. Using nano- and micro-bubbles as contrast imaging agents, this method can help discriminate between healthy and sick tissues. Lipids, proteins, and biodegradable polymers are the main building blocks for bubble formation. Polymer-based bubbles, as opposed to lipids and proteins, exhibit good stability throughout circulation, ensuring biosafety. Due to their numerous functional groups, polymeric backbones also enable drug loading and targeting ligand attachment. Additionally, by enhancing the penetration of blood vessels and cell membranes, the explosive effect of bubbles caused by ultrasound waves can facilitate drug extravasation and cell internalization [194]. For instance, focused ultrasound (FUS) supported by a boron-polymer/microbubble combination (B-MB) was suggested as a treatment to cure brain glioma. B-MBs could access the blood-brain tumor barrier safely while boron levels are increased. For ultrasound imaging and DOX delivery, Chen et al. [195] described theranostic polymer microcapsules made of hydrogen-bonded multilayers of tannic acid and poly(N-vinylpyrrolidone) (PVP). In contrast to commercially available microbubbles, these capsules had good imaging contrast in both brightness and harmonic modes and demonstrated sustained contrast over 6 months. Tuning the ultrasonic irradiation could likewise be used to achieve controlled DOX release. Furthermore, the number of layers, type of polymer (more flexible poly(methacrylic acid) versus relatively rigid tannic acid), and gram atomic mass of the polymer could all be altered to customize the imaging contrast of the microcapsules.

8.14.3.4 THERANOSTICS BASED ON PHOTOACOUSTIC IMAGING

Although ultrasound can distinguish between tissue interfaces by varying the speed of sound, it cannot provide real-time tissue imaging. By combining the spatiotemporal resolution of ultrasound with the spectral contrast of optics, photoacoustic imaging can expand the contrast of ultrasound from the anatomical to the molecular. This method depends on the photoacoustic effect, in which an object absorbs optical energy and releases it as an acoustic vibration. Drug delivery, theranostics, and photoacoustic diagnostics have all benefited from this combination. NIR light-absorbable with hydrophilic and hydrophobic polymers and a NO-releasing donor, DETA NONOate, were used to create photosensitizer-free polymeric nanocapsules.

8.14.3.5 THERANOSTIC BASED ON NUCLEAR IMAGING

Nuclear imaging methods are effective tools for measuring the distribution of radiotracers throughout the entire body while penetrating deep tissue. The majority of radiotracers, however, are tiny compounds that lack tissue selectivity *in vivo* and may be prone to fast renal elimination. Biomedical polymers have been applied to deliver radiotracers and anticancer medications in theranostic systems to increase tissue selectivity. PEG chains, folic acid, and radioactive 131I were used to introduce amine-terminated poly(amidoamine) dendrimers. The created dendrimers allowed for simultaneous *in vivo* radiation and targeted SPECT imaging. The combination of various polymer-based theranostic devices offers highly selective targeting of sick tissues and enables imaging-guided medical therapy. However, none of them have received clinical use approval. Given the significantly enhanced entanglement of the combined two modalities, the main reason is that these systems do not produce well-balanced and adequate therapeutic and diagnostic outcomes, especially for cancer theranostics. The *in vivo* transport process of intravenously (IV) administered theranostic agents also requires completing the CAPIR cascade which includes sequential circulation in the blood compartments, accumulation in the tumor, deep penetration into avascular tumor tissue to reach tumor cells, followed by cellular internalization and intracellular drug release. Although entering cells and penetrating the tumor may not be necessary for the diagnostic modality to function, the two processes may improve the contrast by increasing the tissue concentration. Therefore, in relation to polymer-based theranostic systems, the efficient completion of the CAPIR cascade may not only increase the therapeutic outcomes but also improve the diagnostic quality. The effectiveness of CAPIR completion is determined by two key phases: tumor extravasation and tumor penetration. Traditional polymer-based theranostic systems rely on the tumor's increased permeability and retention

(EPR) effect to extravasate through the highly permeable tumor vasculature and collect in the tumor lesion. However, such passive diffusion-based extravasation is severely constrained by the absence of the EPR effect in human tumors, as well as its heterogeneous and dynamic nature [196]. Fast transcytosis-inducing polymers have recently been reported by Shen et al. as drug-delivery vehicles. These polymers take advantage of the intrinsic active transport (ATP) pathways of endothelial cells and tumor/stromal cells and circumvent the challenges of passive extravasation and penetration of conventional nanomedicines, greatly increasing tumor accumulation and infiltration as well as therapeutic efficacy. They believe that theranostic systems based on polymers could offer improved therapeutic and diagnostic outcomes [197].

8.15 BIOMEDICAL BIOLOGY POLYMERS FOR IMMUNE THERAPY AND ANTI-INFECTION

8.15.1 BIOMEDICAL POLYMERS USED IN IMMUNOTHERAPY AND VACCINE DEVELOPMENT

Many efforts have been made to promote the use of biomedical polymers in cancer immunotherapy due to their favorable properties for drug delivery to tumors. Biomedical polymers have been looked at as effective vaccines and other immunotherapeutic delivery systems, as well as immunologic stimulants, to engage the immune system. Biomedical polymers could be applied as a delivery system for passive transfer or active transfer to administer immune-bioactive drugs while shielding them from the environment, prolonging their stay in circulation, minimizing systemic side effects, and inducing potent immune responses. A biomedical polymer-based nanosized delivery method for vaccines that can imitate pathogens enables effective antigen delivery to APCs. A depot effect is another property of large polymeric carriers, which raises the local concentration of antigens at certain places through timely release without repeated immunizations [198]. More intriguingly, a number of polymers have been created that have immunostimulatory properties and can be recognized by particular receptors expressed on immune cells to activate specific immunological pathways. Immunostimulating polymers that have the tendency to stimulate the immune system may also work effectively as delivery vehicles to trigger synergistic immunological reactions. The most recent developments in biological polymers used for vaccinations and immunotherapy have been reviewed in this section [199, 200].

8.15.1.1 CANCER-RELATED BIOMEDICAL POLYMERS

Cancer vaccinations are made with biomedical polymers. A vaccine is an active immunization technique that stimulates a particular immune response against a particular infection. Adjuvants and suitable antigens are typically used in cancer vaccines to stimulate powerful immune responses. Cancer vaccinations not only treat pre-existing tumors but also work to stop a particular cancer from developing. Cancer vaccines are typically categorized into whole cancer cell vaccines, subunit vaccinations, dendritic cell vaccines, and nucleic acid vaccines (such as mRNA) based on the type of antigen. Delivering antigens and adjuvants to APCs in a manner that encourages T cells to exert cytotoxic antitumor effects following maturation is essential for the promotion of effective cancer vaccines. Recent studies have shown that biomedical polymers that have undergone specific engineering can effectively promote cancer vaccines in a number of ways, including by increasing the stability of antigens by preventing their degradation, incorporating multiple adjuvants and antigens to enhance the immunogenicity, enhancing the uptake and cross-presentation of antigen by APCs, by enabling active targeting of APCs via specific recognition receptors, or by activating endosomal Toll-like receptor. As a result, many biomedical polymers have been researched for the production of vaccines [200]. Effective antigen delivery to APCs is crucial for APC activation and antigen presentation in cancer vaccines. Antigens are successfully delivered to APCs with better antigen presentation using biomedical polymers. For instance, PLGA nanoparticles have been effectively used for the encapsulation of various antigens, such as peptides, proteins, cell lysates, or nucleic acids, realizing delayed release of antigens to trigger more potent immune responses. Additionally, PLGA has the ability to co-deliver antigens and adjuvants into the same APC at the same time, eliciting potent T-cell responses even at very low doses. Lynn and colleagues showed how polymeric carriers could improve the efficacy of cancer vaccinations by creating a library of polymers containing adjuvants with various densities in addition to merely encapsulating antigens and adjuvants. A suitable injection method is yet another important element influencing vaccination efficacy with the addition of prolonged release of antigens. A greater immune response was seen after intradermal vaccination, according to earlier investigations. Zaric et al. recently created a microneedle that has an antigen-loaded with PLGA nanoparticles [201]. Intriguingly, antigen-encapsulated PLGA nanoparticles could be released into the skin layers harboring numerous APCs after the microneedle was dissolved in the skin, thereby activating DCs. Another crucial tactic to trigger potent immune responses is the use of cancer cell membranes as novel antigens, in addition to novel administration methods. Cancer cell membranes were used by Kroll et al. [202]. PLGA nanoparticles coated with adjuvant induce robust and multiantigenic antitumor immune responses. Additionally, Liu's team [203] created a cancer vaccine that was coated with the membrane of a cancer cell. They discovered that following additional mannose alteration, the nanovaccine may target APCs specifically because of the interaction between mannose and its receptor on APCs, resulting in noticeably improved immune responses. A lot of interest has recently been paid to mRNA-based nucleic acid vaccines because of their distinct benefits over traditional immunizations. However, because mRNA is a relatively delicate macromolecule, it cannot easily cross physiological barriers to enter target cells and is not stable in physiological environments. Many biomedical polymers have been applied to deliver mRNA in order to increase the immunogenicity of mRNA vaccines. Nucleic acid distribution has been done using PEI, the most popular cationic polymer [204]. Different modified PEIs have been created for the delivery of mRNA vaccines due to their toxicity. A supramolecularly assembled polymeric nanoparticle based on CpG/polyamidoamine-thioketal-adamantane and poly-[(N-2-hydroxyethyl)-aspartamide]-Pt(IV)/-cyclodextrin was created by Chen's group [298]. They discovered that polymeric nanoparticles could work as a tumor vaccine for stimulation of the immune system and cause the release of tumor-associated antigens following intravenous injection.

8.15.1.2 ICB USES BIOMEDICAL POLYMERS

Although using biomedical polymers to promote immunotherapy for cancer has mainly concentrated on cancer vaccines by effectively delivering antigens and adjuvants, ineffective immune responses still exist due to immune resistance. Immune checkpoint treatment is a typical immune resistance mechanism that is crucial in defending healthy tissues from immune system assault. By suppressing regulatory signals expressed by immune cells or tumor cells, ICB therapy has produced excellent therapeutic effects for the treatment of numerous types of tumors. Anti-cytotoxic T-lymphocyte-associated antigen 4 (anti-CTLA-4) and anti-programmed cell death protein 1 and its ligand (anti-PD-1/anti-PD-L1) antibodies are two examples of different ICB inhibitors [205]. The current ICB still faces a few hurdles, such as enhancing efficacy and lowering immune-related adverse effects, despite the fact that several of these ICB inhibitors have demonstrated exceptional therapeutic results [228]. As a result, numerous methods based on biomedical polymers have been created to increase ICB's effectiveness while minimizing its negative effects. Li et al. [206] synthesized a biocompatible and biodegradable poly (ethylene glycol)-block-poly (D, L-lactide) (PEG-PDLLA) copolymer to encapsulate CTLA-4 siRNA (siCTLA-4) by the use of N-bis(2-hydroxyethyl)-N-methyl-N-(2-cholesteryloxycarbonyl aminoethyl) ammonium bromide (BHEM-Chol). By efficiently suppressing CTLA-4 molecules expressed on the surface of activated T cells, they showed that siCTLA-4-loaded polymeric nanoparticles could boost anticancer immune responses. Along with polymeric nanoparticles, some polymer-based hydrogels have also been created to aid in the delivery of ICB inhibitors. Alginate hydrogels were employed by Li et al. [207] to deliver anti-PD-1 and anti-inflammatory medications locally and concurrently. Strong immune responses and anticancer effects were attained because anti-PD-1 and anti-inflammatory medications were delivered effectively.

8.15.1.3 POLYMERS FOR CYTOKINE TREATMENT IN BIOMEDICINE

Contrary to immunotherapy by ICB, cytokine therapy uses direct immunotherapy to activate various immune cells to destroy cancer cells. The released proteins known as cytokines are crucial in controlling both the innate and adaptive immune systems. Interferons (IFNs), interleukins (ILs), and granulocyte-macrophage colony-stimulating factor (GM-CSF) have been the main cytokines used in cancer immunotherapy to date. There are still several side effects, such as acute inflammation and autoimmune illnesses, even if cytokine therapies have stimulated large antitumor immune responses. Widespread interest has also been shown in the effective delivery of cytokines using biomedical polymers. The FDA has approved the use of IL-2, a growth factor for T cell activation and proliferation, to treat renal cell carcinoma and melanoma. Polymeric nanoparticles have been coated with IL-2 in the past to encourage T-cell activation and invasion. For illustration, Pork and Coworkers [208] used biodegradable polymers to incorporate IL-2 and transforming growth factor (TGF) inhibitors. Naturally increasing the infiltration of natural killer cells and $CD8^+$ T cells, the sustained release of IL-2 and TGF inhibitors significantly slowed the growth of tumors. Additionally, by conjugating IL-2 to the exterior of hydroxyethyl starch nanocapsules, effective T cell targeting relying on IL-2 receptor-mediated internalization was made possible. In addition to IL-2, IL-7 is a key cytokine for T-cell development and proliferation. It can also be combined with antigens in polymeric nanoparticles to elicit potent humoral and cellular immune responses that prevent the formation of tumors. Additionally, IL-12 has been encapsulated in a variety of biodegradable polymer-based nanoparticles, such as PLGA, PLA, polysaccharide chitosan, and PBAE, to encourage the activation of T cells.

8.15.1.4 ADOPTION OF BIOMEDICAL POLYMERS FOR T-CELL THERAPY

Attaching drug-loaded particles to the T cell membrane is a particular potential tactics. For instance, Irvine's team effectively expanded T cells in the tumor by loading cytokines into nanoparticles and conjugating them to T cells. Using such a method, additional molecules that support T cell activation could be delivered. Later on, they combined nanoparticles loaded with NSC-87877 with tumor-specific T cells, obtaining Significantly increased tumor T cell expansion [209].

8.15.2 BIOMEDICAL POLYMERS WITH ANTI-BACTERIAL PROPERTIES

Drug resistance, which lowers the effectiveness of antibacterial medications, is a severe issue brought on by the misuse of antibiotics. As a result, numerous antibacterial materials are being thoroughly researched on a global scale to combat bacterial illnesses. Polymers have drawn more attention recently because of their tremendous flexibility and diversity. Another class of innovative antibacterial polymers with strong antibacterial efficacy and a variety of applications are poly(ionic liquid) and their copolymers. Additionally, the chirality of polymers is a crucial component of their antibacterial activity and has been researched in the creation of bactericidal and anti-adhesive polymers. The aforementioned antibacterial polymers increase bactericidal efficacy and offer strong candidates for the prevention and treatment of infections. Medical device-induced infections are primarily caused by biofilm, which is resistant to standard antibiotic therapy. To cope with biofilm, various high-performance polymeric nanosystems were created. Bactericide-loaded micelles with improved biofilm permeation and acid-triggered charge conversion properties were created by a pH-responsive PBAE-based copolymer. These micelles may aggregate in biofilm and release antibacterial content driven by bacterial enzymes. Clustered polymeric nanoparticles conjugated to azithromycin with charge- and size-adaptive characteristics were created [210]. Based on poly(amidoamine) dendrimers with amino ends, the nanoparticles were able to deconstruct in the acidic biofilm microenvironment to provide improved antibiofilm performances both *in vitro* and *in vivo*. By combining the result of an antibiotic and reactive oxygen species, Rose Bengal polydopamine nanoparticles functionalized with polymyxin B can penetrate biofilm at low pH levels and bind to bacteria to demonstrate bactericidal properties. Drug delivery nanotransporters with thermoresponsive inspiration might provide integrated luminescence monitoring and synergistic chemo-photothermal antibacterial characteristics, which improve antibiotic permeation to kill drug-resistant bacteria. Such polymeric nanomaterials enhance the antibiotics' ability to deliver themselves effectively inside biofilms. Some antibacterial polymers with nanoscale dimensions might also perform exceptionally well. For instance, imidazolium-based block copolypeptides could be prefabricated into nanoparticles with strong antibacterial activity and less hemolysis when made using simple ring-opening polymerization and click chemistry techniques. Additionally, the antibacterial capabilities of polymeric nanomaterials could be increased with the addition of inorganic components. It is encouraging that antibiotic-loaded polymeric vesicles connected to ceria may prevent bacterial infection and get rid of superoxide free radicals at the same time. Polymeric nanoparticles have developed into potent therapeutic agents for treating difficult infections brought on by biofilms. Medical devices that require intervention or implantation are used in various clinical specialties, including orthopedics, dentistry, and plastic surgery. However, the high morbidity of infections brought on by medical devices results in treatment failure and excruciating pain for patients. Different types of polymer coatings are created to prevent or treat medical device-related infections because the side effect of these infections is the attachment and growth of bacteria on the surfaces of medical devices. Numerous medical devices, including titanium implants, can be effectively coated with an antibacterial substance using a catechol-based strategy. The biocompatibility of medical equipment is crucial because they come into direct contact with the human body. The ratio of positively and negatively charged groups in copolymers can be changed to increase the compatibility of quaternary ammonium coating. Medical catheters with a transparent covering made of fluorinated copolymer-based water-separable nanoparticles with hexamethylene biguanide component displayed outstanding antibacterial activity and biocompatibility. Making polymer coatings responsive to internal or external stimuli is another method to increase biocompatibility. A variety of outside factors, including radiation, could cause the molecular structure to change, creating surfaces with switchable bactericidal-release properties [211]. The surface coating might exhibit balanced bactericidal and antifouling capabilities to suit the demand of certain medical equipment by the logical combination of polymers and inorganic components, such as super hydrophilic polymer and gold nanorods. Infection-responsive antibacterial polymeric coatings could be created by taking into account the differences in the microenvironments between healthy and diseased tissues, such as pH levels and enzymes. Because the top layer could collapse in a low pH environment to expose the bactericidal components, hierarchical structures having a bactericidal lower layer and a multi-carboxyl polymeric top layer can be used to create the reactive antibacterial coating [212]. Aminoglycoside antibiotics could be decorated onto the surface of medical implants with high amounts using surface-initiated living polymerization and the Schiff base reaction. Aminoglycoside antibiotics could be loaded onto the surface of medical implants through surface-initiated living polymerization and the Schiff base reaction, and the loaded antibiotics could be released by the triggering of a weakly acidic environment of infection, showing a self-adaptive antibacterial profile. A triblock copolymer with a polyester block, two PPF diol, and diisocyanate are used to create a new unsaturated PU, also known as PPFU. The PPFU can be modified later by grafting compounds like peptides onto the highly reactive unsaturated double bonds, which promotes selective cell adhesion and migration [213].

8.15.3 ANTIVIRUS-USE BIOMEDICAL POLYMERS

The novel coronavirus pneumonia outbreak caused by SARS-CoV-2 has nearly completely spread around the world and caused significant harm to human society. Generally speaking, a virus is a form of non-cellular, tiny, and simple organism that only contains one type of nucleic acid (DNA or RNA), multiplies by replication, and parasitizes other live cells in order to exist. Nearly a third of all deaths worldwide each year are the result of viral infections, which account for 20% of all communicable diseases. Accurate and prompt diagnosis, effective disinfection, effective personal or collective protective equipment, and early and effective therapies are very necessary to restrict the spread of the virus, which eventually necessitates multidisciplinary solutions. The use of polymer materials for anti-virus applications, such as viral detection and diagnosis, disinfection, prevention, and therapy, is the only topic we discuss in this section [214].

8.15.3.1 BIOMEDICAL POLYMERS FOR VIRAL TESTING AND DETECTION

The first issue in preventing the devastation of human society brought on by infectious diseases is to identify an unknown pathogen that resembles an epidemic and can manifest itself in several ways. Traditional immunoassays, such as ELISA and paper-based LFAs, are less sensitive to viruses and can result in false-positive results that are hard to identify from genuine positive samples. To get over these restrictions, optical immunosensors such as colorimetry, fluorescence, and surface plasmon resonance have been created. For instance, polydiacetylene (PDA), a conjugated polymer, exhibits distinct chromatic properties [215].

8.15.3.2 BIOMEDICAL MATERIALS FOR CLEANING

The major method for controlling the spread of the virus is to clean up the environment to stop it from infecting people. In addition to natural polymers and their derivatives, biomedical polymers used for virus disinfection can be primarily synthetic polymers with antiviral properties by incorporating antiviral components (such as metal ions), functional organic chains, and charged groups, such as quaternary ammonium polymers. Grafting with functional groups, ligands, and chains is a particular modifying technique. Chitin, another polymer originating from animals or fungi, can be chemically modified by removing some of its acetyl groups to produce chitosan, an organic polysaccharide that is plentiful in natural resources. Because of the electrostatic interaction between the positively charged polycation of chitosan and the negatively charged surface of the virus, which can prevent virus infection or directly kill the virus by rupturing its protective membrane, chitosan, and its derivatives have already been shown to have direct antiviral activity. Additionally, PEI's anti-virus mechanism for synthetic polymers is similar to chitosan's in that its positive charge can draw in naturally negatively charged viruses. A sustainable and biodegradable antiviral filter membrane built of amyloid nanofibers manufactured from food-grade iron oxide and milk protein nanoparticles has been developed to stop the spread and transmission of viruses through ambient fluids. A very promising method of treating virus-contaminated water is made possible by the filter membrane's superior disinfection results against enveloped, non-enveloped, air- and water-borne viruses like SARS-CoV-2, influenza A (H1N1), and enterovirus 71 (EV71) [216].

8.15.3.3 BIOMEDICAL POLYMERS TO INHIBIT VIRUSES

Physical barriers like masks, gloves, protective suits, and other protective gear shield people from infection. By altering the surface of a material, people can decrease the adherence and survival of viruses and provide the modified material with stronger antiviral properties. The text depicts a variety of antiviral surfaces and coatings, including organic and polymeric, metal, and inorganic nanomaterial-coated surfaces. Additionally, some inorganic materials, including copper, silver, gold, zinc, and TiO_2 nanoparticles, have broad-spectrum antiviral properties. In order to create organic composites with anti-virus impact through surface modification, researchers can embed these nanoparticles into textiles and films [217]. Chitosan, a substance that can destroy viruses, can also be utilized to make anti-virus materials. Chitosan hydrogels were created and used to create antiviral gloves and spray. The materials having a photodynamic effect are another type of antiviral substance that can eradicate surface viruses. Vitamin K-containing nanofiber membranes (VNFMs) were made stable and reusable, and they produced ROS under UV and sunlight with a strong antiviral impact (> 99.9%). A brief (90 min) irradiation exposure time. Antiviral adhesion is an approach for materials design in addition to killing the viruses adhered to the upper layer of the materials. Additionally, it is stated that a hydrophilic negatively charged polyester fabric with a fluorinated polycationic coating is created using a one-step vapor deposition. Additionally, the materials mentioned above can create physical barriers, and several polymers are used inside the human body to protect against viruses. Preventing aerosol-borne viruses from entering mammalian nasal cavities and coming into contact with epithelial cells might be the best way to reduce the transmission of these viruses. This polymer has shown the potential to disrupt COVID-19's attachment to host cells, thus preventing virus penetration into the body. For example, a nanoscale gel that responds to poly(N-vinyl caprolactam) was created, which showed almost no cytotoxic effect on cervix vaginal epithelial cells but can inhibit intracellular HIV-1 replication and has a significant protective effect against HIV infection [218].

8.15.3.4 UTILIZING BIOMEDICAL POLYMERS TO TREAT VIRAL INFECTIONS

The creation of therapeutic medications and vaccines is the final step in the prevention and control of viruses. Polymeric and small molecule-based antiviral medications are both available. Recently, polymers have gained a lot of attention due to their unique properties when compared to small-molecule medications. This is due to the wide range of chemical composition and topological structure tunability of polymers. Additionally, polymers exhibit "multivalence," which allows them to use multiple pendent groups, polymer main chains, or even self-assembled nanostructures to bind to numerous complementary receptors on biological targets at the same time, significantly improving the antiviral activity of the drugs. Therefore, the development of polymeric materials for anti-virus infection is highly promising. Carrageenan, agar, fucoidan, and alginate are examples of marine-derived sulfated polysaccharides that have broad-spectrum antiviral activity against a variety of enveloped and nonenveloped viruses, making them one of the potential sources for the creation of novel antiviral drugs. It has also been investigated how these sulfated polysaccharides prevent various stages of virus infection in host cells such as stopping initial virus entrance and decreasing transcription and translation via controlling host cells' immunological responses. Additionally, it was investigated whether highly sulfated synthetic glycomimetics can act as suppressors of virus binding or infection. According to the findings, both long-chain and short-chain glycol polymers might prevent the spread of the herpes simplex virus (HSV), influenza A virus (IAV), and Merkel cell polyomavirus (MCPyV) [219]. Another potential family of antiviral medications is multivalent binding inhibitors, which prevent viruses from adhering to host cell membranes by imitating mucin for antiviral functions. For instance, the highly secured target of virus attachment ligands (VALs) is heparan sulfate proteoglycan (HSPG). By mimicking the proteoglycan, certain polymers can prevent the interaction and cross-talk between viruses and cells. In order to simulate HSPG, for instance, gold nanoparticles with polymeric sulfonates were created. The long-chain sulfonates in these particles helped to produce strong multivalent binding to VALs and deform the viruses, which ultimately led to the irreversible loss of viral infectivity. Additionally, the nanoparticles barely exhibit cytotoxicity and can be applied to treat infections caused by the respiratory syncytial virus (RSV), dengue virus, HPV, and HSV. A type of mucin called hemagglutinin is crucial to the invasion of the host cells by the virus. To accomplish this, a multivalent polymer-functionalized bacteriophage capsid with hemagglutinin ligand was created, which could mimic the hemagglutinin ligand arrangement on the head of the influenza A virus to match the geometry of the hemagglutinin binding site. It assists in binding to the virus in a specified multivalent mode in this way. In other words, the capsids have the ability to cover the entire virus envelope, preventing the virus from attaching to host cells and thwarting virus infection. In a nutshell, polymer shows great promise as a viral infection medication [220].

8.16 REGENERATIVE BIOMEDICAL POLYMER

8.16.1 BIOMEDICAL POLYMERS FOR TISSUE REGENERATION AND REPAIR

Biomedical polymer materials that, alone or in combination with cells and biological elements, can treat disease and injury by regenerating or repairing functional tissues are referred to as biomedical polymers for tissue cure and regeneration. The design of biomedical polymers for tissue cure and regeneration must take into account a number of key factors. Few foreign-body reactions should be induced, biophysical or biochemical interactions with the surroundings should activate useful functions, and certain cell behaviors should be controlled or regulated. *Ex vivo* and *in situ,* biomedical polymers are two categories that can be applied to categorize materials for tissue cure and regeneration. For a variety of uses, biomedical polymers in different formulations, such as scaffolds, sponges, micro- and nanoparticles, fibers, films, and hydrogels, have been produced. The interactions between cells/tissue and biomaterials are influenced by the stiffness, topology, porosity, and degradation of each physical architectural or chemical component [221]. The MNPs, or micro- and nanoparticles, can target or penetrate tissues and cells, and also promote cell-mediated movement and contact with specific cells. The MNPs-based approaches exhibit promise for biomedical applications, particularly in tissue regeneration and immunomodulation [222]. The hydrogel is a flexible substance that is soft and has great biocompatibility and biodegradability [223]. In contrast to real proteins, synthetic polypeptides, a newly developed class of hydrogel polymers, have a far more biomimetic secondary structure. In the biomedical field, polyurethane (PU), which has great elasticity, biocompatibility, biodegradability, and a customizable chemical structure, is frequently employed. In tissue repair and regeneration, a surface that promotes cell-selective adhesion and directed migration is extremely important, especially for the repair and regeneration of blood vessels, neurons, dermis, cartilage, bone, and other tissues. Physical signals (stiffness, charges, topology, and hydrophilicity), biochemical cues (peptides and antibodies), and physiological cues (chemokines and cytokines) can all be used to construct cell-selective biomaterials. The proliferation, differentiation, adhesion, and migration of the chosen cells can be encouraged by the cell-selective surfaces, which is advantageous for tissue regeneration and repair. Surfaces with a gradient that can mediate cell adhesion and migration are crucial for tissue regeneration and repair [224].

8.16.1.1 INTERACTIONS BETWEEN CELLS AND TISSUES IN THE CONTEXT OF BIOMEDICAL POLYMER BIOMATERIALS AND IMMUNOMODULATION

For tissue repair and regeneration, a variety of biomaterials are implanted, either alone or in conjunction with additional bioactive components that will communicate with the immune system, such as immune cells, cytokines, and chemokines. Including cytokines, chemokines, and different immune cells. Due to their molecular scale contact with cells and their interpretation at the bulk size, polymeric biomaterials, whether derived naturally or synthetically, can affect the immune cell function. Innate immune cells approach the implanted biomaterials one at a time during the acute inflammatory phase. The most significant actors are tissue-resident macrophages, blood-derived monocytes, neutrophils, and monocytes. Neutrophil activation may be impacted by biomaterial surface properties. For instance, when the wettability (hydrophilicity) or surface roughness of titanium implants change, neutrophils respond to the implants differently [225]. In comparison to rough hydrophilic surfaces, smooth or rough hydrophobic surfaces increase the amount of pro-inflammatory cytokines and enzymes secreted by neutrophils. Neutrophils' reactions to polymers are similarly influenced by their roughness. On polymeric cardiovascular implants, the rougher sections of polytetrafluoroethylene and Dacron can cause noticeably more neutrophil death and ROS production than the smoother sections. Similar results have been seen on electrospun scaffolds with small fiber diameters, which cause neutrophils to generate more traps as compared to those with large fiber diameters. As a result, biomedical polymers alter the mechanical, chemical, and morphological characteristics of tissues, causing tissue healing and regeneration.

8.16.1.2 USES FOR BIOMEDICAL POLYMERS IN TISSUE REGENERATION AND REPAIR

In particular, by modifying the microenvironment to encourage the innate ability of tissues with regard to repair and regeneration, polymeric materials have been used in the diagnosis and treatment of many disorders related to skin regeneration and repair.

1. In the early acute inflammatory phase of wound healing, neutrophils and monocytes/M1-like macrophages predominate in the cutaneous wound microenvironment. In diabetic skin wounds, biodegradable polyurethane antimicrobial hydrogel and cryogel exhibit immunomodulation by increasing the M2-macrophage population ratio. Additionally, some amphiphilic NPs are employed to transition macrophages from the M1 to the M2 polarity, as shown by the decreased expression of the M1 marker CD80.
2. Repair of the myocardium. When myocardial infarction (MI) begins, ischemia and hypoxia cause cardiomyocytes to undergo apoptosis. Because of their quick ROS-scavenging abilities, ROS-responsive biomaterials are seen as viable MI therapeutic candidates because they can lessen tissue damage by clearing out extra ROS from inflammatory areas. For instance, to create fibrous patches, 1,6-hexamethylene diisocyanate (HDI), 1,6-thioketal (PTK), and polyurethane (PFTU) were electrospun with gelatin. The substance effectively removes excessive ROS, lowers cell apoptosis, and reduces inflammation, which results in a better recovery of myocardial functions with the least amount of infarction [226].
3. Bone defects: The dual function of antibacterial and osteointegration is in high demand in orthopedics, particularly for individuals who have diabetes and osteoporosis simultaneously. Although single-purpose coatings have been widely developed, there is a lack of a simple and reliable approach for producing therapeutically useful implants with this dual function. Grafted hyperbranched poly(L-lysine) (HBPL) covalently onto titanium substrate, resulting in antibacterial activity against various pathogens and promotion of proliferation, spreading, adhesion, and osteogenic differentiation of MC3T3-E1 cells *in vitro*. The *in vivo* results confirm the good antimicrobial and anti-inflammatory response in a rat model, thereby resulting in better osteointegration than the control group.

4. Treatment of osteoarthritis (OA): ROS generation in the osteoarthritis microenvironment is a unique and potential medication release trigger. A novel polyurethane with ROS-responsive thioketal groups was created and later formed into dexamethasone-loaded nanoparticles (NPs). In the monosodium iodoacetate-induced OA model, our nanoplatform efficiently reduces the ROS level and improves the phenotypic of M2 macrophages.
5. Bowel disorders and biomaterials. In contrast, gastroenterology receives little emphasis when discussing biomaterials. Delivering targeted protein to the colon can help treat inflammatory illnesses including irritable bowel syndrome. The alginate/chitosan microcapsule loading IL-1 receptor antagonist can inhibit the highly expressed IL-1 cytokine in the inflamed colonic mucosa.
6. A decrease in pneumonia: In order to reduce inflammation and associated consequences in pneumonia, targeting macrophages, dendritic cells, and B-cells with NPs carrying proteins, peptides, and anti-inflammatory medicines has proven to be a successful technique. The biodegradable polymers and lipid-based nanomicelles possess amazing properties, such as the ability to form aerosol readily, biocompatibility, the ability to target particular regions of the lung cell population, stability, and the ability to release encapsulated medications in a particular way. To lessen the severity of the cytokine storm syndrome, mice with acute lung injury (ALI) were given nasal administration of hesperidin/chitosan NPs [227].
7. Regeneration of periodontal tissue. Collagen is mixed with cytokines and other biomaterials to create a special property for an immune response that enhances the bioactivity of periodontal regenerations. The immune system can be strengthened with chitosan as an alternative to encapsulate medications or growth factors for periodontal regeneration.
8. Regeneration and repair of the cornea. Research on corneal wound treatment has recently focused on properties of the cornea, such as immune response, preventing angiogenesis, and modifying cell signaling. Investigations on the immunocompatibility of various poly(glycerol sebacate) (PGS)-PCL nanofibrous scaffold compositions were conducted. Human corneal endothelial cells and human conjunctival epithelial cells are immediately viable in these scaffolds due to their well-defined mechanical properties. The scaffolds with a greater PGS/PCL ratio exhibit the best cell organization and cyto- and immuno-compatibility, while none of the PGS/PCL

combinations have any impact on granulocytes, such as naive and activated peripheral blood mononuclear cells (PBMCs) [228].

9. When neural stem/progenitor cells (NSPC) are transported using a hydrogel containing repeating units of the tetrapeptide, Arg-Ala-Asp-Ala, poly (RADA), with a C-terminal IKVAV sequence, it promotes better tissue healing in the rat brain than saline alone after a resection-type injury, compared with NSPC transplantation when PCL is absent. In conclusion, the modulation of different immune cells and/or biomolecules by polymeric biomaterials such as scaffolds, hydrogels, fibers, and NPs may improve tissue cure and regeneration. These biomaterials could be altered physically or chemically to hasten the healing processes of different tissues.

8.16.2 POLYMERS USED IN BIOMEDICINE TO REGENERATE THE HEART

8.16.2.1 A POLYELECTROLYTE LAYER WITH BIOACTIVE AGENTS

Cardiovascular stents have probably been used for the past 20 years to treat cardiovascular diseases (CVDs), and they have established themselves as the gold standard globally. Although it has been demonstrated that drug-eluting stents that release antiproliferative medications are effective at preventing restenosis, patients are at high risk for late stent thrombosis (ST) because of incomplete endothelium regeneration. In this situation, the new coating generation must improve endothelium renewal while keeping restenosis inhibition. Biomedical polymers, like polyelectrolytes and amphiphilic block copolymers, offer a viable option for creating medicinal coatings that can control cell responses like adhesion, migration, and bioactive polyelectrolyte coating. In order to functionalize medical devices, chemical grafting is widely used to stabilize tiny bioactive molecules (such as medications, antibodies, and peptides, etc.) over polymer covering. Functional groups are prevalent in polyelectrolytes, which have the ability to be intrinsically functionalized. In order to treat cardiovascular disease, bioactive substances have also been successfully converted into polyelectrolyte coating [229].

8.16.2.2 A POLYELECTROLYTE LAYER WITH CERTAIN MECHANICAL CHARACTERISTICS

All eukaryotic cells' growth, functionality, and homeostasis are significantly influenced by the extracellular matrix (ECM). Numerous studies have suggested that the extracellular microenvironment's biophysical cues, such as topographical features and stiffness, are essential for controlling cell proliferation, differentiation, migration, and other processes. It is well-recognized that stiffness, a crucial mechanical component of the ECM, has a significant impact on cellular behaviors. Numerous strategies have been developed recently to achieve bio-inspired stiffness and control cell behaviors. Ji et al. have successfully modulated the substrate stiffness of a coating made of PrS/DNA polyelectrolytes. According to the study, the coating reduced both SMC and EC cell attachment by reducing stiffness. It's interesting to note that SMCs were more sensitive to the stiffness lowering than ECs. Another crucial biophysical cue for controlling cellular behavior is the topographical characteristics of the polyelectrolyte coating. A versatile technique that has been used to research cell adhesion, differentiation, and spread behaviors is alternately depositing on a patterned substrate to create a patterned polyelectrolyte LbL coating. Through the use of pH, ionic strength, molecular weight, and other factors, the roughness of the polyelectrolyte coating can be altered. In light of this, the coating serves as a useful representation of stochastic topographic cues for examining cell behaviors [230].

8.16.2.3 A COVERING MADE OF SYNTHETIC POLYMER THAT IS POROUS

The rapid advancement in interface science over the past two decades has led to a notable improvement in cardiovascular stents for reduced hyperplasia and improved endothelium regeneration. These studies' fundamental approaches typically call for combining stents with biofunctional species such as peptides, proteins, DNAs/RNAs, etc. However, there are only a few different types of synthetic polymers that can currently be used as coating materials for cardiovascular stents. These polymers are typically dissolved in organic solvents during production. Therefore, despite the enormous success in concept fields, it is still difficult to combine biofunctional species with coating materials in actual applications. More recently, porous materials with linked pore structures have offered a different method of delivering bioactive chemicals. The simplicity, gentleness, and overall independence of the solute in the adsorbed solution were the distinguishing characteristics of this capillarity-based wicking process. The same group has described a self-healing spongy coating platform based on the crosslinked poly(-caprolactone)-poly(ethylene glycol)-poly(-caprolactone) (PCL-PEG-PCL, PCEC) triblock amphiphilic copolymer in order to modify the releasing profile of encapsulated drugs. The PEG chain gave the coating material the ability to swell in water, which caused the freeze-drying process to create porous structures. The PCEC networks conducted an ultrafast activation of polymer chain mobility by altering the ratio of PEG and PCL segments in the networks, realizing the self-healing behavior of the porous structure [231].

8.16.3 HEMOSTASIS AND WOUND CLOSING BIOMEDICAL POLYMERS

Sutures have traditionally been the first option for hemostasis and wound closure during surgery; nevertheless, they increase patient trauma and induce secondary tissue damage. Due to their significant biocompatibility, biodegradability, and simplicity of application, fast-acting and strong bio-adhesives are anticipated to be possible substitutes for sutures for tissue sealing and hemostasis. The significance of bio-adhesives in the sector of biomedicine has recently come to light, motivating researchers all over the world to enhance their structure and functionality and use them in different fields, including stomach/intestinal tissue repair, hemostasis, and myocardial infarction treatment. The advancements in the creation of glue- and tape-type bioadhesives and also the challenges associated with their clinical translation are highlighted. As a result, the present state of research on bio-adhesives for tissue sealing and hemostasis is reviewed. The abolishment of moisture from the tissue surface and obtaining close adherence are both facilitated by *in situ* curing of glue-type adhesives. They are therefore frequently employed for quick hemostasis and the adhesive filling of tissue defects. Additionally, glue-type adhesives can be safely used in minimally invasive surgery due to their injectability. Aldehydes, isocyanates, and N-hydroxy succinimide (NHS)-activated adhesive groups are frequently combined with curing processes like light, temperature, and self-assembly to create injectable glue-type bioadhesives. By mixing methacrylylated gelatin, o-nitrobenzyl modified hyaluronic acid, and a photoinitiator, for instance, Zhu, Zhang, Ouyang, and colleagues [232] created a photo-reactive injectable sticky hydrogel. To create a gel that could cling to and plug bleeding arteries and hearts under ultraviolet irradiation to achieve hemostasis, the hydrogel precursor underwent fast crosslinking. However, there are still many issues with glue-type adhesives that must be resolved in the ensuing research, such as non-reversibility, relatively poor adhesive properties, and lengthy curing times that can last minutes, hours, or days. Adhesive tape, myocardial patches, bandages, antifouling coatings, and wearable electronics are all examples of medical technology. The tape-type bio-adhesives demonstrate an instant bonding ability than the glue-type bio-adhesives since they don't need further stimulation or prolonged curing. In order for these materials to acquire good adhesion onto the surface of tissues covered in blood, bodily fluids, or perspiration, effective drainage of liquid is essential because they cannot be cured *in situ* to eliminate moisture from the tissue surface. Patches for tissue regeneration and the formation of cardiac patches, although rapid adherence could not be possible due to the slow diffusion and tangling of polymer chains. Zhao's team [233] combined a biopolymer with NHS-modified poly(acrylic acid) to create dry double-sided tape (DST). Due to the fast absorption of interface moisture, it displayed instant and excellent adhesion to a variety of wet tissues. This caused quick and temporary physical crosslinking with the adhered tissue's surface via the formation of hydrogen bonds and electrostatic interactions. The adhesion stability was further increased by subsequent covalent crosslinking of NHS with amine groups on the surface of the tissue. The DST could be employed as a strain sensor carrier to immediately seal ruptured intestines, stomach tissues, lung lobes, or trachea. Additionally, bio-inspired adhesives are continually being created. They used a gradient electrostatic complex technique and one-sided dipping to create the Janus sticky hydrogel. The Janus adhesive hydrogel's two faces showed remarkably different sticky and non-adhesive properties, and the phase separation brought on by electrostatic interaction enhanced the hydrophobicity and drainage capacity of the attached side. In place of surgical sutures, this Janus hydrogel was utilized to treat stomach perforations. The adhesive hydrogel's sticky side was securely attached to the stomach tissue. In order to seal the perforation site and speed up wound healing, the adhesive side of the adhesive hydrogel securely bonded to the stomach tissue. The other side helped to avoid postoperative adhesions. Additionally, the carboxyl adhesive's wet tissue adhesion strength was significantly increased by adding the hydrophobic hyperbranched structure, which not only provided excellent drainage characteristics but also remarkably increased the carboxyl content on the adhesive's surface. In addition to oysters, bio-adhesives made from topological microstructures inspired by nature, such as hexagonal papillae, suction patches, and spider silk fibers, have been frequently described. These adhesives have been utilized in wearable electronics and cardiac patches, and they have significantly aided advancement in the pertinent disciplines [234]. The ultimate objective of bio-adhesives is to be translated into the clinic, despite the fact that their significance and potential have been consistently researched. Therefore, in addition to concentrating on how well bio-adhesives adhere, researchers also need to take into account how well they cure, how quickly they degrade, and what effects their degraded components have on the human body. In order to realize clinical applications for bio-adhesives, research still needs to be done.

8.17 BIOACTIVE BIOMEDICAL POLYMERS FOR CANCER PRECESSION THERAPY

8.17.1 BIOACTIVE PEPTIDES FOR THE TREATMENT OF CANCER

Peptides are structures with less than 50 amino acids that are covalently linked together to carry out a variety of biological processes in the human body. Due to their superior biological activities, including targeting ability, stimuli-responsiveness, and therapeutic power, as well as their good biocompatibility and biosafety, peptides have been extensively used for cancer therapy. The peptides might either be in a monomer or assembled state to produce their anti-cancer properties [235].

8.17.1.1 PEPTIDE MONOMER FOR THE TREATMENT OF CANCER

Several peptide-based medications for cancer treatment have received FDA or European Medicines Agency (EMA) approval. Additionally, 1167 peptide-based drug studies for the treatment and cure of cancer are still in the pre-clinical and clinical stages. One could receive peptide-based medicines for treating cancer. Different sources could be used to obtain peptide-based medicines for cancer treatment. Natural hormones or their analogs, such as gonadotropin-releasing hormone (GnRH) analogs, have been used directly in the treatment of prostate, breast, and other cancers. The natural environment, such as terrestrial and marine ecosystems, could also be used to screen peptide-based medications. The screening of peptide-based pharmaceuticals for cancer therapy was made possible by the invention of high throughput screening techniques, such as phage display and one-bead-one-compound. The anti-cancer actions of the peptide would be achieved by activating or deactivating the protein and associated signaling pathways, such as blocking the secretion of testosterone. To precisely deliver drugs by locating the tumor, it is common knowledge that the microenvironment of a tumor differs from the milieu of a healthy body. In freshly generated tumor blood vessels, ECM, stromal cell membranes, and cancer cell membranes, numerous components are overexpressed or even selectively expressed, providing peptide binding sites. For instance, overexpressed proteins such as CD13 (aminopeptidase N) and CD105 (endoglin) that are targeted are seen in the new blood vessels of tumors. In addition, fibronectin, hyaluronic acid, and other parts of the ECM in tumors may serve as peptide-binding targets. The membrane proteins on stromal cells, such as bombesin receptors, and also the membrane proteins on tumor cells, such as the integrin family and others, are overexpressed and serve as targets. The features of the microenvironment in various tumor types influence the choice of targets [236]. Following targeting, controlled drug release is crucial for cancer chemotherapy. It has been discovered that the tumor site has several overexpressed enzymes that can enzymolyze particular peptide sequences, such as protein phosphatase, lactate dehydrogenase, and matrix metalloproteases. Consequently, a typical peptide drug conjugate, a targeting sequence, an enzyme cleavable sequence, in order to achieve accurate drug administration, targeting sequence, enzyme cleavable sequence, and drug conjugate, a type of typical peptide drug conjugate, or PDC, are therefore constructed. The PDC with reduced molecular weight has a stronger penetrating capacity when compared to antibody-drug conjugates, demonstrating substantial advantages for drug delivery of solid tumors.

8.17.1.2 SELF-ASSEMBLED PEPTIDE FOR THE TREATMENT OF CANCER

Because of the conjunction with surgery and interstitial chemotherapy for cancer, the self-assembled peptide-based hydrogel has been enormously used. For long-term chemotherapy, the hydrogel's disintegration could release the medication. The carriers for nanostructures could be given intravenously. With continuous drug release, the nanostructures with drug content passively and actively accumulate in the tumor site. Lower the drug release associated with highly organized nanostructures, such as greater amounts of β-sheet structures [237]. Due to unwanted deconstruction or additional aggregation, the pre-formed self-assembly of peptides may operate less well. Furthermore, it is common knowledge that nanostructures with larger sizes show long-term retention, whereas those with smaller sizes would deeply penetrate tumor tissues. The great performance of drug delivery requires programmed self-assembly. In order to accurately modify peptide self-assembly *in situ in vivo*, the idea of "*in vivo* self-assembly" is presented. At low pH levels, certain amino acids, including histidine, arginine, and lysine, may be protonated. This can upset the hydrophilic-hydrophobic equilibrium of self-assembling peptides and cause *in situ* self-assembly in tumors. *In situ-generated* peptide assemblies in an acidic tumor environment may serve as a drug store for loading and dispensing medications. For instance, in lysosome acidity, the *in situ* self-assembly process and subsequent assembly structures were able to disrupt the lysosome and directly kill cancer cells. Through the β-sheet, the peptide self-assembled into fibers on the cell membrane sequence, which was set off by binding to some particular cancer cell receptors. The receptor-related pathway was blocked by the peptide fibers, which also prevented tumor growth. Enzyme-induced self-assembly (EISA), which is caused by enzyme cleavage, can occur when an enzyme is overexpressed in the ECM and cancer cells. The drug-conjugated peptide residue self-assembled into a hydrogel and carried out prolonged drug release. It was possible to create peptide hydrogel *in situ* by injecting the peptide solution intratumorally rather than implanting it through surgery [238].

8.17.2 BIOMEDICAL POLYMERS BASED ON BIOMEMBRANES

To address the varied needs for *in vivo* biomedical applications, great efforts have been made to give polymeric materials, in particular polymeric nanoparticles (PNs), more and better functionality. Unfortunately, growing functionalization manipulation complexity seems to be keeping these PNs from being clinically applicable because mass production is so challenging. Furthermore, under difficult *in vivo* conditions, these artificially created functions are still far from satisfactory. Due to the lengthy evolution of naturally existing biofunctions, researchers have begun to look to nature for guidance. The cell membrane coating technique has recently come up as cutting-edge nanotechnology, allowing PNs to share the complete range of cellular biological activities, and was motivated by the fact that the membrane-membrane interactions between cells play considerably essential roles in many of these processes. The need to solve the problems with cell-based treatments, such as their challenging production and storage processes, and their safety concerns, such as the carcinogenicity of cancer cells and the potential for tumor stem cells to promote metastasis. The physical extrusion approach is mostly used to cover the cell membrane on PNs [238]. Additionally, attempts have been made to control membrane-PN adhesion, such as by increasing correlation through the interaction of a ligand grafted onto PNs and a transmembrane receptor expressed on cell membranes. Dynamically changing affinity between cell membranes and PNs core was achieved by taking advantage of the proton sponge effect, which is frequently exploited for the endosomal escape of PNs. As a result, the PN core sheds the outer membranes for enhanced uptake by tumors. Red blood cells (RBCs), B cells, cancer cells, immune cells, leukocytes, stem cells, platelets, and even bacteria are all used as sources of coated membranes. In general, the inherent roles of source cells in the host body are strongly tied to the biofunctions supplied by cell membranes. The variety of the cell bank enables the creation of a toolbox of biofunctions originating from different cell membranes, providing PNs with a potent armory for equipping them with naturally derived biofunctions as needed. Up to this point, the recorded biofunctions of cell membranes have been concentrated on several features, such as longer circulation duration, targeting cells/tissues, and immunological activation.

8.17.2.1 BIOMEDICAL POLYMERS WITH A BIO MEMBRANE FOUNDATION ARE EMPLOYED FOR EXTENDED CIRCULATION TIMES.

The enhancement of the *in vivo* effectiveness of transported cargoes (such as medicines, imaging agents, etc.) as well as the likelihood that they will arrive at the intended places depends on the prolonged circulation length of foreign PNs. This goal can be attained by coating RBC membrane onto PNs for biomimetic camouflage to avoid complement system, reticuloendothelial system, and immune system monitoring since RBC membranes contain "do not eat me" signals like CD47. In light of the immunogenicity of PEG segments and the time-consuming PEGylation method, this covert strategy appears to be superior to the well-known PEGylation. Notably, it has been demonstrated that nearly all cell lines' membranes contribute in some capacity. Based on the idea that these cells might survive and multiply in the host body, the results seem reasonable. Actually, even cancer cells produce a lot of "don't eat me" signals in order to avoid being eliminated by the immune system.

8.17.2.2 TARGETING BIO TAXIS WITH BIOMEDICAL POLYMERS BASED ON BIO MEMBRANES

The dynamic lateral fluidity of cell membranes enables multivalent contacts of one or more ligand-receptor pairs between membranes, enhancing the targeted cells' ability to recognize membrane-coated PNs. In light of this, site-specific transport via the cell membrane driving biotargeting may be preferable to the traditional targeting. A good candidate for targeted therapy for thrombotic illnesses including ischemic stroke and cardiac ischemia-reperfusion injury is the platelet membrane-coated PNs, which, like their parent cells, demonstrated a significant tendency to aggregate in wounded blood arteries. In order to provide a fresh target for the tumor-tropic attract the membrane of platelet-coated nanomaterials, blood vessels in tumors were purposefully destroyed [238]. Additionally, randomly migrating tumor cells (CTCs) could survive and spread to new metastatic niches by attracting platelets to surround them through specific biomolecular binding like P-Selectin and CD44 receptors. The platelet membrane-coated PNs were directed at the CTC capture for cancer diagnostics and tumor metastasis inhibition based on this property [239]. Since cell membranes contain self-adhesion proteins, cancer cell membrane-coated nanomaterials demonstrated an intriguing ability to target the homotypic. The cancer cell membrane-coated nanomaterials showed an intriguing capacity to target homotypic cancer cells because cell membranes contain self-adhesion proteins. Even when heterologous tumors are present, this self-recognition was shown to be extremely specific in targeting homologous tumors *in vivo*. Thus, the strategy of coating cancer cells with a membrane has been mostly applied to allow PNs to target tumors that originated from the same source of cancer cells. Because mesenchymal stem cells from the human umbilical cord are highly migratory and drawn to inflamed areas, the coating of PNs with these cells increased their propensity to gravitate towards tumors.

8.17.2.3 CANCER IMMUNOTHERAPY USES BIOMEDICAL POLYMERS BASED ON BIOMEMBRANES

The poor results of cancer immunotherapy are caused by a variety of immunosuppressive mechanisms that are produced alongside tumor progression. Although this effect seems insufficient, theoretically, cancer cell membranes could provide sources of neoantigens for the creation of vaccines. Immunotherapy would benefit from the increased expression of immune-associated chemokines on membranes, the nanoscale size, and the non-living nature of PNs due to the cell membranes extracted from the *in vitro* primed immune cells to get across immunosuppressive barriers. The interleukin-2 (IL-2)-loaded PLGA nanoparticles were coated with the cell membranes extracted from ovarian cancer cell lysate-primed DCs, which preserved the antigen-presentation components (such as MHC, CD86, and CD40) on their surface, to mimic DCs' antigen-presenting ability to stimulate T cells for therapeutic and preventative effects against cancers. When specific proteins (such as RANKL or DNAM-1) are present, such as on NK cell membranes coated on photodynamic therapy (PDT)-capable PNs, they can reverse the intratumoral immunosuppressive microenvironment and act as an immune inducer to stimulate the immune response brought on by PDT-induced immunogenic cell death for activating APCs [240]. Additionally, it has been demonstrated that the beta cell membrane coated on nanofiber scaffolds stimulates the growth of additional beta cells because these interactions are necessary for both cells to survive. Artificially modifying cell membranes by physical, chemical, and biological methods paves the way for the advancement of biomembrane coating nanotechnology. Direct chemical alteration of cell membranes is not a preferred method to maintain the membrane's biological activity. The combination of their separate biofunctions existing on the upper layer of PNs is made possible by the hybrid coating made of numerous membranes from several source cells. For instance, merging the membranes of bacteria and cancer cells could produce the combined biofunctions of malignant antigens and natural adjuvants, creating an immunotherapeutic formulation that targets cancer. The application of cell membranes as a coating provides a platform technology to easily equip PNs with biologically sophisticated capabilities on their surface. Clarification of cellular membrane-membrane contact processes, the identification of additional membrane biofunctions, and knowledge of the function of biomembrane-PN interactions in biomedical applications are all necessary for this field to advance in the future.

8.17.3 BIOMEDICAL POLYMERS DERIVED FROM MICROORGANISMS FOR CANCER TREATMENT

Living microorganisms (including bacteria, fungi, viruses, and micro-algae) and their natural components have recently come to light for their unrivaled capacity to treat tumors due to their unique targeting, immune-evoking, simple functionalization, and natural therapeutic features. In order to create cell-membrane immunotherapy, NK-NPs were able to induce pro-inflammatory M1-macrophage polarization in tumors. DAMPs (CRT exposure, ATP secretion, and HMGB1 release) could be produced by dying tumor cells as an effect of NK-NPs. They are used to develop living microbe-involved therapeutic formulas that integrate with therapeutic bacteria and their components using a variety of integration strategies, including physical assembly, chemical conjugation, and biological recombinant expression.

Additionally, microbial components that contain proteins, peptides, polysaccharides, and lipids could be used to create powerful therapeutic nanomedicines from natural biomedical polymers that are derived from living things. Active therapeutic medicines based on life-unit-involved polymers (LIPs) established a new paradigm in cancer treatment, which was encouraged by the improved therapeutic outcomes of malignancies reported in numerous literature [241]. Particularly, biomedical polymers stand out as the best materials due to their excellent biocompatibility and simplicity of modification. For instance, highly cytoprotective potentiality can be obtained to protect against external aggressors by hierarchically encapsulating the microbes with active polymers through self-crosslinking and self-polymerization, which is also advantageous to speed up microorganism-based disease therapy. According to one study, coating live attenuated bacterial surfaces with synthetic nanoparticles made from self-assembling cationic polymers and plasmid DNA allows bacteria to successfully evade phagosomes and increase gastric acid tolerance after oral administration due to the protective effect of the coated nanoparticle layer, which significantly facilitates bacterial penetration into the blood circulation and supports cancer immunotherapy. This approach built a typical paradigm for the organic integration of prebiotics and probiotics to globally alter gut microbiota, which is a safe and effective treatment for gastrointestinal cancer. The growth of pro-tumoral *Fusobacterium nucleatum* (*F. nucleatum*) was also evidently inhibited by chemically conjugating chemotherapeutic drug-loaded dextran nanoparticles with the targeted phages of *F. nucleatum* after oral administration due to the overall modulation of gut microbiota. In addition to altering individual microbes, the use of biomedical polymers to encapsulate several microorganisms was investigated for the curing of diseases. To load lytic phages and maintain phage activity, hollow PLGA polymeric microspheres were created and delivered to the deep lung. *Pseudomonas aeruginosa* (*P. aeruginosa*)-caused lung infections and infection-related cystic fibrosis can finally be decreased by dry powder inhalation. The notion of cascaded bacterial metabolism could be expanded to an anticancer application by merging several microorganisms with multifunctional biomedical polymers in light of the exceptional qualities of this inventive bacterial micro-ecosystem. Due to their high stability, inherent bioaffinity, ease of extraction and modification, and higher biosecurity, microorganism-derived polymeric components (such as bacterial surface layer proteins and biopolymers, virus-like particles (VLPs), and fungi-related polysaccharides) also demonstrate superior properties in biomedical applications [241, 242].

8.17.4 BIOMEDICAL POLYMERS FOR CANCER TREATMENT USING CELLS

Currently, cancers are treated using therapeutic methods that primarily use cells from the innate and adaptive immune systems [242, 243]. Although cell therapy has made unbelievable progress, it still faces obstacles in proving its clinical treatment efficacy and tolerable toxicity. However, cellular carriers encounter issues with efficiency and safety. At the same time, these cells can also act as carriers to deliver anticancer drugs to tumors with high specificity and efficiency. Therefore, to improve the viability, functionality, and sustainability of cells, polymeric carriers containing bioactive drugs have been developed. To increase the spatiotemporal precision of drug delivery to tumors, polymeric materials are also employed to encapsulate pharmaceuticals before connecting them to cellular carriers. Researchers have used polymeric materials to create hydrogels, microneedles, and micro/nanoparticles that work with cells to achieve these two objectives [243]. Polymer-based hydrogels are extensively researched because of their porous networks, swelling capacity, acceptable mechanical strength, mild gelation conditions, and excellent biocompatibility, they have the potential to be a cancer therapeutic. Immune cells, such as T cells, NK cells, dendritic cells, and chimeric antigen receptor (CAR) T-cells, could be loaded into environmentally sensitive smart hydrogels in cell-based immunity therapy. These kinds of smart hydrogels have a number of benefits for enhancing cell viability and improving the tumor microenvironment, which significantly enhances the immune response to tumors. N-acetylglucosamine and glucuronic acid are repeating disaccharides that make up the linear, non-sulfated glycosaminoglycan known as hyaluronic acid. As a non-immunogenic, hydrophilic, biodegradable, and structurally adaptable polymer, HA is frequently used to create hydrogels [161]. The HA hydrogel also made it possible to simultaneously load other bioactive medications, such as an antibody against programmed death-ligand 1 (PD-L1) that was coupled to platelets and interleukin-15, which was packed in nanoparticulate particles. By reversing the immunosuppressive milieu and supporting cell function, the release of these cargos prevented tumor spread and recurrence. A T-cell stimulating cytokine-loaded PEG-chitosan hydrogel, an injectable polyisocyanopeptides hydrogel, and a GFOGER-modified alginate hydrogel implant have also been manufactured to enhance T cell penetration, enhance T cells' sustainability and proliferation capability, and ultimately achieve a high anticancer activity. A different type of biomedical platform for cell administration in cancer therapy is the polymeric microneedle (MN) array patch [160]. MN can deliver the cargos to the dermis through minimally invasive, painless stratum corneum piercing [161, 162]. The MN array patch has a number of benefits, including simple structural manipulation, circumvention of the first-pass action of ingested drugs, and simple retrieval. The polyester known as PLGA, which is biodegradable and biocompatible, has attracted considerable attention in biomedical applications. In the body, PLGA hydrolyzes to produce the harmless natural metabolites lactic acid and glycolic acid [163]. Additionally, PLGA has a strong enough mechanical property to make the microneedles that can be inserted into tissues. Thus, PLGA could be used to create MN array patches. Cellular therapy can use microparticles and nanoparticles, which are common carriers in the sector of cancer treatment [159]. Particulate medications, such as chemotherapeutic or immunomodulating agents, may attach to or be absorbed by cells in cell-mediated anticancer therapy to enhance their pharmacokinetics and capacity to target particular organs. As a result, the drug-loaded particles can aggregate specifically in the tumor or other target areas, considerably increasing therapeutic effectiveness and lowering systemic off-target toxicity. Chemotherapy medications directly attack cancer cells when supplied by cells. Drugs for recruiting immune cells, immunomodulating tumor microenvironment, or improving cellular physiological function can also be put into nanoparticles. Nanoparticles on erythrocytes can specifically aggregate in the spleen to provide a potent immunological response based on an erythrocyte-driven immune targeting process. Ovalbumin was consequently coupled to a polystyrene nanoparticle. In order to transport antigens to the spleen, ovalbumin is interconnected to polystyrene nanoparticles and then adhered to erythrocytes, strongly activating an adaptive immune response. Additionally, a polyelectrolyte multilayer particle made of poly(vinyl alcohol) and PLGA was created to deliver interferon (IFN) to the macrophage it attached to. The particles' disk-like structure could stop the macrophage carrier from engulfing the drug backpack during endocytosis [93]. This method has demonstrated the potential to polarize M2-type macrophages that are linked with tumors to the M1 phenotype, changing the tumor microenvironment to one that is pro-inflammatory and encouraging an anticancer immune response to prevent metastasis and restrict tumor growth [168]. For medication delivery to the lung, erythrocytes have also been fitted with PLGA nanoparticles that have been loaded with chemokines [78].

8.18 CONCLUSION

Polymers are extremely important in many areas of daily life, including the medical industry. For the planned uses, such as the creation of pharmaceutical formulations, nanoparticles, drug conjugates, or tissue engineering reasons, choosing and designing the appropriate polymer is of utmost significance. In general, polymers have to be biocompatible, economical, nontoxic, and versatile.

These polymers may be employed in a broad variety of applications in almost all medical sectors due to their diverse polymer classes' capacity to adapt formulations by altering molecular weight, the degree of crystallization, cross-linking degree, copolymers, and producing polymer blends. When creating pharmaceutical products, polymeric materials are often used as excipients to preserve the medicine from deterioration or to regulate drug release, among other uses. In order to obtain and enhance the therapeutic impact of some medications, smart polymers have shown significant promise in the development of more intelligent drug delivery systems. Smart polymers do this by displaying changes in their physiochemical characteristics in response to external or endogenous stimuli. These polymers have demonstrated promising outcomes in the treatment of cancer and drug targeting. Polymeric materials have been used to create several biosensors for a variety of ailments. Conversely, conducting polymers are being studied for a variety of biological applications, including the creation of controlled drug release, artificial muscles, the stimulation of nerve regeneration, and brain recording.

KEYWORDS

- **biomedical**
- **drug delivery**
- **natural polymer**
- **polycaprolactone**
- **polyglycolic acid**
- **polylactic glycolic acid**
- **polymers**

REFERENCES

1. Mangal, S., Larson, I., Meiser, F., & Morton, D. A. (2015). Particle engineering of polymers into multifunctional interactive excipients. In V. K. Thakur & M. K. Thakur (Eds.), *Handbook of Polymers for Pharmaceutical Technologies* (pp. 1–31).
2. Abdul Amer, Z. J., Kareem Ahmed, J., & Fahim Abbas, S. (2014). Chitosan/PMMA bioblend for drug release applications. *International Journal of Engineering and Technology, 4*, 318–324.
3. Shinde, P., Patil, M., & Patil, A. (2017). Methods, types and applications of pharmaceutical polymers. *World Journal of Pharmacy and Pharmaceutical Sciences, 6*, 784–797.
4. Ribeiro, C., Sencadas, V., Correia, D. M., & Lanceros-Méndez, S. (2015). Piezoelectric polymers as biomaterials for tissue engineering applications. *Colloids and Surfaces B: Biointerfaces, 136*, 46–55.
5. Doppalapudi, S., Jain, A., Khan, W., & Domb, A. J. (2014). Biodegradable polymers-an overview. *Polymers for Advanced Technologies, 25*, 427–435.
6. Debotton, N., & Dahan, A. (2017). Applications of polymers as pharmaceutical excipients in solid oral dosage forms. *Medical Research Reviews, 29*, 52–97.
7. Khar, R. K., Vyas, S. P., Ahmad, F. J., & Jain, G. K. (Eds.). (2013). *Lachman/Lieberman's: The Theory and Practice of Industrial Pharmacy* (4th ed., pp. 403–448, 576–596).
8. Tomar, S., Singh, L., & Sharma, V. (2016). Miraculous adjuvants: The pharmaceutical polymers. *International Research Journal of Pharmacy, 7*(7), 405–515.
9. Jain, N. K. (2008). Pharmaceutical Product Development (1st ed.). *CBS Publishers & Distributors, 5*(5), 5–90.
10. Ahluwalia, V. K., & Mishra, A. (2008). *A Textbook of Polymer Science (*1st ed.). Ane Books Pvt. Ltd, *79*(9), 89–91.
11. Jyotirmoy, D., Das, M., & Das, A. (2017). Excellency of natural polymer in drug delivery system: A review. *International Journal of Pharmaceutical and Biological Science Archive, 5*(1), 17–22.
12. Sadhar, G. (2016). Natural polymers-based drug delivery systems. *World Journal of Pharmacy and Pharmaceutical Sciences, 5*(4), 805–816.
13. Gandhi, K. J., Deshmane, S. V., & Biyani, K. R. (2012). Polymers in pharmaceutical drug delivery system: A review. *International Journal of Pharmaceutical Sciences Review and Research, 14*(2), 57–66.
14. Kaushik, K., Sharma, R. B., & Agarwal, S. (2016). Natural polymers and their applications. *International Journal of Pharmaceutical Sciences Review and Research, 37*(2), 30–35.
15. Hofmann, W. (1968). *Vulkanizatsiia I vulkanizuiushchie agenty.* Moscow, *5*(5), 8–16.
16. Eirich, F. R. (Ed.). (1978). *Science and Technology of Rubber.* Academic Press. New York, *5*(48), 56–81.
17. Singh, D. P. (2020). Research in polymeric materials 'Current status and future potential.' In *National Conference in Advances of Applied Sciences and Mathematics, NCASM-20*, Chitkara University, Rajpura, India (pp. 24–25).
18. Mustafa, N. S., Omer, M. A. A., Garlnabi, M. E. M., & Ismail, H. A. (2016). Reviewing of general types of polymers, their properties and application in medical field. *International Journal of Science and Research (IJSR), 5*(8), 212–219.

19. Deb, P. K., Kokaz, S. F., Abed, S. N., Paradkar, A., & Tekade, R. K. (n.d.). Pharmaceutical and biomedical applications of polymers. *In Basic Fundamentals of Drug Delivery* (Chapter 6), 405–502.
20. Brady, J., Drig, T., Lee, P. I., & Li, J. X. (2016). Polymer properties and characterization. In *Developing Solid Oral Dosage Forms: Pharmaceutical Theory and Practice* (2nd ed.), Thakur Publication, 89–105.
21. Srinivasan, M., Rajabi, M., & Mousa, S. (2015). Multifunctional nanomaterials and their applications in drug delivery and cancer therapy. *Nanomaterials, 5*(9), 1690–1703. doi: 10.3390/nano5041690.
22. Kharkwal, H., Malhotra, B., & Janaswamy, S. (2017). Natural polymers for drug delivery: An introduction. In H. Kharkwal & S. Janaswamy (Eds.), *Natural Polymers for Drug Delivery* (pp. 1–9).
23. Parisi Ortensia Ilaria, Curcio, M., & Puoci, F. (2015). Polymer chemistry and synthetic polymers. In F. Puoci (Ed.), *Advanced Polymers in Medicine* (pp. 1–33).
24. Olatunji, O. (2016). Classification of natural polymers. In O. Olatunji (Ed.), *Natural Polymers* (pp. 1–17).
25. Saldivar-Guerra, E., & Vivaldo-Lima, E. (2013). Introduction to polymers and polymer types. In *Handbook of Polymer Synthesis, Characterization, and Processing*, Jyoti Publication, 59–85.
26. Ali Shah, A., Hasan, F., Shah, Z., Kanwal, N., & Zeb, S. (2013). Biodegradation of natural and synthetic rubbers: A review. *International Biodeterioration & Biodegradation, 83*(83), 145–157. doi: 10.1016/j.ibiod.2013.05.017.
27. Hamielec, A. E., & Penlidis, A. (1989). *Comprehensive Polymer Science and Supplements, 5*(8), 89–105.
28. Hong, K., & Mays, J. W. (2001). *Encyclopedia of Materials: Science and Technology 5*(9), 181–202.
29. Bamford, C. H., Barb, W. G., Jenkins, A. D., & Onyon, P. F. (1958). The Kinetics of Vinyl Polymerization by Radical Mechanism. *Butterworths Scientific Publications, 8*(2), 22–52.
30. North, A. M. (1966). The Kinetics of Free Radical Polymerization. Pergamon Press, 5(2), 9–31.
31. Mishra, B., & Khare, P. K. (2015). Review on Polymers. *Journal of Emerging Technologies and Innovative Research, 2*(2), 258–264.
32. Turnhaut, J. V. (1999). *Thermally stimulated discharge of electrets*. In G. M. Sessler (Ed.), *Electrets* (Vol. 1). Laplacian Press.
33. Bhajantri, R. B., Ravindrachary, V., Harisha, A., Crasta, V., Nayak, S. P., & Poojary, B. (2006). Synthesis and Characterization of Fe_3O_4-oleate/Poly (vinyl alcohol) Nanocomposites for Electrical Applications. *Polymer, 47*(3591), 3591.
34. Zidan, H. M. (2003). Electron spin resonance and ultraviolet spectral analysis of UV-irradiated PVA films filled with MnCl2 and CrF3. *Journal of Applied Polymer Science, 88*(1), 104–111.
35. Şirin, A., Zayim, E. Ö., & Sesalan, B. Ş. (2017). The optical properties of a novel metal-free phthalocyanine. *Chemical Papers. 71*, 2107–2115.
36. Devi, C. U., Sharma, A. K., & Rao, V. N. (2002). Electrical and optical properties of pure and silver nitrate-doped polyvinyl alcohol films. *Materials Letters. 56*(3), 167–174.
37. Muhammad, F., Yahya, M. Y., Rebar, T., Abdulwahid, Omed Gh. Abdullah, Shujahadeen B. Aziz, Sarkawt A. Hussein, & Fahmi (2016). *J Mater Sci: Mater Electron. 27*, 12112–12118.

38. Migahed, M. D., Ahmed, M. T., & Kotp, A. E. (2005). Computation of TSDC Peak Parameters in Amorphous Polymers using Vogel-Tammann-Fulcher Relaxation Model. *Journal of Macromolecular Science, Part B: Physics, 44*(1), 43–53.
39. Lewandowshi, A. C. (1994). Analytical description of thermally stimulated luminescence and conductivity without the quasiequilibrium approximation. *Physical Review B, 49*(12), 8029.
40. Thielen, A. A. (1994). Characterization of polyester films used in capacitors. I. Transient and steady-state conductivity. *Journal of Applied Physics, 76*(8), 4689–4695.
41. Christodoulides, C. (1998). Modelling of thermally stimulated depolarization current peaks obtained by global and thermal cleaning experiments. *Journal of Physics D: Applied Physics. 31*(20), 2898.
42. Khalil, K.M., El-Khatib, R. M., Ali, T. T., Mahmoud, H. A., & Elsamahy, A. A. (2013). Titania nanoparticles by acidic peptization of xerogel formed by hydrolysis of titanium (IV) isopropoxide under atmospheric humidity conditions. *Powder Technology. 245*, 156–162.
43. Mustafa, N. S., Omer, M. A. A., Garlnabi, M. E. M., & Ismail, H. A. (2016). Reviewing of general polymer types, properties and application in medical field. *International Journal of Science and Research (IJSR), 5*(8), 212–221.
44. Van Krevelen, D. W., & Te Nijenhuis K. (2009). Properties of polymers: their correlation with chemical structure; their numerical estimation and prediction from additive group contributions. Elsevier; Feb 9.
45. Jenni, A., Zurbriggen, R., Holzer, L., & Herwegh, M. (2005). Changes in microstructures and physical properties of polymer-modified mortars during wet storage. *Cement and Concrete Research*, 35, 79–90.
46. Kumari, P., & Kumar, A. (2019). A review on properties, classification and application of Polymer. *Journal of Advanced and Scholarly Researches in Allied Education, 16*(2), 1122–1129.
47. Ravindranath, K., & Rao, M. C. (2013). Physical properties and application of conducting polymers: An overview. *International Journal of Applied Pharmacy and Biological Chemistry (IJAPBC), 2*(1), 190–200.
48. Sen, D., Mohite, B., & Kayande, N. (2019). Review on polymer. *International Journal of Polymer Science and Materials (IJPSM), 4*(10), 1–15.
49. Perez Rail, L. (Ed.). Classification, types and chemistry of polymer. *Pharmaceutical Analytical Chemistry: Open Access, 7*, 1–2.
50. Chakraborty, I. (Lecture note). Polymer chemistry: Part-1: History and classification of polymer (pp. 1–21).
51. Priya, V. S. V., Roy, H. K., Jyothi, N., & Prasanthi, N. L. (2016). Polymers in drug delivery technology, types of polymers and application. *Scholars Academic Journal of Pharmacy, 3*(5), 305–308.
52. Odian, G. (2004). *Principles of Polymerization* (4th ed.). Wiley Interscience.
53. Narayan, R., & Pettigrew, C. (1999). ASTM Standardization News, December, 36–42.
54. Kolybaba, M., Tabil, L. G., Panigrahi, S., Crerar, W. J., Powell, J., & Wang, B. (2003). CSAE. InASAE Annual Intersectional Meeting. Fargo (ND), The Society for Engineering in Agricultural, Food and Biological Systems 2003, 8–18.
55. Gjullin, C. M. (1947). Effect of clothing color on the rate of attack of Aedes mosquitoes. *Journal of Economic Entomology, 40*, 326–327.

56. Gillies, M. T. (1980). The role of carbon dioxide in host-finding by mosquitoes (Diptera: Culicidae)—A review. *Bulletin of Entomological Research, 70*, 525–532.
57. Mukherjee, S., & Gowen, A. (2015). A review of recent trends in polymer characterization using non-destructive vibrational spectroscopic modalities and chemical imaging. *Analytica Chimica Acta, 895*, 12–34.
58. Pethrick, R. A., & Dawkins, J. W. (1999). Modern techniques for polymer characterization. Wiley.
59. Murthy, S. (2004). Recent developments in polymer characterization using X-ray diffraction. *The Rigaku Journal, 21*(1), 15–24.
60. Mulla, S. M., Phale, P. S., & Saraf, M. R. (2012). Use of X-ray diffraction technique for polymer characterization and studying the effect of optical accessories. Automotive Materials Lab, ARAI: Pune, India. 2012, 6–20.
61. Nicholas, P. C. (1996). *Polymer Characterization.* 17–24, Noyes Publications.
62. Arza, S. (2008). Wiley. Characterization and analysis of polymers. John Wiley & Sons; 2008, 243–246.
63. Zhuang, J., Gordon, M. R., Ventura, J., Li, L., & Thayumanavan, S. (2013). Multi-stimuli responsive macromolecules and their assemblies. *Chemical Society Reviews, 42*, 20–22.
64. Vercruysse, J., Delaet, U., Van Assche, I., Cappuyns, P., Arata, F., Caporicci, G., et al. (2013). Stability and repeatability of a continuous twin screw granulation and drying system. *European Journal of Pharmaceutics and Biopharmaceutics, 85*, 1031–1038.
65. Kestur, U., & Desai, D. (2017). Excipients for conventional oral solid dosage forms. In Koo, O. M. Y. (Ed.), *Pharmaceutical Excipients Properties, Functionality, and Applications in Research and Industry* (pp. 51–95).
66. Hong, Y., Liu, G., & Gu, Z. (2016). Recent advances of starch-based excipients used in extended-release tablets: a review. *Drug Delivery, 23*, 12–20.
67. Karolewicz, B. (2016). A review of polymers as multifunctional excipients in drug dosage form technology. *Saudi Pharmaceutical Journal, 24*, 525–536.
68. Hoag, S. W. (2017). Capsules dosage form: formulation and manufacturing considerations. In *Developing Solid Oral Dosage Forms: Pharmaceutical Theory and Practice* (2nd ed.). 92–150. Elsevier Inc.
69. Gullapalli, R. P., & Mazzitelli, C. L. (2017). Gelatin and non-gelatin capsule dosage forms. *Journal of Pharmaceutical Sciences, 106*(6), 1453–1465.
70. Al-Tabakha, M. M. (2010). HPMC capsules: current status and future prospects. *Journal of Pharmacy & Pharmaceutical Sciences, 13*, 428–442.
71. Dias, F., & Duarte, C. (2013). Cellulose and its derivatives are used in the pharmaceutical compounding practice. In T. Van de Ven & L. Godbout (Eds.), *Cellulose – Medical, Pharmaceutical and Electronic Applications* (pp. 141–162).
72. Kulkarni, V. S., & Shaw, C. (2016). Use of polymers and thickeners in semisolid and liquid formulations. *Essentials of Chemical Formulation of Semisolid and Liquid Dosages, 21*, 33–69.
73. Malviya, R., Srivastava, P., & Kulkarni, G. T. (2011). Applications of mucilages in drug delivery—a review. *Advances in Biological Research (Rennes), 5*, 1–7.
74. Gowdhaman, P., Antonyraj, K., & Annamalai, V. (2015). An effective approach on physical and dielectric properties of PZT-PVDF composites. *International Journal of Advanced Science and Research, 1*, 322–328.

75. Reddy, K. V. R., Reddy, D. B. V., & Shruti, P. (2015). Principles of drug release in various dosage forms. *World Journal of Pharmacy Research, 3*, 426–439.
76. Yadav, B. K., Gidwani, B., & Vyas, A. (2015). Rosin: recent advances and potential applications in novel drug delivery system. *Journal of Bioactive and Compatible Polymers, 31*, 111–126.
77. Madhuri, B., Gawali, V. B., & Mahesh, B. (2012). Ocular inserts: a rate-controlled drug delivery system a review. *International Journal of Pharmaceutical Erudition, 2*, 49–63.
78. Zhao, Z., Ukidve, A., Krishnan, V., Fehnel, A., Pan, D. C., Gao, Y., Kim, J., Evans, M. A., Mandal, A., Guo, J., Muzykantov, V. R., & Mitragotri, S. (2021). Systemic tumour suppression via the preferential accumulation of erythrocyte-anchored chemokine-encapsulating nanoparticles in lung metastases. *Nat Biomed Eng, 5*, 441–454.
79. Karthikeyan, D., Bhowmick, M., Pandey, V., Nandhakumar, J., Sengottuvelu, S., Sonkar, S., et al. (2008). The concept of ocular inserts as drug delivery systems: an overview. *Asian Journal of Pharmaceutics, 2*, 192.
80. Rathore, K. S., & Nema, R. K. (2009). Review on ocular inserts. *International Journal of PharmTech Research, 1*, 164–169.
81. Pramod, S. K., & Navnath, K. A. (2017). Ocular inserts: an overview. *World Journal of Pharmacy and Pharmaceutical Sciences, 6*, 141–148.
82. Benagiano, G., Gabelnick, H., & Brosens, I. (2015). Long-acting hormonal contraception. *Women's Health, 11*, 749–757.
83. Mansour, D. J., & Percy, L. (2017). Denying contraceptive choice to women with a previous ectopic pregnancy. *Journal of Family Planning and Reproductive Health Care, 43*, 232–233.
84. Cook, M. T., & Brown, M. B. (2018). Polymeric gels for intravaginal drug delivery. *Journal of Controlled Release, 270*, 145–157.
85. Sahoo, C. K., Nayak, P. K., Sarangi, D. K., & Sahoo, T. K. (2013). Intra vaginal drug delivery system: an overview. *American Journal of Advanced Drug Delivery, 1*, 43–55.
86. Ghori, M. U., & Conway, B. R. (2015). Hydrophilic matrices for oral control drug delivery. *American Journal of Pharmacological Sciences, 3*, 103–109.
87. Siepmann, J., & Siepmann, F. (2012). Swelling controlled drug delivery systems. In *Fundamentals and Applications of Controlled Release Drug Delivery* (pp. 153–170).
88. Khan, H., Shukla, R. N., & Bajpai, A. K. (2016). Genipin-modified gelatin nanocarriers as swelling controlled drug delivery system for in vitro release of cytarabine. *Materials Science and Engineering C, 61*, 457–465.
89. Larson, N., & Ghandehari, H. (2012). Polymeric conjugates for drug delivery. *Chemistry of Materials, 24*, 840–853.
90. Patil, J. P., Mahajan, H. S., & Patel, R. C. (2015). A review on polymer-drug conjugate—what, why, and how? *International Journal of Pharmaceutical Sciences and Research, 6*, 4611–4621.
91. Pang, X., Yang, X., & Zhai, G. (2014). Polymer-drug conjugates: recent progress on administration routes. *Expert Opinion on Drug Delivery, 11*, 1075–1086.
92. Duro-Castano, A., Movellan, J., & Vicent, M. J. (2015). Smart branched polymer-drug conjugates as nano-sized drug delivery systems. *Biomaterials Science, 3*, 1321–1334.
93. Ukidve, A., Zhao, Z., Fehnel, A., Krishnan, V., Pan, D. C., Gao, Y., Mandal, A., Muzykantov, V., & Mitragotri, S. (2020). Erythrocyte-driven immunization via biomimicry of their natural antigen-presenting function. *Proc Natl Acad Sci USA, 117*, 17727–17736.

94. Feng, Q., & Tong, R. (2016). Anticancer nanoparticulate polymer-drug conjugate. *Bioengineering & Translational Medicine, 1*, 277–296.
95. Vilar, G., Tulla-Puche, J., & Albericio, F. (2012). Polymers and drug delivery systems. *Current Drug Delivery, 9*, 1–28.
96. Hana, K., Etrych, T., Rıhova, B., Kostka, L., Starovoytova, L., Kovar, M., et al. (2013). HPMA copolymer conjugates of DOX and mitomycin C for combination therapy: physicochemical characterization, cytotoxic effects, combination index analysis, and anti-tumor efficacy. *Macromolecular Bioscience, 13*, 1648–1660.
97. Dipti, K., Tiwari, A. K., Chuttani, K., Kumar, R., Mishra, K., Singh, H., et al. (2011). Research article polyethyleneglycolylated isoniazid conjugate for reduced toxicity and sustained release. *Therapeutic Delivery, 2*, 205–212.
98. Nishiyama, K., Kitamura, N., Sakane, T., Hibi, T., & Yamamoto, A. (2011). Development of polyethylene glycol conjugated alendronate, a novel nitrogen-containing bisphosphonate derivative: evaluation of absorption, safety, and effects after intrapulmonary administration in rats. *Journal of Pharmaceutical Sciences, 100*, 3783–3792.
99. Crucho, C. I. C., & Barros, M. T. (2017). Polymeric nanoparticles: a study on the preparation variables and characterization methods. *Materials Science and Engineering C, 5*, 1–53.
100. Rao, J. P., & Geckeler, K. E. (2011). Polymer nanoparticles: preparation techniques and size-control parameters. *Progress in Polymer Science, 36*, 887–913.
101. El-Say, K. M., & El-Sawy, H. S. (2017). Polymeric nanoparticles: a promising platform for drug delivery. *International Journal of Pharmaceutics, 528*, 675–691.
102. Nagavarma, B. V. N., Yadav, H. K. S., Ayaz, A., Vasudha, L. S., & Shivakumar, H. G. (2012). Different techniques for preparation of polymeric nanoparticles: a review. *Asian Journal of Pharmaceutical and Clinical Research, 5*, 16–23.
103. Patel, T., Zhou, J., Piepmeier, J. M., & Saltzman, W. M. (2012). Polymeric nanoparticles for drug delivery to the central nervous system. *Advanced Drug Delivery Reviews, 64*, 701–705.
104. Attili-Qadri, S., Karra, N., Nemirovski, A., Schwob, O., Talmon, Y., & Nassar, T. (2013). The oral delivery system prolongs blood circulation of docetaxel nanocapsules via lymphatic absorption. *Proceedings of the National Academy of Sciences, 110*, 17498–17503.
105. Nanjwade, B. K., Singh, J., Parikh, K. A., & Manvi, F. V. (2009). Preparation and evaluation of carboplatin biodegradable polymeric nanoparticles. *International Journal of Pharmaceutics, 385*, 176–180.
106. Kant, G., Gupta, P., Gupta, U. D., Jain, N. K., & Agrawal, G. P. (2010). Gelatin nanocarriers as potential vectors for effective management of tuberculosis. *International Journal of Pharmaceutics, 385*, 143–149.
107. Alibolandi, M., Sadeghi, F., Abnous, K., Atyabi, F., Ramezani, M., & Hadizadeh, F. (2015). The chemotherapeutic potential of doxorubicin-loaded PEG-b-PLGA nano polymersomes in mouse breast cancer model. *European Journal of Pharmaceutics and Biopharmaceutics, 94*, 521–531.
108. Li, L., Lin, S. L., Deng, L., & Liu, Z. G. (2013). Potential use of chitosan nanoparticles for oral delivery of DNA vaccine in black seabream *Acanthopagrus schlegelii* Bleeker to protect from Vibrio parahaemolyticus. *Journal of Fish Diseases, 36*, 987–995.

109. Han, J., Zhao, D., Li, D., Wang, X., Jin, Z., & Zhao, K. (2018). Polymer-based nanomaterials and applications for vaccines and drugs. *Polymers, 10*, 1–14.
110. Morrison, P. W., & Khutoryanskiy, V. V. (2014). Advances in ophthalmic drug delivery. *Therapeutic Delivery, 5*, 1297–1315.
111. Sikandar, M. K., Sharma, P. K., & Visht, S. (2011). Ocular drug delivery system: an overview. *International Journal of Pharmaceutical Sciences and Research, 2*, 1168–1175.
112. Boateng, J., Okeke, O., & Khan, S. (2015). Polysaccharide-based formulations for mucosal drug delivery: a review. *Current Pharmaceutical Design, 21*, 4798–4821.
113. Rey-Rico, A., & Cucchiarini, M. (2018). PEO-PPO-PEO tri-block copolymers for gene delivery applications in human regenerative medicine—An overview. *International Journal of Molecular Sciences, 19*, 1–15.
114. Isaacman, M. J., Barron, K. A., & Theogarajan, L. S. (2012). Clickable amphiphilic triblock copolymers. *Journal of Polymer Science Part A: Polymer Chemistry, 50*, 2319-2329.
115. Taribagil, R. R., Hillmyer, M. A., & Lodge, T. P. (2010). Hydrogels from ABA and ABC triblock polymers. *Macromolecules, 43*, 5396–5404.
116. Behbahani, M., Abandansari, H. S., Salarian, M., Babapour, M., Bagheri, A., & Nabid, M. R. (2014). Synthesis and application of a thermosensitive tri-block copolymer as an efficient sample treatment technique for preconcentration and ultra-trace detection of lead ions. *Microchimica Acta, 181*, 129–137.
117. Ulery, B. D., Nair, L. S., & Laurencin, C. T. (2011). Biomedical applications of biodegradable polymers. *Journal of Polymer Science Part B: Polymer Physics, 49*, 832–864.
118. James, H. P., John, R., Alex, A., & Anoop, K. R. (2014). Smart polymers for the controlled delivery of drugs: concise overview. *Acta Pharmaceutica Sinica B*, *2*(4), 202–250.
119. Gandhi, A., Paul, A., Sen, S. O., & Sen, K. K. (2015). Studies on thermoresponsive polymers: phase behavior, drug delivery, and biomedical applications. *Asian Journal of Pharmaceutical Sciences, 10*, 99–107.
120. Khan, S., Ullah, A., Ullah, K., & Rehman, N. U. (2016). Insight into hydrogels. *Designed Monomers and Polymers, 19*, 456–478.
121. Wang, K., Strandman, S., & Zhu, X. X. (2017). A mini-review: shape memory polymers for biomedical applications. *Frontiers of Chemical Science and Engineering, 11*, 1–11.
122. Serrano, M. C., & Ameer, G. A. (2012). Recent insights into the biomedical applications of shape-memory polymers. *Macromolecular Bioscience, 12*, 1156–1171.
123. Baker, R. M., Tseng, L. F., Iannolo, M. T., Oest, M. E., & Henderson, J. H. (2016). Self-deploying shape memory polymer scaffolds for grafting and stabilizing complex bone defects: a mouse femoral segmental defect study. *Biomaterials, 76*, 388–398.
124. Landsman, T. L., Bush, R. L., Glowczwski, A., Horn, J., Jessen, S. L., Ungchusri, E., et al. (2016). Design and verification of a shape memory polymer peripheral occlusion device. *Journal of the Mechanical Behavior of Biomedical Materials, 63*, 195–206.
125. Nakamura, H., Etrych, T., Chytil, P., Ohkubo, M., Fang, J., & Ulbrich, K., et al. (2014). Two-step mechanisms of tumor-selective delivery of N-(2-hydroxypropyl) methacrylamide copolymer conjugated with pirarubicin via an acid cleavable linkage. *Journal of Controlled Release, 174*, 81–87.

126. Zardad, A. Z., Choonara, Y. E., du Toit, L. C., Kumar, P., Mabrouk, M., & Demarco Kondiah, P. P., et al. (2016). A review of thermo- and ultrasound-responsive polymeric systems for delivery of chemotherapeutic agents. *Polymers, 8*, 1–22.
127. Bhatnagar, S., & Venuganti, V. V. K. (2015). Cancer targeting: responsive polymers for stimuli-sensitive drug delivery. *Journal of Nanoscience and Nanotechnology, 15*, 1925–1945.
128. Guo, J., Skinner, G. W., Harcum, W. W., & Barnum, P. E. (1998). Pharmaceutical applications of naturally occurring water-soluble polymers. *Pharmaceutical Science & Technology Today, 1*, 254–261.
129. Zheng, W., Sauer, D., & McGinity, J. W. (2005). Influence of hydroxyethylcellulose on the drug release properties of theophylline pellets coated with Eudragit RS 30 D. *European Journal of Pharmaceutics and Biopharmaceutics, 59*, 147–154.
130. Khoder, M., Schropp, V., Zeitler, S., Pereira, B., Habashy, R., Royall, P. G., Wang, J. T., & Alhnan, M. A. (2020). A novel natural GRAS-grade enteric coating for pharmaceutical and nutraceutical products. *International Journal of Pharmaceutics, 584*, 119–392.
131. Yang, R. K., Fuisz, R. C., Myers, G. L., & Fuisz, J. M. (2018). Uniform films for rapid dissolve dosage form incorporating taste-masking compositions. *US Patent 9,931,305 B2*, 2018.
132. Van Ngo, H., Nguyen, P. K., Van Vo, T., Duan, W., Tran, V. T., Tran, P. H., & Tran, T. T. (2016). Hydrophilic-hydrophobic polymer blend for modulation of crystalline changes and molecular interactions in solid dispersion. *International Journal of Pharmaceutics, 513*, 148–152.
133. Makadia, H. K., & Siegel, S. J. (2011). Poly lactic-co-glycolic acid (PLGA) as biodegradable controlled drug delivery carrier. *Polymers, 3*, 1377–1397.
134. Teo, A. J. T., Mishra, A., Park, I., Kim, Y. J., Park, W. T., & Yoon, Y. J. (2016). Polymeric biomaterials for medical implants and devices. *ACS Biomaterials Science & Engineering, 2*, 454–472.
135. Gunatillake, P. A., & Adhikari, R. (2016). Nondegradable synthetic polymers for medical devices and implants. In *Biosynthetic Polymers for Medical Applications*. 302–324. Elsevier Ltd.
136. Manavitehrani, I., Fathi, A., Badr, H., Daly, S., Shirazi, A. N., & Dehghani, F. (2016). Biomedical applications of biodegradable polyesters. *Polymers, 8*, 1–22.
137. Shastri, V. P. (2003). Non-degradable biocompatible polymers in medicine: past, present, and future. *Current Pharmaceutical Biotechnology, 4*(5), 331–337.
138. Eschbach, L. (2000). Nonresorbable polymers in bone surgery. *Injury, 31*(4), S-D22–S-D27.
139. Shelton, J. R., Lando, J. B., & Agostini, D. E. (1971). Synthesis and characterization of poly (β-hydroxybutyrate). *Journal of Polymer Science Part B: Polymer Letters, 9*, 173–178.
140. Nair, L. S., & Laurencin, C. T. (2007) Biodegradable polymers as biomaterials. *Progress in Polymer Science, 32*, 762–798.
141. Feng, J., Zhuo, R. X., & Zhang, X. Z. (2012). Construction of functional aliphatic polycarbonates for biomedical applications.*Prog Polym Sci, 37*, 211–236.
142. He, C., Zhuang, X., Tang, Z., Tian, H., & Chen, X. (2012). pH-Responsive poly (ethylene glycol)/poly (L-lactide) supramolecular micelles based on host–guest interaction. *Adv Healthcare Mater, 1*, 48–78.

143. Rubin, L. R. (1983). Polyethylene as a bone and cartilage substitute: A 32 year retrospective. In L. R. Rubin (Ed.), *Biomaterials in Reconstructive Surgery* (pp. 474–493). C V Mosby.
144. Dong, C. X., Zhu, S. J., Mizuno, M., & Hashimoto, M. (2011). Fatigue behavior of HDPE composite reinforced with silane modified TiO_2. *J. Mater. Sci. Technol., 27*(1), 659–677.
145. Pinto, S., Alves, P., Matos, C. M., Santos, A. C., Rodrigues, L. R., Teixeira, J. A., & Gil, M. H. (2010). Poly (dimethyl siloxane) surface modification by Low pressure plasma to improve its characteristics towards biomedical applications. *Colloids Surf. B Biointerfaces, 81*(1), 20–26.
146. Owen, M. J. (1981). Why Silicones behave funny. *Chemtech, 11*, 288.
147. Wang, L., Sun, B., Ziemer, K. S., Barabino, G. A., & Carrier, R. L. (2010). Chemical and physical modifications to poly(dimethylsiloxane) surfaces affect adhesion of Caco-2 cells. *J. Biomed. Mater. Res. A, 93*(4), 1260–1271.
148. Li, B., Chen, J., & Wang, H. (2006). RGD peptide-conjugated Poly(dimethylsiloxane) promotes adhesion, proliferation, and collagen secretion of human fibroblasts. *J. Biomed. Mater. Res. A, 79A*, 989–998.
149. Basu, A., & Domb, A. J. (2018). Recent advances in polyanhydride based biomaterials. *Adv Mater, 30*, 1706815.
150. Rothemund, S., & Teasdale, I. (2016). Preparation of polyphosphazenes: a tutorial review. *Chem Soc Rev, 45*, 5200–5215.
151. Allcock, H. R., & Morozowich, N. L. (2012). Comparison of the synthesis and bioerodible properties of N-linked versus O-linked amino acid substituted polyphosphazenes. *Polym Chem, 3*, 578–590.
152. Lakshmi, S., Katti, D. S., & Laurencin, C. T. (2003). Iodegradable polyphosphazenes for drug delivery applications. *Adv Drug Deliver Rev, 55*, 467–482.
153. Greco, R. S. (Ed.). (1994). Implantation Biology: The host response and Biomedical Devices. CRC Press. 45–64.
154. Chlupac, J., Filova, E., & Bacakova, L. (2009). Blood vessel replacement: 50 years of development and tissue engineering paradigms in vascular surgery. *Physiol. Res., 58*(2), S119–S13.
155. Rahimy, M. H., Peyman, G. A., Chin, S. Y., Golshani, R., Aras, C., Borhani, H., & Thompson, H. (1994). Polysulfone capillary fiber for intraocular Drug delivery: In vitro and in vivo evaluations. *J. Drug Target., 2*(4), 2455–2480.
156. Zhu, J., & Marchant, R. E. (2011). Design properties of hydrogel tissue engineering scaffolds. *Expert Rev. Med. Devices, 8*(5), 607–626.
157. Harris, J. M., (ed.) (1992). Poly (ethylene glycol) chemistry: biotechnical and biomedical applications. Springer Science & Business Media; 81–90.
158. Harris, J. M., Zalipsky, S., editors. (1997). Poly (ethylene glycol) chemistry and biological applications. *American Chemical Society;* 87–95.
159. Liu, X., Wang, D., Zhang, P., & Li, Y. (2019). Recent advances in nanosized drug delivery systems for overcoming the barriers to anti-PD immunotherapy of cancer. *Nano Today, 29,* 100801.
160. Makvandi, P., Jamaledin, R., Chen, G., Baghbantaraghdari, Z., Zare, E. N., Di Natale, C., Onesto, V., Vecchione, R., Lee, J., Tay, F. R., Netti, P., Mattoli, V., Jaklenec, A., Gu, Z., & Langer, R. (2021). Stimuli-responsive transdermal microneedle patches. *Mater Today, 47,* 206–222.

161. Lee, J. Y., Chung, S. J., Cho, H. J., & Kim, D. D. (2016). Iodinated hyaluronic acid oligomer-based nanoassemblies for tumor-targeted drug delivery and cancer imaging. *Biomaterials, 85,* 218–231.
162. Ulery, B. D., Nair, L. S., & Laurencin, C. T. (2011). Biomedical applications of biodegradable polymers. *J Polym Sci B Polym Phys, 49,* 832–864.
163. Wang, Y. C., Yuan, Y. Y., Du, J. Z., Yang, X. Z., & Wang, J. (2009). Recent progress in polyphosphoesters: from controlled synthesis to biomedical applications. *Macromol Biosci, 9,* 1154–1164.
164. Chen, W. H., Liao, W. C., Sohn, Y. S., Fadeev, M., Cecconello, A., Nechushtai, R., & Willner, I. (2018). Stimuli-responsive nucleic acid-based polyacrylamide hydrogel-coated metal–organic framework nanoparticles for controlled drug release. *Adv Funct Mater, 28,* 1705137.
165. Du, X. J., Wang, J. L., Liu, W. W., Yang, J. X., Sun, C. Y., Sun, R., & Wang, J. (2015). Regulating the surface poly (ethylene glycol) density of polymeric nanoparticles and evaluating its role in drug delivery in vivo. *Biomaterials, 69,* 1–11.
166. Wang, J. L., Du, X. J., Yang, J. X., Shen, S., Li, H. J., Luo, Y. L., & Wang, J. (2018). The effect of surface poly (ethylene glycol) length on in vivo drug delivery behaviors of polymeric nanoparticles. *Biomaterials, 182,* 104–113.
167. Du, J. Z., Li, H. J., & Wang, J. (2018). Tumor-acidity-cleavable maleic acid amide (TACMAA), a powerful tool for designing smart nanoparticles to overcome delivery barriers in cancer nanomedicine. *Acc Chem Res, 51,* 2848–2856.
168. Sun, Q., Zhou, Z., Qiu, N., & Shen, Y. (2017). Rational design of cancer nanomedicine: nanoproperty integration and synchronization. *Adv Mater, 29,* 1606628.
169. Xiao, Z., Su, Z., Han, S., Huang, J., Lin, L., & Shuai, X. (2020). Dual pH-sensitive nanodrug blocks PD-1 immune checkpoint and uses T cells to deliver NF-κB inhibitor for antitumor immunotherapy. *Sci Adv, 6,* eaay7785.
170. Hu, Q., Sun, W., Qian, C., Bomba, H. N., Xin, H., & Gu, Z. (2017). Relay Drug Delivery for Amplifying Targeting Signal and Enhancing Anticancer Efficacy. *Adv Mater, 29,* 1605803.
171. Wang, Z., Song, L., Liu, Q., Tian, R., Shang, Y., Liu, F., & Ding, B. (2021). A tubular DNA nanodevice as a siRNA/chemo-drug co-delivery vehicle for combined cancer therapy. *Angew Chem Int Ed, 60,* 2594–2598.
172. Gabrielson, N. P., Lu, H., Yin, L., Li, D., Wang, F., & Cheng, J. (2012). Reactive and bioactive cationic α-helical polypeptide template for nonviral gene delivery. *Angew Chem Int Ed, 51,* 1143–1147.
173. Fang, H., Guo, Z., Chen, J., Lin, L., Hu, Y., Li, Y., & Chen, X. (2021). Combination of epigenetic regulation with gene therapy-mediated immune checkpoint blockade induces anti-tumour effects and immune response in vivo. *Nat Commun, 12,* 6742.
174. Wang, M., Liu, H., Li, L., & Cheng, Y. (2014). A fluorinated dendrimer achieves excellent gene transfection efficacy at extremely low nitrogen to phosphorus ratios. *Nat Commun, 5,* 3053.
175. Zhang, Z., Qiu, N., Wu, S., Liu, X., Zhou, Z., Tang, J., & Shen, Y. (2021). Dose-independent transfection of hydrophobized polyplexes. *Adv Mater, 33,* 2102219.
176. Uchida, H., Itaka, K., Nomoto, T., Ishii, T., Suma, T., Ikegami, M., & Kataoka, K. (2014). Supramolecular chiroptical switches. *J Am Chem Soc, 136,* 12396–12400.

177. Haabeth, O. A. W., Blake, T. R., McKinlay, C. J., Waymouth, R. M., Wender, P. A., & Levy, R. (2018). Polymers for cytosolic protein delivery. *Proc Natl Acad Sci USA, 115,* E915.
178. Lv, J., Fan, Q., Wang, H., & Cheng, Y. (2019). Polymers for cytosolic protein delivery. *Biomaterials, 218,* 119358.
179. Su, Z., Xiao, Z., Wang, Y., Huang, J., An, Y., Wang, X., & Shuai, X. (2020). Codelivery of anti-PD-1 antibody and paclitaxel with matrix metalloproteinase and pH dual-sensitive micelles for enhanced tumor chemoimmunotherapy. *Small, 16,* 1906832.
180. Liang, S., Liu, Y., Jin, X., Liu, G., Wen, J., Zhang, L., & Lu, Y. (2016). Phosphorylcholine polymer nanocapsules prolong the circulation time and reduce the immunogenicity of therapeutic proteins. *Nano Res, 9,* 1022–1031.
181. Liu, Y., Du, J., Yan, M., Lau, M. Y., Hu, J., Han, H., & Lu, Y. (2013). Biomimetic enzyme nanocomplexes and their use as antidotes and preventive measures for alcohol intoxication. *Nat Nanotech, 8,* 187–192.
182. Postupalenko, V., Sibler, A. P., Desplancq, D., Nominé, Y., Spehner, D., Schultz, P., Weiss, E., & Zuber, G. (2014). Intracellular delivery of functionally active proteins using self-assembling pyridylthiourea-polyethylenimine. *J Control Release, 178,* 86–94.
183. Li, X., Cai, X., Zhang, Z., Ding, Y., Ma, R., Huang, F., Liu, Y., Liu, J., & Shi, L. (2020). Mimetic heat shock protein mediated immune process to enhance cancer immunotherapy. *Nano Lett, 20,* 4454–4463.
184. Zhen, X., & Jiang, X. (2020). Polymer-based activatable optical probes for tumor fluorescence and photoacoustic imaging. *Wiley Interdisciplinary Reviews: Nanomed Nanobiotechnol, 12,* e1593.
185. Wang, Y., Zhou, K., Huang, G., Hensley, C., Huang, X., Ma, X., Zhao, T., Sumer, B. D., DeBerardinis, R. J., & Gao, J. (2014). A nanoparticle-based strategy for the imaging of a broad range of tumours by nonlinear amplification of microenvironment signals. *Nat Mater, 13,* 204–212.
186. Zheng, X., Wang, X., Mao, H., Wu, W., Liu, B., & Jiang, X. (2015). Hypoxia-specific ultrasensitive detection of tumours and cancer cells in vivo. *Nat Commun, 6,* 5834.
187. Loeb, G. E., Bak, M. J., Salcman, M., & Schmidt, E. M. (1977). Parylene as a chronically stable, reproducible microelectrode insulator. *IEEE Trans Biomed Eng, BME-24,* 121–128.
188. Yang, Z., Deng, J., Sun, X., Li, H., & Peng, H. (2014). Polymer and its advantages. *Adv Mater, 26,* 2643–2647.
189. Inoue, A., Yuk, H., Lu, B., & Zhao, X. (2020). Strong adhesion of wet conducting polymers on diverse substrates. *Sci Adv, 6,* eaay5394.
190. Zhai, D., Liu, B., Shi, Y., Pan, L., Wang, Y., Li, W., Zhang, R., & Yu, G. (2013). Highly sensitive glucose sensor based on Pt nanoparticle/polyaniline hydrogel heterostructures. *ACS Nano, 7,* 3540–3546.
191. Tang, C., Xie, S., Wang, M., Feng, J., Han, Z., Wu, X., Wang, L., Chen, C., Wang, J., Jiang, L., Chen, P., Sun, X., & Peng, H. (2020). A fiber-shaped neural probe with alterable elastic moduli for direct implantation and stable electronic–brain interfaces. *J Mater Chem B, 8,* 4387–4394.
192. Hong, G., Antaris, A. L., & Dai, H. (2017). Near-infrared fluorophores for biomedical imaging. *Nat Biomed Eng, 1,* 0010.

193. Miao, Q., Xie, C., Zhen, X., Lyu, Y., Duan, H., Liu, X., Jokerst, J. V., & Pu, K. (2017). Molecular afterglow imaging with bright, biodegradable polymer nanoparticles. *Nat Biotechnol, 35,* 1102–1110.
194. Ye, M., Qian, Y., Tang, J., Hu, H., Sui, M., & Shen, Y. (2013). Targeted biodegradable dendritic MRI contrast agent for enhanced tumor imaging. *J Control Release, 169,* 239–245.
195. Duan, L., Yang, L., Jin, J., Yang, F., Liu, D., Hu, K., Wang, Q., Yue, Y., & Gu, N. (2020). Micro/nano-bubble-assisted ultrasound to enhance the EPR effect and potential theranostic applications. *Theranostics, 10,* 462–483.
196. Chen, J., Ratnayaka, S., Alford, A., Kozlovskaya, V., Liu, F., Xue, B., Hoyt, K., & Kharlampieva, E. (2017). Micro/nano-bubble-assisted ultrasound to enhance the EPR effect and potential theranostic applications. *ACS Nano, 11,* 3135–3146.
197. Danhier, F. (2016). To exploit the tumor microenvironment: Since the EPR effect fails in the clinic, what is the future of nanomedicine? *J Control Release, 244,* 108–121.
198. Zhou, Q., Dong, C., Fan, W., Jiang, H., Xiang, J., Qiu, N., Piao, Y., Xie, T., Luo, Y., Li, Z., Liu, F., & Shen, Y. (2020). Tumor extravasation and infiltration as barriers of nanomedicine for high efficacy: The current status and transcytosis strategy. *Biomaterials, 240,* 119902.
199. Leleux, J., & Roy, K. (2013). Biophysical attributes of CpG presentation control TLR9 signaling to differentially polarize systemic immune responses. *Adv Healthcare Mater, 2,* 72–94.
200. Zhang, Y., Ma, S., Liu, X., Xu, Y., Zhao, J., Si, X., Li, H., Huang, Z., Wang, Z., Tang, Z., Song, W., & Chen, X. (2021). Supramolecular assembled programmable nanomedicine as in situ cancer vaccine for cancer immunotherapy. *Adv Mater, 33,* 2007293.
201. Liu, W. L., Zou, M. Z., Liu, T., Zeng, J. Y., Li, X., Yu, W. Y., Li, C. X., Ye, J. J., Song, W., Feng, J., & Zhang, X. Z. (2019). Cytomembrane nanovaccines show therapeutic effects by mimicking tumor cells and antigen presenting cells. *Nat Commun, 10,* 3199.
202. Zaric, M., Lyubomska, O., Touzelet, O., Poux, C., Al-Zahrani, S., Fay, F., Kroll, A. V., Fang, R. H., Jiang, Y., Zhou, J., Wei, X., Yu, C. L., Gao, J., Luk, B. T., Dehaini, D., Gao, W., & Zhang, L. (2017). Nanoparticulate delivery of cancer cell membrane elicits multiantigenic antitumor immunity. *Adv Mater, 29,* 1703969.
203. Yang, R., Xu, J., Xu, L., Sun, X., Chen, Q., Zhao, Y., Peng, R., & Liu, Z. (2018). Cancer cell membrane-coated adjuvant nanoparticles with mannose modification for effective anticancer vaccination. *ACS Nano, 12,* 5121–5129.
204. Lai, W. F. (2011). Biomaterials, biological molecules, and polymers in developing vaccines. *Expert Rev Med Devices, 8,* 173–185.
205. Polla Ravi, S., Shamiya, Y., Chakraborty, A., Elias, C., & Paul, A. (2021). Restoring anti-tumor functions of T cells via nanoparticle-mediated immune checkpoint modulation. *Trends Pharmacol Sci, 42,* 813–828.
206. Li, S. Y., Liu, Y., Xu, C. F., Shen, S., Sun, R., Du, X. J., Xia, J. X., Zhu, Y. H., & Wang, J. (2016). Hydrogel dual delivered celecoxib and anti-PD-1 synergistically improve antitumor immunity. *J Control Release, 231,* 17–28.
207. Li, Y., Fang, M., Zhang, J., Wang, J., Song, Y., Shi, J., Li, W., Wu, G., Ren, J., Wang, Z., Zou, W., & Wang, L. (2016). . Combination delivery of TGF-β inhibitor and IL-2 by nanoscale liposomal polymeric gels enhances tumour immunotherapy. *OncoImmunology, 5,* e1074374.

208. Park, J., Wrzesinski, S. H., Stern, E., Look, M., Criscione, J., Ragheb, R., Jay, S. M., Demento, S. L., Agawu, A., Licona Limon, P., Ferrandino, A. F., Gonzalez, D., Habermann, A., Flavell, R. A., & Fahmy, T. M. (2012). Synapse-directed delivery of immunomodulators using T-cell-conjugated nanoparticles. *Nat Mater, 11,* 895–905.
209. Stephan, M. T., Stephan, S. B., Bak, P., Chen, J., & Irvine, D. J. (2012). Self-targeting, zwitterionic micellar dispersants enhance antibiotic killing of infectious biofilms—An intravital imaging study in mice. *Biomaterials, 33,* 5776–5787.
210. Tian, S., Su, L., Liu, Y., Cao, J., Yang, G., Ren, Y., Huang, F., Liu, J., An, Y., Van der Mei, H. C., Busscher, H. J., & Shi, L. (2020). Multifunctional antibacterial materials comprising water dispersible random copolymers containing a fluorinated block and their application in catheters. *Sci Adv, 6,* eabb1112.
211. Li, W., Zhang, H., Li, X., Yu, H., Che, C., Luan, S., Ren, Y., Li, S., Liu, P., & Yu, X. (2020). Mimetic heat shock protein mediated immune process to enhance cancer immunotherapy. *ACS Appl Mater Interfaces, 12,* 7617–7630.
212. Zhao, Y. Q., Sun, Y., Zhang, Y., Ding, X., Zhao, N., Yu, B., Zhao, H., Duan, S., & Xu, F. J. (2020). *ACS Nano, 14,* 2265–2275.
213. Sun, Y., Zhao, Y. Q., Zeng, Q., Wu, Y. W., Hu, Y., Duan, S., Tang, Z., & Xu, F. J. (2019). Dual-functional implants with antibacterial and osteointegration-promoting performances. *ACS Appl Mater Interfaces, 11,* 36449–36457.
214. Tang, Z., Kong, N., Zhang, X., Liu, Y., Hu, P., Mou, S., Liljeström, P., Shi, J., Tan, W., Kim, J. S., Cao, Y., Langer, R., Leong, K. W., Farokhzad, O. C., & Tao, W. (2020). A materials-science perspective on tackling COVID-19. *Nat Rev Mater, 5,* 847–860.
215. Kim, E., Lim, E. K., Park, G., Park, C., Lim, J. W., Lee, H., Na, W., Yeom, M., Kim, J., Song, D., & Haam, S. (2021). Advanced Nanomaterials for Preparedness Against (Re-) Emerging Viral Diseases. *Adv Mater, 33,* 2005927.
216. Palika, A., Armanious, A., Rahimi, A., Medaglia, C., Gasbarri, M., Handschin, S., Rossi, A., Pohl, M. O., Busnadiego, I., Gübeli, C., Anjanappa, R. B., Bolisetty, S., Peydayesh, M., Stertz, S., Hale, B. G., Tapparel, C., Stellacci, F., & Mezzenga, R. (2021). Anjanappa RB. An antiviral trap made of protein nanofibrils and iron oxyhydroxide nanoparticles. *Nat Nanotechnol, 16,* 918–925.
217. Imani, S. M., Ladouceur, L., Marshall, T., Maclachlan, R., Soleymani, L., & Didar, T. F. (2020). Antimicrobial nanomaterials and coatings: Current mechanisms and future perspectives to control the spread of viruses including SARS-CoV-2. *ACS Nano, 14,* 12341–12369.
218. Pyrć, K., Milewska, A., Duran, E. B., Botwina, P., Dabrowska, A., Jedrysik, M., Benedyk, M., Lopes, R., Arenas-Pinto, A., Badr, M., Mellor, R., Kalber, T. L., Fernandez-Reyes, D., Schätzlein, A. G., & Uchegbu, I. F. (2021). Mellor R. SARS-CoV-2 inhibition using a mucoadhesive, amphiphilic chitosan that may serve as an anti-viral nasal spray. *Sci Rep, 11,* 20012.
219. Soria-Martinez, L., Bauer, S., Giesler, M., Schelhaas, S., Materlik, J., Janus, K., Pierzyna, P., Becker, M., Snyder, N. L., Hartmann, L., & Schelhaas, M. (2020). Prophylactic antiviral activity of sulfated glycomimetic oligomers and polymers. *J Am Chem Soc, 142,* 5252–5265.
220. Lauster, D., Klenk, S., Ludwig, K., Nojoumi, S., Behren, S., Adam, L., Stadtmüller, M., Saenger, S., Zimmler, S., Hönzke, K., & Yao L. (2020). Phage capsid nanoparticles with defined ligand arrangement block influenza virus entry. *Nat Nanotechnol, 15,* 373–379.
221. Gaharwar, A. K., Singh, I., & Khademhosseini, A. (2020). *Nat Rev Mater, 5,* 686–705.

222. Spicer, C. D. (2020). Hydrogel scaffolds for tissue engineering: the importance of polymer choice. *Polym Chem, 11,* 184–219.
223. Xie, J., Shen, K., Zheng, H., Yao, Y., Chen, Y., & Gao, C. (2021). Grafting of CAG peptides and (polyethylene glycol) on unsaturated polyurethane films to promote selective adhesion and migration of urethral epithelial cells. *J Mater Chem B, 9,* 6201–6211.
224. Zhang, H., Zheng, X., Ahmed, W., Yao, Y., Bai, J., Chen, Y., & Gao, C. (2018). Design and applications of cell-selective surfaces and interfaces. *Biomacromolecules, 19,* 1746–1763.
225. Abaricia, J. O., Farzad, N., Heath, T. J., Simmons, J., Morandini, L., & Olivares-Navarrete, R. (2021). Olivares-Navarrete R. Control of innate immune response by biomaterial surface topography, energy, and stiffness. *Acta Biomater, 133,* 58–73.
226. Xie, J., Yao, Y., Wang, S., Fan, L., Ding, J., Gao, Y., Li, S., Shen, L., Zhu, Y., & Gao, C. (2021). Co-delivery of azithromycin and ibuprofen by ROS-responsive polymer nanoparticles synergistically attenuates the acute lung injury. *Adv Healthcare Mater, 11,* 2101855.
227. Jin, H., Zhao, Z., Lan, Q., Zhou, H., Mai, Z., Wang, Y., Ding, X., Zhang, W., Pi, J., Evans, C. E., & Liu, X. (2021). Nasal delivery of hesperidin/chitosan nanoparticles suppresses cytokine storm syndrome in a mouse model of acute lung injury. *Front Pharmacol, 11.*
228. Salehi, S., Czugala, M., Stafiej, P., Fathi, M., Bahners, T., Gutmann, J. S., Singer, B. B., & Fuchsluger, T. A. (2017). Poly (glycerol sebacate)-poly (ε-caprolactone) blend nanofibrous scaffold as intrinsic bio-and immunocompatible system for corneal repair. *Acta Biomater, 50,* 370–380.
229. Thierry, B., Kujawa, P., Tkaczyk, C., Winnik, F. M., Bilodeau, L., & Tabrizian, M. (2005). Delivery platform for hydrophobic drugs: prodrug approach combined with self-assembled multilayers. *J Am Chem Soc, 127,* 1626–1627.
230. Han, C., Luo, X., Zou, D., Li, J., Zhang, K., Yang, P., & Huang, N. (2019). Nature-inspired extracellular matrix coating produced by micro-patterned smooth muscle and endothelial cells endows cardiovascular materials with better biocompatibility. *Biomater Sci, 7,* 2686–2701.
231. Wang, J., Chen, X. C., Xue, Y. F., Hu, M., Wang, Y. B., Ren, K. F., & Ji, J. (2019). Thermo-triggered ultrafast self-healing of microporous coating for on-demand encapsulation of biomacromolecules. *Biomaterials, 192,* 15–25.
232. Hong, Y., Zhou, F., Hua, Y., Zhang, X., Ni, C., Pan, D., Zhang, Y., Jiang, D., Yang, L., Lin, Q., Zou, Y., Yu, D., Arnot, D. E., Zou, X., & Zhu, L. (2019). A strongly adhesive hemostatic hydrogel for the repair of arterial and heart bleeds. *Nat Commun, 10,* 2060.
233. Yuk, H., Varela, C. E., Nabzdyk, C. S., Mao, X., Padera, R. F., Roche, E. T., & Zhao, X. (2019). Dry double-sided tape for adhesion of wet tissues and devices. *Nature, 575,* 169–174.
234. Kim, D. W., Baik, S., Min, H., Chun, S., Lee, H. J., Kim, K. H., Lee, J. Y., & Pang, C. (2019). Highly permeable skin patch with conductive hierarchical architectures inspired by amphibians and octopi for omnidirectionally enhanced wet adhesion. *Adv Funct Mater, 29,* 1807614.
235. Zhang, C., Wu, W., Li, R., Qiu, W., Zhuang, Z., Cheng, S., & Zhang, X. (2018). Peptide-based multifunctional nanomaterials for tumor imaging and therapy. *Adv Funct Mater, 28,* 1804492.

236. Ciobanasu, C. (2021). Peptides-based therapy and diagnosis. Strategies for non-invasive therapies in cancer. *J Drug Target, 29,* 1063–1079.
237. Matson, J. B., Newcomb, C. J., Bitton, R., & Stupp, S. I. (2012). Nanostructure-templated control of drug release from peptide amphiphile nanofiber gels. *Soft Matter, 8,* 3586–3595.
238. Hu, J., Zhang, M., Zheng, D., Hong, S., Feng, J., & Zhang, X. Z. (2018). miRNA delivery by nanosystems: state of the art and perspectives. Pharmaceutics. *Biomacromolecules, 19,* 2043–2058.
239. Hu, Q., Sun, W., Qian, C., Wang, C., Bomba, H. N., & Gu, Z. (2015). Anticancer platelet-mimicking nanovehicles. Advanced Materials (Deerfield Beach, Fla.). *Adv Mater, 27,* 7043–7050.
240. Deng, G., Sun, Z., Li, S., Peng, X., Li, W., Zhou, L., Ma, Y., Gong, P., & Cai, L. (2018). Cell-membrane immunotherapy based on natural killer cell membrane coated nanoparticles for the effective inhibition of primary and abscopal tumor growth. *ACS Nano, 12,* 12096–12108.
241. Liu, Z., Guo, K., Zhao, N., & Xu, F. J. (2019). Polysaccharides-based nanohybrids: Promising candidates for biomedical materials. *Sci China Mater, 62,* 1831–1836.
242. Zhang, X., Wei, W., & Tan, R. (2015). Symbionts, a promising source of bioactive natural products. *Sci China Chem, 58,* 1097–1109.
243. Mitrousis, N., Fokina, A., & Shoichet, M. S. (2018). . Biomaterials for cell transplantation. *Nat Rev Mater, 3*(11), 441–456.

CHAPTER 9

Plastic Degradation by Using Microbes: A Sustainable Approach to Tackle Environmental Threats

K. M. ANIL KUMAR,[1] E. SUMITHA,[2] M. B. SHIVASWAMY,[3] H. S. NAGENDRA PRASAD,[3] R. KAVYA,[3] and B. S. MADHUKAR[3]

[1]*Department of Environmental Science, JSS Academy of Higher Education and Research, Mysore, Karnataka, India*

[2]*Department of Microbiology, JSS Academy of Higher Education and Research, Mysore, Karnataka, India*

[3]*Department of Chemistry, Sri Jayachamarajendra College of Engineering, JSS Science and Technology University, Mysore, Karnataka, India*

ABSTRACT

Synthetic plastics are essential in our current lifestyle and impact the environment and human health. Plastics are high molecular weight organic materials. Existing lines of evidence suggest that petroleum-derived polymers are extremely resistant to natural biodegradation pathways. Some microorganisms possess the unique property to degrade plastics, which have been isolated and characterized extensively. Factors that account for the degradation of the polymer include molecular weight, the complexity of the chemical structure, and the variations in crystalline nature. The present study was carried out to degrade the polymers into their monomers. Characteristics including the nature of the polymer and environmental conditions were found to hamper the rate of degradation. Various approaches exist to monitor the rate of biological degradation, with the need of the hour being to generate biobased plastic material that

Sustainability in Polymer Technology and Plastics Engineering: Concepts, Strategies, and Opportunities. Tamara Tatrishvili, Neha Kanwar Rawat, Swati Gokul Talele, and A. K. Haghi (Eds.)

can degrade efficiently. Studies have also shown that these recycled products can generate petrochemical products, which will in turn meet the demand for organic fuels. In the present study, we attempted to degrade the plastics using fungal strains (*Aspergillus Niger, Rhizopus* sp., *Penicillium* sp., *Trichoderma* sp., and *Fusarium* sp.), and bacterial strains (*Bacillus cereus, Klebsiella* sp., *Salmonella typhi, Staphylococcus aureus*, and *Escherichia coli*). The degradation efficiency of these microorganisms on these polymers was determined by the parameters (1) Mean weight loss and (2) Chemical and structural modification of LDPE sheets. Further, SEM and FTIR characterization studies were performed to confirm the same. The outcome of the work could gain insights into understanding the genes responsible for degrading the plastic, which in turn can be accelerated by the recombinant DNA technology approach.

9.1 INTRODUCTION

High molecular weight organic polymers are derived from various hydrocarbons and petroleum derivatives. These polymers are referred to as plastics [1]. The term "plastic" is derived from the Greek word "Plastikos" which means "that which can be molded into various shapes." Plastics are polymers that begin to move when heated and may thus be cast into molds. Plastic is made up of the elements chlorine, oxygen, hydrogen, carbon, silicon, and nitrogen. Polyethylene accounts for 64% of all plastic and has the general formula C_nH2_n [2]. Synthetic plastics manufacture is one of the world's fastest-expanding industries. Because of their distinct qualities, plastics outperform other materials. Plastics have both advantages and downsides. Plastics are sturdy, long-lasting, and lightweight. On the other hand, they are damaging to the natural environment, being resistant to deterioration and causing pollution. Plastics represent a major hazard to our world by collecting in vast amounts [3, 4]. Plastics are used for packaging and a variety of other applications such as agricultural film production, diaper packaging, and fishing nets. Plastics play an essential role in every area of the global economy. Plastics are in great demand in rapidly increasing industries such as agriculture, architecture and construction, health, and consumer products, and no one can function without them. Plastics, the backbone of many businesses, are utilized in the production of a variety of products used in our everyday lives, such as defense materials, sanitary wares, tiles, plastic bottles, fake leather, and several other home items. Plastics are also utilized in food, pharmaceutical, detergent, and cosmetic packaging [5]. Microplastics (MPs) are plastic particles with a diameter of 5 mm that are produced during plastic breakdown and may have ecotoxicological consequences. Fibrous MPs may be breathed, stay in the lungs, and, when combined with other pollutants such as dyes and plasticizers, can cause health consequences such as carcinogenicity and mutagenicity [6–8]. Histopathology study revealed deformation of epithelial cells in the midgut area following chronic and acute exposures to polystyrene microplastics at 1 and 100 mg/L, respectively [9]. Plastics can decay in the environment through four different mechanisms: hydrolytic degradation, photodegradation, thermo-oxidative degradation, and biodegradation [10].

Microorganisms have the capacity to adapt to practically any environment and have the ability to destroy a variety of chemicals, including microplastics [11]. The employment of microorganisms to destroy microplastics would improve biodegradation while causing no damage to the environment [12, 13]. As a result, it is a viable and ecologically safe technique for facilitating natural bioremediation and influencing the cleansing of natural ecosystems without having negative consequences. However, few functional microorganisms have been identified so far, the interactions between microbes and microplastics have yet to be explored, and there is still a dearth of understanding regarding microplastic elimination. Therefore, it is critical to summarize and analyze existing research in order to establish how to get more functional microorganisms and increase their microplastic-degrading activity, as well as to encourage a better understanding of how bacteria metabolize and utilize microplastics [14, 15]. The current study summarizes recent breakthroughs in the microbial breakdown of synthetic plastics such as polyethylene (PE), polystyrene, polypropylene, polyvinyl chloride, polyethylene terephthalate, and polyurethane, as well as an overview of the enzymes involved in biodegradation. Therefore, the study adds to current information in the field of microbial and/or enzymatic degradation of synthetic polymers. Microbial and enzymatic polymer degradation is a potential technique for converting plastic waste into carbon dioxide, polymer monomers, and maybe value-added chemicals.

9.2 CLASSIFICATION OF PLASTICS BASED ON THERMAL PROPERTIES

9.2.1 THERMO-PLASTICS

Thermoplastics, when simply put, are resins with the ability to transform into liquids upon heating and into rigid materials upon cooling. These plastics, typically referred to as common plastics, have molecular weights ranging from 20,000 to 5,00,000 amu (atomic mass units). Atoms and molecules are joined end to end in carbon chains to form macromolecules arranged linearly in the form of chains. Opening of the double bond, which is necessary to create linear macromolecules, and the reaction is carried out by the use of free radicals. Polypropylene and polyethylene are examples of addition polymerization, a form of polymerization. Some examples are Polyvinyl chloride (PVC), polytetrafluoroethylene (PTFE), and polystyrene (PS) [16].

9.2.2 THERMOSETTING PLASTICS

The chemical interaction of two or more components under ambient conditions, or when caused by radiation or heat, results in the formation of thermosetting polymers, which are networks with a high degree of cross-linking. The process of thermoset production is irreversible. Typically, thermosets are rigid, hard, and have high strength and stiffness, hence used in structural applications. When heated, they have a tendency to be more temperature resistant and, unlike thermoplastics, do not creep or deform at higher degrees [17].

9.3 BIODEGRADATION

When microorganisms utilize an organic material as a source of carbon and energy, it results in biodegradation, which is the decay or disintegration of the material. The recalcitrant compounds that are difficult to degrade could become pollutants in the environment. This encourages the process of biodegradation. Biodegradation occurs under aerobic conditions as well as anaerobic conditions. Oxygen acts as an electron acceptor in aerobic conditions, and nitrate, sulfate, or any other compound acts as an electron acceptor under anaerobic conditions. Water and soil environments can be treated using biodegradation [18].

9.3.1 AEROBIC RESPIRATION

As mentioned earlier, it is the process where huge carbon compounds are broken down into smaller compounds by microorganisms using oxygen as an electron acceptor [19] (Figure 9.1).

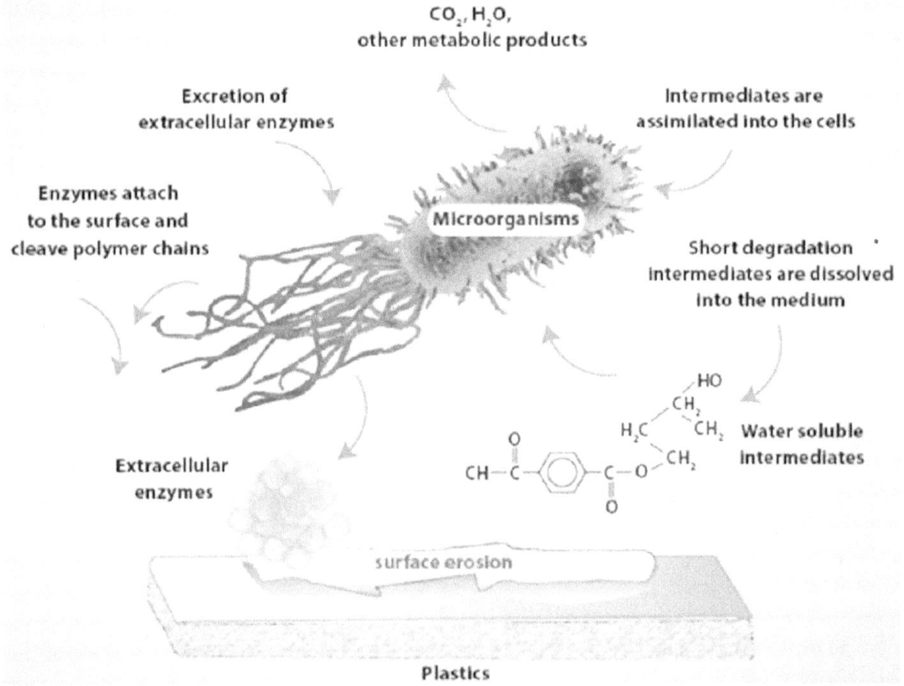

FIGURE 9.1 Organic compound + oxygen → water + carbon dioxide + Carbon residual.

9.3.2 *ANAEROBIC RESPIRATION*

It is a process where biodegradation occurs without the help of oxygen. However, the natural decrease of pollutants at hazardous waste sites depends heavily on oxygen. In place of oxygen, anaerobic bacteria employ electron acceptors such as nitrate, iron, sulfate, manganese, and CO_2 to break down big organic molecules into smaller compounds [20].

Organic compound → methane + carbon dioxide + water + Carbon residual

Many polymers are large in size and water-insoluble, hence preventing them from entering the cells of microorganisms through their cell walls. By secreting extracellular enzymes, microorganisms are able to utilize these polymers as a source of energy. These enzymes function outside of the bacterial cells to depolymerize the polymers. Enzymes play a part in the intracellular and extracellular biodegradation of polymers. The processes involved in the biological deterioration of polymers are depolymerization and mineralization. Exoenzymes, which are enzymes secreted outside of the cell, disassemble big polymers to create small, water-soluble molecules. These compounds are used as an energy source because they can cross semi-permeable bacterial membranes. Depolymerization, the process by which big polymers are broken down, and mineralization, the process by which end products such as water, carbon dioxide, and other inorganic species are produced, are two different processes [21] (Figure 9.2).

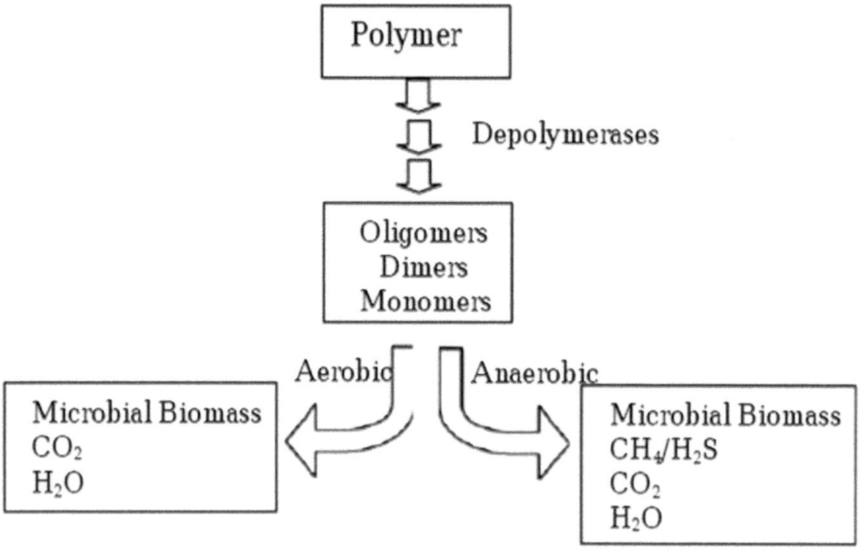

FIGURE 9.2 Schematic diagram of polymer degradation under aerobic and anaerobic conditions.

Source: Reprinted with permission from Ref.[33]. Copyright © 2003 Elsevier Ltd.

9.4 DEGRADATION OF POLYMERS

The structural characteristics of polymers, the presence of a degradative microbial population, and the environmental factors that favor microbial growth all affect how easily they can be broken down. Bacteria, fungi, and other heterotrophic microbes may be able to feed on polymers. Molecular weight, crystallinity, and physical shapes all affect the biodegradability of a polymer. In general, a polymer's capacity to be broken down by micro-organisms decreases as its molecular weight rises. The repeating units of a polymer, however, can be broken down and mineralized considerably more quickly as monomers, dimers, and oligomers. Because bacteria need the substrate to be taken up through the cellular membrane and then further degraded by cellular enzymes. High molecular weights lead to a decrease in solubility, making them less ideal for microbial attack. It should be noted that the degradation of polymers may be aided by simultaneous abiological and biological processes. Biodeterioration, bio-fragmentation, mineralization, and assimilation are various forms of microbial polymer degradation [22].

9.4.1 BIODETERIORATION

Polymers, in addition to inorganic and organic substances, undergo a variety of modification processes as a result of microbial growth and metabolism. Biodeterioration, according to [22], is "any undesirable change in a material's properties caused by the vital activities of organisms." It is noticeable in artwork made of paper, wood, stone, vegetal/animal fibers, and parchment as well as in monuments and wall paintings. To determine the type of interaction that is taking place inside the substance and to evaluate the cause-and-effect of the biodeterioration action of a specific identified biological agent, it is necessary to comprehend the physiological and morphological characteristics of the biodeteriogens. During this process, the polymer is modified in such a way as to form a colony and spread as a biofilm [23] (Figure 9.3).

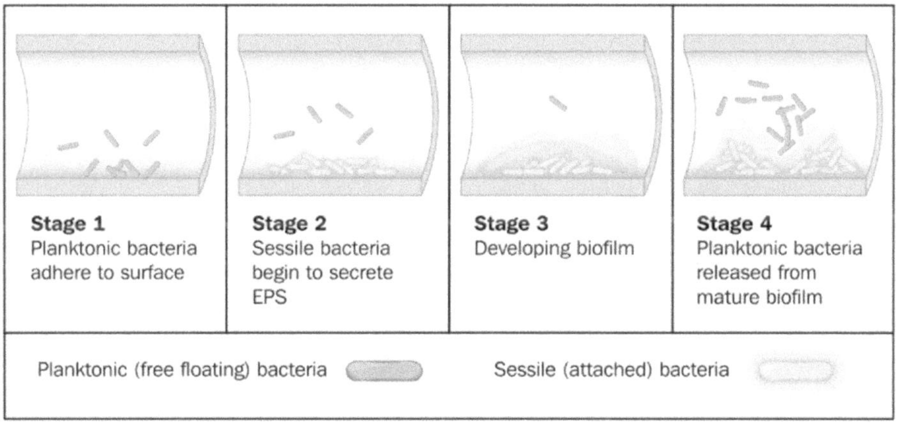

FIGURE 9.3 Biofilm formation by bacteria in the lumen of a tube.

Image Credit: Blamb/ Shutterstock. https://www.news-medical.net/life-sciences/Biofilm-Analysis-Tools-and-Techniques.aspx

Plastisphere is a thin biofilm layer formed by microbes that develop on plastic. *Cyanobacteria*, bacteria, fungi, and protozoa make up the plastisphere. The extracellular enzymes released by the bacteria in the plastisphere accelerate the physical and chemical degradation of plastics. Plastics degrade physically and chemically, becoming benzyl alcohol, carboxylic acid, benzoic acid, benzaldehyde, and other compounds that serve as a source of energy for bacteria. Microbes release a range of secondary metabolites into the ecosystem [23].

9.4.2 BIO-FRAGMENTATION

The continuation of the biodeterioration step is biofragmentation. Polymers are converted into oligomers and monomers during biofragmentation. Adsorption, desorption, and degradation of microplastic are caused by bacterial biofilm development on the plastic surface. In marine ecology, microplastics are viewed as an ecological niche for microorganisms. Exopolysaccharides (EPS) produced by bacteria cause biofilm to adhere to a plastic surface. The polymer is degraded by extracellular enzymes produced by bacteria in the biofilm. The ability of oxygenases, which are primarily found in bacteria, to break down oxygen molecules when they are introduced to carbon chains results in the formation of less damaging alcohol and peroxyl compounds. Additionally, lipases, esterases, or endopeptidases for amide groups catalyze the transition of carboxylic groups [25].

9.4.3 MINERALIZATION

Cell membranes allow plastic polymers that are created during the bio-fragmentation process to enter microbial cells. Large monomers are unable to enter cells and must remain outside. Small monomers that are transported inside the cells are oxidized and utilized to produce energy. In the end, this energy is used to produce biomass [26].

9.4.4 ASSIMILATION

Atoms are incorporated into the microbial cells during the assimilation step for complete breakdown. Secondary metabolites are transferred to other bacteria or carried outside of the cells where they are further degraded and used. During the breakdown of metabolites, the oxidized products, including CO_2, N_2, H_2O, and CH_4, are released [27] (Figure 9.4).

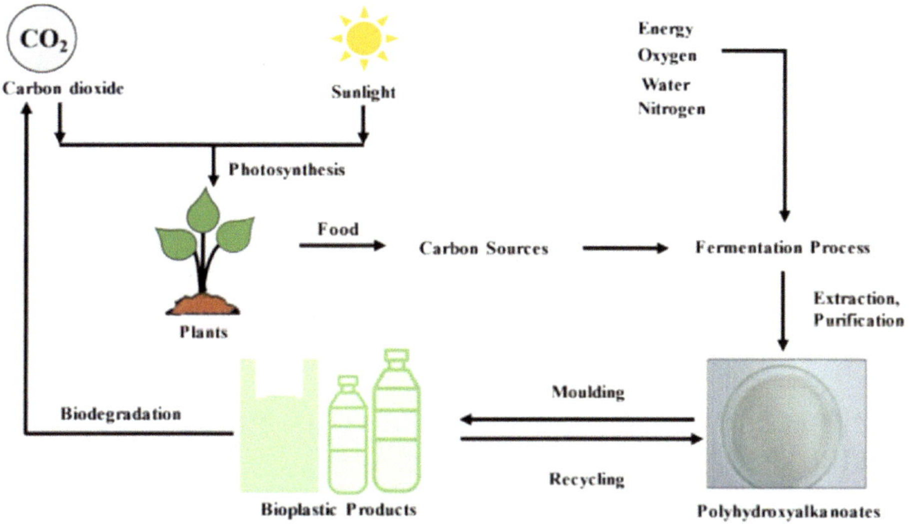

FIGURE 9.4 Schematic illustration of plastic biodegradation as a complex, step-wise pathway. First, during biodeterioration (1), abiotic and biotic factors, like light and secreted hydrophobins, modify the polymer surface and increase its surface area. Next, during bio fragmentation; (2), organisms secrete enzymes that break the polymer down into its monomer components. Finally, during assimilation; (3), the organism metabolizes the monomers into biomass, carbon dioxide, and water.

9.5 HOW ARE THE DEGRADABLE PLASTICS CLASSIFIED BASED ON PATHWAY OF DEGRADATION?

The plastics that are degradable are again classified into two, based on their mode of degradation pathway: Oxo-biodegradable plastics and hydro-biodegradable plastics.

9.5.1 OXO BIODEGRADABLE PLASTICS

Oxo-degradable is a separate category. It is a standard plastic that has been combined with an additive to mimic biodegradation rather than bioplastic or biodegradable plastic. As opposed to biodegradable and compostable plastics, which degrade at the molecular or polymer level, oxo-degradable polymers swiftly fragment into smaller and smaller fragments, known as microplastics [28].

9.5.2 HYDRO BIODEGRADABLE PLASTICS

They are plastics whose biodegradation is sparked by a chemical event called hydrolysis, in which water combines with one substance to create another. These plastics are produced from renewable resources. They have oil and starch in them.

In some cases, hydro-biodegradable plastics have been produced using genetically engineered crops. These polymers are produced using materials including sugar cane, corn, wheat, and petroleum. Compared to oxo-biodegradable polymers, they disintegrate and biodegrade more quickly [29].

9.6 MECHANISMS OF PLASTIC DEGRADATION IN THE ENVIRONMENT

This degrading behavior has been noted for polymeric systems containing polylactides, whose processes of degradation are well documented in the literature. Surface erosion or bulk degradation mechanisms can both degrade polymers. Hydrolysis of chemical bonds in the polymer chain at the material's core [30]. The microbial enzymes break down the polymer chains into water-soluble monomers and oligomers, which are then readily absorbed and metabolized by microbial cells, causing the biodegradation of plastics. The employment of microorganisms and enzymes in the biodegradation of polymers is categorized [31] polymers as the main source of the organic chemical. The *Aspergillus* and *Pseudomonas* species identified from the plastic waste site had minimal degradation potential. Beginning with microbial adhesion to its surface, polythene degrades. It has been discovered that a number of actinobacteria and fungi that degrade wood, including *Streptomyces badius*, *Streptomyces setonii*, and *Streptomyces viridosporus*, create extracellular enzymes (peroxidases, lactases, and oxidases) that cause polythene to break down [32].

9.6.1 HDPE AND LDPE MYCO DEGRADATION ANALYSIS

Each of the 500-ml autoclaved flasks in the five sets received 100 ml of Czapekdox broth supplemented with sterilized HDPE, LDPE, and sugar before the possible pure fungal isolates were added. The five inoculation sets were incubated for a total of 0, 15, 30, 45, 60, and 90 days in a static environment at room temperature of 28°C [24].

9.7 CONCLUSION

It is evident that we cannot meet our daily needs without plastic, but given its negative effects, it is necessary to establish effective procedures for its safe disposal and investigate alternatives such as starch-based and mixed plastics. Despite numerous papers highlighting the promise of plastic-degrading microorganisms, none of them have been discovered to have any real-world applications. Hence, there is a critical need to screen effective species and create technology that can effectively break down plastic without harming the environment. Due to their prospective uses in the sectors of environmental preservation and the maintenance of physical health, biodegradable polymers have attracted significantly greater interest in recent years. There are several potential bacteria in our environment that can break down polyester-type biodegradable polymers.

KEYWORDS

- **bacterial strains**
- **biobased plastic**
- **biological degradation**
- **exopolysaccharides**
- **fungal strains**
- **microplastics**
- **polytetrafluoroethylene**
- **polyvinyl chloride**
- **synthetic plastics**

REFERENCES

1. Elahi, A., Qamar, S. A., & Fouda, A. (2021). Plastics degradation by microbes: A sustainable approach. *Journal of King Saud University-Science, 33*(6), 101538.
2. Kale, S. K., Deshmukh, A. G., Dudhare, M. S., & Patil, V. B. (2015). Microbial degradation of plastic: A review. *Journal of Biochemical Technology, 6*(2), 952–961.
3. Yang, Z., Lü, F., Zhang, H., Wang, W., Shao, L., Ye, J., & He, P. (2021). Is incineration the terminator of plastics and microplastics? *Journal of Hazardous Materials, 401,* 123429.

4. Al-Thawadi, S. (2020). Microplastics and nano plastics in aquatic environments: Challenges and threats to aquatic organisms. *Arabian Journal for Science and Engineering, 45*(6), 4419–4440.
5. Piergiovanni, L., & Limbo, S. (2016). *Food Packaging Materials* (pp. 33–49). Basel, Switzerland: Springer.
6. Zhang, C., Chen, X., Wang, J., & Tan, L. (2017). Toxic effects of microplastic on marine microalgae *Skeletonema costatum*: Interactions between microplastic and algae. *Environmental Pollution, 220,* 1282–1288.
7. Gasperi, J., Wright, S. L., Dris, R., Collard, F., Mandin, C., Guerrouache, M., & Tassin, B. (2018). Microplastics in air: Are we breathing it in? *Current Opinion in Environmental Science & Health, 1,* 1–5.
8. Wang, X., Bolan, N., Tsang, D. C., Sarkar, B., Bradney, L., & Li, Y. (2021). A review of microplastics aggregation in aquatic environment: Influence factors, analytical methods, and environmental implications. *Journal of Hazardous Materials, 402,* 123496.
9. Suman, T. H., Jia, P. -P., Li, W. -G., Junaid, M., Xin, G. -Y., Wang, Y., & Pei, D. -S. (2020). Acute and chronic effects of polystyrene microplastics on brine shrimp: First evidence highlighting the molecular mechanism through transcriptome analysis. *Journal of Hazardous Materials, 400,* 123220.
10. Nuzzi, L. C., Cerrato, F. E., Erikson, C. R., Webb, M. L., Rosen, H., Walsh, E. M., & Labow, B. I. (2013). Psychosocial impact of adolescent gynecomastia: A prospective case-control study. *Plastic and Reconstructive Surgery, 131*(4), 890–896.
11. Brooks, A. N., Turkarslan, S., Beer, K. D., Lo, F. Y., & Baliga, N. S. (2011). Adaptation of cells to new environments. *Wiley Interdisciplinary Reviews: Systems Biology and Medicine, 3,* 544–561.
12. Sen, S. K., & Raut, S. (2015). Microbial degradation of low-density polyethylene (LDPE): A review. *Journal of Environmental Chemical Engineering, 3*(1), 462–473.
13. Qi, X., Ren, Y., & Wang, X. (2017). New advances in the biodegradation of Poly (lactic) acid. *International Biodeterioration & Biodegradation, 117,* 215–223.
14. Shah, Z., Krumholz, L., Aktas, D. F., Hasan, F., Khattak, M., & Shah, A. A. (2013). Degradation of polyester polyurethane by a newly isolated soil bacterium, *Bacillus subtilis* strain MZA-75. *Biodegradation, 24,* 865–877.
15. Yuan, J., Ma, J., Sun, Y., Zhou, T., Zhao, Y., & Yu, F. (2020). Microbial degradation and other environmental aspects of microplastics/plastics. *Science of the Total Environment, 715,* 136968.
16. Mirji, R., & Lobo, B. (2017). Computation of the mass attenuation coefficient of polymeric materials at specific gamma photon energies. *Radiation Physics and Chemistry, 135,* 32–44.
17. Chattopadhyay, I. (2022). Role of microbiome and biofilm in environmental plastic degradation. *Biocatalysis and Agricultural Biotechnology, 39,* 102263.
18. Poznyak, T., Oria, J. I. C., & Poznyak, A. (2018). Ozonation and biodegradation in environmental engineering: Dynamic neural network approach. *Elsevier.* 1st Edition – November 7, 2018; p. 18.
19. Bouwer, E. J., & Zehnder, A. J. (1993). Bioremediation of organic compounds—putting microbial metabolism to work. *Trends in Biotechnology, 11*(8), 360–367.
20. Eskander, S., & Saleh, H. E. D. (2017). Biodegradation: Process mechanism. *Environmental Science & Engineering, 8*(8), 1–31.

21. Gautam, R., Bassi, A. S., & Yanful, E. K. (2007). A review of biodegradation of synthetic plastic and foams. *Applied Biochemistry and Biotechnology, 141,* 85–108.
22. Banerjee, A., Chatterjee, K., & Madras, G. (2014). Enzymatic degradation of polymers: A brief review. *Materials Science and Technology, 30*(5), 567–573.
23. Allsopp, D., Seal, K. J., & Gaylarde, C. C. (2004). Introduction to biodeterioration. Cambridge University Press. https://catdir.loc.gov/catdir/samples/cam041/2003060350.pdf.
24. Hadar, Y., & Sivan, A. (2004). Colonization, biofilm formation and biodegradation of polyethylene by a strain of *Rhodococcus ruber*. *Applied Microbiology and Biotechnology, 65,* 97–104.
25. Davis, E. K., & Raja, S. (2020). Sources and impact of microplastic pollution in Indian aquatic ecosystem: A review. *Sustainable Mining,* ISSN: 0973–4929, *15*(Special Issue 1), 1–9.
26. Dussud, C., & Ghiglione, J. F. (2014). Bacterial degradation of synthetic plastics. In *CIESM workshop monogr* (Vol. 46, pp. 49–54).
27. Tokiwa, Y., Calabia, B. P., Ugwu, C. U., & Aiba, S. (2009). Biodegradability of plastics. *International Journal of Molecular Sciences, 10*(9), 3722–3742.
28. Tsuji, H., & Miyauchi, S. (2001). Enzymatic hydrolysis of poly (lactide)s: Effects of molecular weight, L-lactide content, and enantiomeric and diastereoisomeric polymer blending. *Biomacromolecules, 2*(2), 597–604.
29. Iwata, T., & Doi, Y. (1998). Morphology and enzymatic degradation of poly (L-lactic acid) single crystals. *Macromolecules, 31*(8), 2461–2467.
30. Li, S. (1999). Hydrolytic degradation characteristics of aliphatic polyesters derived from lactic and glycolic acids. *Journal of Biomedical Materials Research: An Official Journal of The Society for Biomaterials, The Japanese Society for Biomaterials, and The Australian Society for Biomaterials, 48*(3), 342–353.
31. Premraj R, Dobley M. (2005). Biodegradation of polymers. *Indian Journal of Biotechnology, 4,* 186–193.
32. Thamizhmarai, T., & Kannahi, M. (2018). Biodegradation of low-density polyethylene by *Aspergillus* sp. and *Pseudomonas* sp. isolated from plastic dumped site — A SEM analysis. *International Journal of Pharmaceutical Sciences Review and Research, 50*(1), 182–187.
33. Gu, Ji-Dong & Ford, Timothy & B, Mitton & Mitchell, Ralph. (2000). Microbial degradation of polymeric materials.

CHAPTER 10

Polymeric Nanoparticles and Nanofibers for Drug Delivery Applications: Current Successes in Achieving the Sustainable Development Goals

ANIKET A. NERKAR,[1] SHWETA H. SHAHARE,[1] SWATI G. TALELE,[1] LAXMIKANT B. BORSE,[1] and HITESH V. SHAHARE[2]

[1]*Sandip Institute of Pharmaceutical Sciences, Nashik, Maharashtra, India*

[2]*SNJBs Shriman Sureshdada Jain College of Pharmacy, Chandwad, Nashik, Maharashtra, India*

ABSTRACT

The utilization of nanoparticles and nanofibers for local medication delivery is described in this review article for the first time in detail. Researchers have looked into using nanotechnology, particularly nanoparticles and nanofibers, as drug delivery systems. To be able to increase the active component's flow into and through the skin, this method concentrates the active ingredient in the carrier. In order to distribute hydrophobic and hydrophilic medications and release them in a regulated manner over an extended period of time, nanoparticles and nanofibers can both be used. The examples given clearly demonstrate the significant influence this field of study has had and will continue to have on both clinical results and the creation of new products. An accessible surface for medication delivery is human skin. The typical adult body type has skin covering roughly 2 m^2 of its surface and holds about one-third of the blood that circulates through the body. In recent decades, the development of controlled drug delivery has become very important in the pharmaceutical industry. It is known that the human skin surface

Sustainability in Polymer Technology and Plastics Engineering: Concepts, Strategies, and Opportunities. Tamara Tatrishvili, Neha Kanwar Rawat, Swati Gokul Talele, and A. K. Haghi (Eds.)

contains an average of 10–70 hair follicles and 200–250 sweat glands per square centimeter of the skin surface. It is one of the more accessible body parts in the human body. There has been great interest in the skin as a site of drug application for both local and systemic effects. However, the skin, especially the stratum corneum, provides a powerful barrier to drug penetration, limiting topical and transdermal bioavailability. Techniques to enhance skin penetration have been created to enhance bioavailability and increase the number of medications for which transdermal and topical administration are practical choices. In recent decades, the number of drugs prescribed in patches has barely increased, and the composition of patch systems has changed very little. Changes were primarily limited to improvements in the materials used.

10.1 INTRODUCTION

10.1.1 STRUCTURE OF SKIN

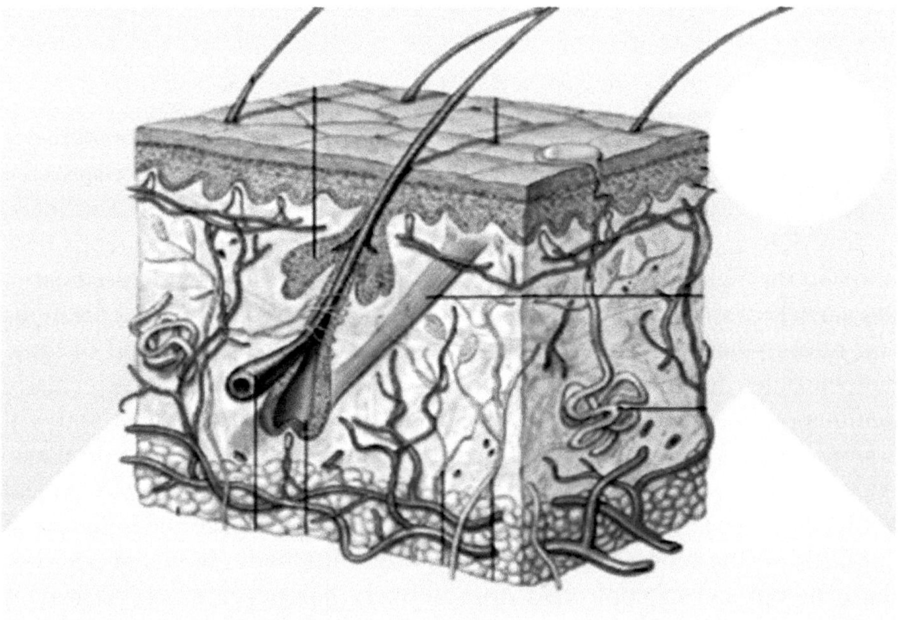

FIGURE 10.1 Structure of skin.

The large surface area of skin (approximately 20 square feet) makes it a potentially interesting route for drug delivery. Human skin equivalents with lipid and permeability profiles similar to human skin have been developed as substitutes for human cadavers and animal skin to characterize drug permeation *in vitro*. Depending on the drug's properties and method of administration, the drug remains localized or penetrates the dermis where, if hydrophilic, it rapidly enters the bloodstream using capillaries. The capillary bed extends into the upper layers of the dermis just below the dermo-epidermal junction. Local drug delivery is primarily aimed at a local effect and may (i) eliminate the need for drug therapy administered systemically and (ii) reach the target site (skin), minimizing the total dose required and reducing targeted side effects. Some examples of topical dermatological products include skin antifungals, sunscreens, keratolytics, local anesthetics, antiseptics, and anti-inflammatory agents for skin conditions like psoriasis. Transdermal drug delivery, on the other hand, is intended for systemic drug action, and the skin is only the gateway through which the drug enters the body. For example, one uses transdermal patches to deliver nicotine for smoking cessation, buprenorphine and fentanyl for chronic pain, and hyosine (also known as scopolamine) for motion sickness. Medication penetration via SCs in topical and transdermal drug delivery is governed by Fick's second law [1] (Figure 10.1).

Chemical enhancers like phospholipids, alkanes, terpenes, sulfoxides, amides, amines, pyrrolidones, polyalcohols, alcohols, esters, surfactants, and fatty acids [1, 2] are utilized to improve topical drug penetration. Unfortunately, chemical penetration enhancers tend to irritate the skin due to their destructive effects on the lipid structure of the skin, which is higher than the skin can tolerate [4]. For this reason, chemical penetration boosters haven't proven very useful.

Physical enhancers, such as skin hydration by occlusion, sonophoresis, laser and/or thermal ablation, electroporation, radiofrequency therapy, iontophoresis, magnetophoresis, high-pressure jets, and microneedle arrays, use physical methods to momentarily compromise the integrity of the skin's barrier function [4, 5]. Some of these techniques have translated into FDA-approved products for both transdermal and topical drug delivery, but physical enhancers are instructive and over time damage the skin's barrier properties, leading to tingling, and irritation, and also causing a burning sensation.

As an alternative, scientists have looked at the use of nanotechnology, in particular nanoparticles and nanofibers, as drug delivery methods for topically applied and transdermally applied drugs. This approach increases drug concentration (Cv) in the vehicle or formulation, resulting in an increase in drug flux. In order to distribute hydrophilic and hydrophobic medications and release them in a regulated manner over an extended period of time, nanoparticles and nanofibers can both be employed. An overview of nanoparticles and nanofibers created for topical medication delivery applications is given in this paper. Although nanoparticle and nanofiber drug delivery methods have been evaluated independently in the past, to the authors' knowledge [6, 7], this is the first in-depth examination of both types of systems. Collectively, these systems have a significant effect on the treatment of many common skin diseases and collectively generate significant commercial revenues [5]. The examples of topical or transdermal medication delivery systems using nanoparticles and nanofibers provided here show that this field of study has and will continue to have a substantial influence on the commercialization of these technologies as well as clinical outcomes.

10.2 NANOPARTICLES FOR TOPICAL THERAPY

Nanoparticles (sometimes called nanocarriers) have been examined for a variety of biomedical applications for over a decade. The most commonly used nanoparticles for transdermal and/or topical drug delivery are polymeric nanoparticles, nanoemulsions, lipid-based nanoparticles (liposomes and solid lipid nanoparticles), metallic nanoparticles, and dendrimers.

10.2.1 NATURAL POLYMER NANOPARTICLES

Although an extensive variety of natural polymers have been employed for nanoparticle fabrication, chitosan-based nanoparticles have been the most extensively investigated for localized drug delivery applications [8]. Additionally, several publications have reported the use of alginate-albumin nanoparticles for topical drug delivery. This is also briefly explained in this section. Chitosan is a linear polysaccharide derived from chitin discovered in the crustacean exoskeletons. The amino clusters of chitosan are readily protonated under acidic to neutral conditions, resulting in a net positive charge. This makes chitosan water-soluble and serves as a bioadhesive as positively charged chitosan can bind to negatively charged mucoproteins. This interaction helps increase the circulation time of chitosan-conjugated drugs, resulting in better drug bioavailability [9]. Chitosan nanoparticles are characterized by (i) ionic cross-linking; (ii) covalent cross-linking [10]; (iii) precipitation [11]; (iv) polymerization [12] or (v) self-assembly process [13], with sizes 20 to 800 nm depending on the method of preparation. Chitosan-drug complexes are commonly formed by electrostatic interactions between cationic chitosan and anionic drugs [14, 15]. For example, Chen et al. prepared chitosan nanoparticles conjugated with antisense oligonucleotides, generating particles with a size of 102.6 ± 12.0 nm, zeta potential 1.45 ± 1.75 mV, and encapsulation efficiency 87.6 ± 3.5%. Kim et al. encapsulated retinol in chitosan nanoparticles, increasing the solubility of retinol by more than 1,600-fold. The polymer degradation process can be used to modulate the drug distribution using chitosan nanoparticles. This may be adjusted by changing the deacetylation level and chitosan's molecular weight. As a result, it has been demonstrated that chitosan nanoparticles offer continuous release for days to months [16]. Polymer degradation enables control over the drug delivery process in chitosan nanoparticles. By adjusting the deacetylation level and chitosan's molecular weight, this may be tailored. Chitosan nanoparticles have therefore been demonstrated to offer persistent release for days to months [17].

Alginate-derived nanoparticles have not been thoroughly investigated as local medication delivery systems, although two reports have recently appeared in the literature. Ibrahim et al. made brimonidine-loaded alginate nanoparticles and demonstrated that they provided sustained release *in vitro* without initial burst and improved intraocular pressure-lowering effect *in vivo* compared to commercial brimonidine tartrate eye drops [18]. Friedman et al. investigated the use of nanoparticles synthesized from chitosan and alginate to transport antimicrobial agents against Propionibacterium acnes for the topical treatment of acne and compared it with benzoyl peroxide alone *in vitro* [19]. It has shown excellent antibacterial activity. Albumin is a non-toxic, biocompatible, and biodegradable carrier protein used for drug delivery. Alginate-derived nanoparticles as delivery systems for local drug delivery have not received much research. Nevertheless, we found only one report in the literature on the use of albumin nanoparticles for local drug delivery. Das et al. displayed that aspirin-loaded albumin nanoparticles produced sustained release for 72 hours *in vitro* and showed promise as a topical treatment for diabetic retinopathy [20].

10.2.2 SYNTHETIC POLYMER NANOPARTICLES

Natural polymers may not be consistent from batch to batch, which makes it challenging to produce particles with reproducible properties and drug release profiles [21]. Synthetic polymers, on the other hand, can be delivered with high purity and batch-to-batch reproducibility to produce nanoparticles with a more consistent drug release profile [21].

10.2.3 TYROSINE-DERIVED POLYMER NANOPARTICLES (TYROSPHERES)

Tyrosine-derived polymeric nanospheres (abbreviated as TyroSpheres) are fully degradable and composed of poly(ethylene glycol), oligomers of desaminotyrosyl tyrosine alkyl esters, and several different natural dicarboxylic acids such as succinic acid. They are synthesized from a family of ABA-type triblock copolymers, including adipic acid, suberic acid, or sebacic acid [22]. In aqueous conditions, these copolymers produced from tyrosine spontaneously self-assemble to create nanospheres with hydrodynamic diameters of around 70 nm [22]. These nanospheres bind firmly to hydrophobic compounds to form complexes. This is demonstrated by the model compound DAF and the clinically used antineoplastic drug paclitaxel, a potent chemotherapeutic agent [23]. In contrast, TyroSpheres do not bind strongly to hydrophilic compounds such as fluorescein [24, 25]. *In vitro* and *in vivo*, gel formulations of TyroSpheres loaded with Nile Red (a hydrophobic model drug) had significantly improved skin permeation compared to aqueous dispersions of TyroSpheres loaded with Nile Red (1.4 to 1.8 times) [26]. Since these TyroSpheres can only bind and distribute hydrophobic, water-insoluble medicines, such as curcumin (unpublished data), cyclosporin A [89], sildenafil [27], Rolipram [28], and paclitaxel [25, 29], they have been extensively researched for this purpose. These nanospheres are also only stable for around 6 months before degrading, most likely due to the hydrolysis of the ester bonds [23]. Drug characteristics, binding and loading effectiveness, and solution pH all affect how quickly drugs are released from TyroSpheres [30]. An example is cyclosporin release has been demonstrated to be inversely proportional to the amount of drug loaded into TyroSpheres [89].

An *in vitro* cytotoxicity assay in serum-free medium showed no significant reduction in cell metabolic activity at concentrations up to 2 mg/mL, indicating that empty TyroSpheres are non-cytotoxic [22]. Additional research has revealed that TyroSpheres can effectively transfer active substances to the skin without degrading the morphology of the skin [26]. His TyroSpheres loaded with paclitaxel have been demonstrated to efficiently deliver paclitaxel to tumor cells as determined by an *in vitro* cell killing assay [30]. Additionally, an *in vivo* study was conducted to compare the toxicity and antitumor efficacy of intravenously administered paclitaxel using TyroSpheres or commercially available Cremophor® EL in two mouse breast cancer xenograft models [25]. Animals treated with paclitaxel-loaded Cremophor® EL showed significant weight loss and did not recover after treatment, whereas animals treated with paclitaxel-loaded TyroSpheres attained normal weight. Furthermore, the anti-neoplastic effects of the two treatments were comparable. In addition, a greater volume of distribution and a longer half-life were measured in animals treated with his TyroSpheres loaded with paclitaxel. These data displayed that TyroSpheres can be employed as a more effective paclitaxel delivery vehicle with fewer side effects in the treatment of breast cancer. Furthermore, by modifying the triblock copolymer, ligands can be covalently attached to TyroSpheres, allowing drug-loaded TyroSpheres to be targeted to specific recipient cells, thereby targeting non-target cells. May reduce unwanted side effects [29]. TyroSpheres also provide adaptability in storage and handling properties, as they can be developed as dry formulations [27, 28], liquid formulations [24], or incorporated into gels to prevent sagging during topical application [26]. Nanospheres derived from tyrosine offer assurance to topical delivery of lipophilic drugs and personal care products to the skin – may be desirable for the treatment of dermatological conditions such as skin cancer, psoriasis, eczema, microbial infections, sunscreens, and cosmetics.

10.2.4 POLY (LACTIC-CO-GLYCOLIC ACID) NANOPARTICLES

Poly (lactic-co-glycolic acid) (PLGA) is biocompatible and biodegradable, widely used in FDA-approved medical devices and pharmaceutical formulations, and commercially available in a variety of molecular weights and copolymer ratios. As a result, it has undergone substantial research for applications involving topical medication administration. One can encapsulate an extensive range of drugs, from hydrophobic to lipophilic, and can be surface-modified for site-specific drug delivery. Additionally, PLGA can be developed as nanoparticles using several different methods: solvent displacement, solvent diffusion, emulsification evaporation, and phase inversion. Sizes range from 10 to 1,000 nm. PLGA nanoparticles are incorporated into gels using hydroxypropyl methylcellulose (HPMC) or Carbopol® to increase skin contact time [31] and reduce water loss by increasing skin further improving drug penetration into deeper layers of the skin [32].

10.3 LIPID-BASED NANOPARTICLES

10.3.1 SOLID LIPID NANOPARTICLES (SLN)

SLNs are made from solid lipids with an average diameter of 50–1000 nm [33] and are fabricated using three main components: lipids, surfactants, and water. The types of lipids used to generate SLNs are very broad and include triglycerides (such as tristearin), partial glycerides, fatty acids (such as stearic acid), steroids (such as cholesterol), and waxes (such as cetyl palmitate) [34]. An active agent is usually required to stabilize the lipid dispersion. SLNs were developed in the 1990s and are still viewed as promising drug delivery systems, especially for applications requiring sustained release. SLNs can be prepared without organic solvents. Furthermore, they are biodegradable, biocompatible, physicochemically stable, have low toxicity, and can be produced on a large industrial scale at a relatively low cost. Modified SLNs loaded with retinyl palmitate displayed that these negatively charged SLNs boost the skin distribution of retinyl palmitate by 4.8-fold compared to neutral SLNs [35].

10.3.2 LIPOSOME

Liposomes are lipid-based spherical vesicles, which are the main components of cell membranes. Liposomes have at least two lipid membranes and can be loaded with drugs. Liposomes have an average diameter of 50–300 nm [36]. To date, approximately 600 clinical studies have involved drug-delivery systems made of lipids [37]. Overall, lipids reduce interactivity between drugs and serum proteins, thereby increasing the circulation time of drugs in the body, preventing rapid clearance, and enabling drug delivery to target sites [37]. Liposomes are characterized by (i) thin film hydration [38]; (ii) reverse phase evaporation [39]; (iii) solvent injection [40]; (iv) detergent depletion [41]; (v) supercritical fluid [42]; (vi) size reduction sonication [43]; (vii) high-pressure homogenization [43]; and (viii) low-pressure extrusion [44]. Liposomes can provide delayed and triggered release of payloads [45].

10.4 METALLIC AND SILICA NANOPARTICLES

Metal nanoparticles in general have drawn a lot of interest due to their effective use. Examples include silver nanoparticles (AgNPs), which exhibit antiseptic and antibacterial activity [46]. It has been documented that SCs of undamaged mouse skin may disperse AuNPs with a diameter of less than 5 nm [47]. Another illustration is the use of metal nanoparticles in sunscreens and cosmetics to block UV rays, such as zinc oxide (ZnO) and titanium dioxide (TiO_2) nanoparticles [48]. Cerium oxide nanoparticles have gained interest recently for use in the treatment of wounds [49]. The size of metal nanoparticles can be optimized to be large enough to reflect and refract high-energy, very short wavelengths (~290–400 nm) of UV A and B, but too low wavelengths too small to reflect visible wavelengths (~400–700 nm). In this way, they remain invisible on the skin [50].

10.5 DENDRIMER

Dendrimers are a class of highly branched, monodisperse, synthetic macromolecules with well-defined compositions and structures. The most commonly cited dendrimers used in nanomedicine are polyamidoamine (PAMAM), poly(L-lysine) (PLL) backbone dendrimers, polyester (PGLSA-OH), polypropylimine (PPI), poly (2,2-bis (hydroxyl methyl) propionic acid) backbone dendrimers (bis-MPA), and aminobis(methylenephosphonic acid) backbone dendrimers. Dendrimer-based NPs have been comprehensively studied due to their scientific and technical interest in potential applications in a wide range of fields such as catalysis, optics, electronics, and biomedical applications [51]. Dendrimers' adjustable surface chemistry permits them to be assembled onto different nanoparticles [52]. For example, magnetic iron oxide nanoparticles have been stabilized with dendrimers [52]. Dendrimer nanoparticles are typically less than 5 nm in size and can be prepared by rapid reduction and nucleation chemistry. It is displayed that dendrimers can be utilized to enhance the skin permeability of hydrophilic drugs. Dendrimer molecular weight, size, surface charge, and hydrophobicity are important parameters that determine local drug delivery and skin penetration [53].

10.6 TOPICAL APPLICATION OF NANOFIBERS

10.6.1 ACNE

Acne is related to an increased rate of sebum secretion and manifests as mild to severe inflammatory lesions on body sites containing numerous sebaceous glands, such as the face, back, and chest [54]. Acne is classified into three types mild, moderate, or severe, based on the type and number of lesions. Treatment options vary according to the severity and stage of the illness. In general, topical treatments are preferred for mild and moderate acne, while systemic treatments are used in severe cases [55]. Antibiotics, retinoids (vitamin A), and combinations of both can be used as topical therapies. Alternative treatments (long-term use of antibiotics) can lead to bacterial resistance. Nanoparticles are being created to enable progressive and sustained administration of these medications to the skin to address these issues. Equal or increased efficacy with a simultaneous reduction in local side effects is an important criterion for evaluating the benefits of nanoparticulate formulations of acne drugs [56]. The medications used to treat acne when nanoparticle systems are applied topically have shown effectiveness, tolerability, compliance, and esthetic acceptance [57, 58].

10.6.2 INFECTION

The rate of emergence of infectious diseases and drug-resistant pathogens is a serious concern. Despite increasing knowledge of microbial etiology and a broad repertoire of antimicrobial therapies, morbidity and mortality from microbial infections remain high. New tactics and antimicrobial drugs are thus clinically necessary to treat infections.

Nanoparticle systems have been used as antibacterial and anti-infective agents, according to a recent study [59]. For example, AgNPs have been studied as antibacterial agents and it has been shown that AgNPs with a diameter of less than 20 nm can bind to sulfur-containing proteins in bacterial cell membranes. This increases the permeability of bacterial membranes, leading to bacterial death [60]. The overall surface area of nanoparticles is connected to their antimicrobial capabilities, with smaller particles having a larger surface area to volume ratio and greater antimicrobial activity [61]. Complexes of AgNPs and polymer gels have been demonstrated to have increased bactericidal activity compared to high concentrations of AgNPs alone [62]. AuNPs have been used to coat various surfaces such as wound dressings, implants, and glass surfaces to maintain an antimicrobial surface [63]. AuNPs have near-infrared light absorption and can destroy bacteria by photothermal heating [64]. AuNPs perforate the cell wall, causing leakage of cell contents and cell death. AuNPs may also bind to bacterial DNA and inhibit DNA unfolding and transcription [65]. Stable conjugates of AuNPs coated with antibiotics have shown bactericidal activity [66].

Other metal oxide nanoparticles have furthermore been looked into for antimicrobial activity. For example, halogen-treated magnesium oxide nanoparticles have displayed antibacterial activity against bacteria and spores [67]. Copper oxide (CuO) nanoparticles have high antibacterial activity against *Bacillus subtilis* [68]. This was attributed to the abundance of amine and carboxyl groups on the cell surface of *Bacillus subtilis*, which has a high affinity for Cu. Cu ions can also interact with DNA molecules, intercalate into nucleic acid strands, and interfere with biochemical processes [68, 69]. Titanium dioxide (TiO_2) nanoparticles are also being researched to fight infections. It has been suggested that UV-irradiated nanostructured TiO_2 surfaces can be used as effective disinfectants to remove pathogenic microorganisms on food contact surfaces [70]. The major drawback of using TiO_2 in this method is that UV light is required to activate the photocatalyst that initiates the killing of bacteria and viruses. Another potential mechanism is that the released Zn^{2+} ions can directly damage the cell membrane and interact with intracellular contents [71].

10.7 ELECTRO-SPUN NANOFIBERS FOR DRUG DELIVERY

10.7.1 INTRODUCTION TO ELECTROSPINNING

A typical setup [72] includes at least a syringe pump, polymer solutions with or without drugs, a high-voltage power supply, a source electrode, and a collector electrode. During the electrospinning process, by applying an electric field across the viscous polymer solution, charges are induced within the polymer and the needle. When the charge repulsion overcomes the surface tension, the tip of the needle (called the 'Taylor cone') jets, producing a very thin, elongated droplet with a very large surface area. Since the solvent in these droplets is rapidly evaporating, nano-sized fibers form and are attracted to an oppositely charged or grounded collector, depositing a solid mat of woven fabric of fibers. Collectors can be fixed or rotating and/or translating. In the electrospinning process, fiber formation depends on various solution and process parameters, as reported by Bhardwaj et al. [73]. These include polymer concentration, polymer molecular weight, solution viscosity, surface tension, polymer conductivity, solvent, electric field strength/voltage, feed flow rate, type of collector, and distance between the needle tip and collector.

10.7.2 ELECTRO-SPUN NANOFIBERS AS A TOPICAL DRUG DELIVERY SYSTEMS

As drug delivery devices, electrospun fiber mats provide a number of appealing characteristics: (i) using an electrospinning process to produce nano-sized particles from various solutions of both natural (chitosan, fibronectin, gelatin, collagen, silk, ethyl cellulose, etc.) and synthetic polymers (PLA, PGA, PLGA, tyrosine-derived, etc.) [73]; (ii) Fiber mats' high surface-to-volume ratio makes it possible to distribute both hydrophilic and hydrophobic medications effectively [74]; (iii) the drug release profile can be tailored for specific clinical applications by adjusting various parameters such as: for example, drug to polymer ratio, fiber diameter, morphology, and/or porosity [75]. Nanofiber functionalization (e.g., surface graft polymerization) is another approach to modulate drug release [7]; (iv) Electrospun fiber mats can offer sustained release, reducing the frequency of topical application and increasing patient compliance; (v) electrospun fiber mats have more porosity with interconnectivity [76, 77]; (vi) nanofiber meshes are soft and suitable for topical applications; (vii) The use of fiber mats in wound dressings as a component of drug-eluting wound therapy methods is possible. Although electrospinning has evolved into an industrial-scale process for manufacturing HEPA filters and insulating materials, the precise reproducibility required for manufacturing medical devices remains a challenge. Modifications like bubble electrospinning [78] and porous hollow tubes [79] have been investigated to scale up the electrospinning process.

10.8 POTENTIAL CLINICAL APPLICATION

10.8.1 INFECTED WOUNDS

Infected skin wounds take longer to heal, lengthen hospital stays, and can be fatal. Topical antibiotics are beneficial in treating infections. Various antibiotics have been impregnated into the electrospun fibers and shown to have bactericidal activity *in vitro*. The antibiotic cefazolin can be loaded onto PLGA nanofibers using a standard electrospinning setup [80]. Verreck et al. employed itraconazole, a hydrophobic antifungal medication, as a topical delivery method using non-biodegradable polyurethane-based nanofibers. This system provides ultra-fast release of payload (< 24 hours), Square root of time indicating diffuse emission [81]. Kim et al. show sustained release of cefoxitin sodium, a hydrophilic antibiotic, for up to 1 week from electrospun copolymers of PLGA and polyethylene glycol (PEG) in an *in vitro*

Staphylococcus aureus inhibition assay confirmed the biological activity of the drug [90]. Huang et al. used coaxial electrospinning, the antibacterial agent gentamicin sulfate was incorporated into polyurethane fibers and released following zero-order kinetics for up to 7 days *in vitro* [82]. Mupirocin is an antibiotic commonly used to treat skin infections, usually prescribed as a 2% cream or ointment used three times per day for up to 10 days. Both single and double spinneret setups were used to fabricate poly (L-lactic acid) (PLLA) fiber mats containing lidocaine hydrochloride and mupirocin (anesthetic), and the setup significantly affected the burst release profile of mupirocin [83].

10.8.2 CHRONIC WOUNDS/WOUND HEALING

Treatment of cutaneous wounds such as burns, traumatic skin injuries, and chronic wounds is currently an important clinical challenge. In general, normal wound healing can be separated into three time-dependent phases: (i) Inflammation that begins immediately after injury. (ii) The formation of new tissue overlaps with the first stage. (iii) Tissue remodeling can lead to scar tissue formation and remodeling for more than a year [84]. In deep burns, normal wound healing is usually not possible because there is no cell source for regeneration. In chronic wounds, the inflammatory phase is often too long, resulting in the wound remaining open for months to years. In these types of wounds, dressings have an important role in (i) protecting the wound; (ii) keeping the wound surroundings wet; (iii) protecting the wound from contamination with exogenous microorganisms; and (iv) accelerating wound closure. An ideal wound dressing should act as a physical separation between the wound and the outside environment but should be oxygen-permeable. Electrospun fiber mats are (i) highly porous membrane structures that facilitate the exchange of liquids and gases with the environment; (ii) a wound must have well-connected pores to prevent bacterial infiltration; (iii) a large surface area; (iv) it can provide localized release of drugs to the skin. Mats made of electrospun fiber can be utilized for drug delivery vehicles to aid wound healing. Sutures have the potential to cause infection, inflammation, or pain.

10.8.3 ANESTHETICS

During surgical procedures, local anesthetics like bupivacaine and lidocaine hydrochloride are frequently used to treat discomfort. These typically need several injections to penetrate tissue. By employing bupivacaine, Weldon et al. created electrospun fibers that offered long-lasting local analgesia [85]. Local analgesia was observed in 9 out of 10 rats treated with bupivacaine-loaded fibers, with onset on day 1 and a maximum noticed on day 3, followed by a loss of effect between 7 and 9 days. This rat model was used to illustrate the effectiveness of the treatment. In order to release gentamycin vancomycin (antibiotics) and lidocaine, Chen et al. created a sandwich-structure electrospun fiber mat with two outside layers made of PLGA/collagen fibers and an inside layer made of PLGA fibers holding three medicines [86].

10.9 FORMULATION OF NANOFIBERS AND NANOPARTICLES FOR TOPICAL APPLICATION IN THE CLINIC

Vehicles for local drug delivery (such as nanofibers and nanoparticles) can deliver therapeutic effects to the target site directly and reduce potentially unwanted systemic side effects. For topical use in the clinic, nanoparticles can be prepared in a variety of methods, including solution/liquid formulations, dry formulations, and viscous formulations (i.e., creams, lotions, gels, ointments). The suspension media for nanoparticles is often selected with a particular therapeutic use in mind. This medium is biocompatible and should be used to facilitate percutaneous absorption. Additionally, a suspension medium can change how quickly medications are released from nanoparticles [87]. A recent review highlights considerations used in developing topical skin formulations [88]. Due to its flexible handling characteristics, the bat's application directly to the wound is a drug-eluting dressing. The bat can be covered with other clinically used non-adhesive dressings to keep the atmosphere around a wound wet. A recent review has focused on the design of electrospun fiber mats for tissue engineering, drug delivery, and wound healing applications [3].

10.10 CONCLUSION

A wealth of research on nanoparticles has shown that nanoparticles have the potential to efficiently transport medication via the skin barrier. Solid lipid nanoparticles and liposomes offer potential value as local drug delivery systems alongside polymer- and metal-based nanoparticles. Electrospun nanofibers are also promising in the field of wound healing and antimicrobial dressing. Translating a very large global research effort into a clinically used product seems slow. Quantitative research that can link exposure to nanoparticle therapeutic and penetration benefits needs to be prioritized more. Further quantitative studies are needed in the area of nanoparticle toxicity. Moreover, comparisons between healthy and diseased skin related to nanoparticle penetration efficiency are missing. Little is understood about the transport mechanisms of nanoparticles through the three major skin layers. Another observation is that few studies have used commercially relevant controls. For example, the efficacy of nanoparticle formulations is often compared to no-treatment or drug-free formulations, rather than using the current clinical "gold standard" as a reference point. As a result, there are many interesting unanswered questions and technical challenges, providing important opportunities for further research studies.

KEYWORDS

- **chitosan**
- **hydroxyl propyl methyl cellulose**
- **poly (lactic-co-glycolic acid)**
- **poly(L-lysine)**
- **silver nanoparticles**
- **solid lipid nanoparticles**
- **titanium dioxide**

REFERENCES

1. Alvarez-Roman, R., Naik, A., Kalia, Y. N., Guy, R. H., & Fessi, H. (2004). Skin penetration and distribution of polymeric nanoparticles. *Journal of Controlled Release, 99,* 53–62.
2. Bhandari, D. X. L. Krishna Hari, Newa, M., Yoon, S. II, Kim, J. S., Kim, D. D., Kim, J. -A., Yoo, B. K., Woo, J. S., Lyoo, W. S., Lee, J. H., Choi, H. G., & Yong, C. S. (2008). Evaluation of skin permeation and accumulation profiles of a highly lipophilic fatty ester. *Archives of Pharmacal Research, 31,* 242–249.
3. Goh, Y. F., Shakir, I., & Hussain, R. (2013). Electrospun fibers for tissue engineering, drug delivery, and wound dressing. *Journal of Materials Science, 48*(8), 3027–3054.
4. Naik, L. A. R. M. P. Aarti, Potts, R. O., & Guy, R. H. (1995). Mechanism of oleic acid-induced skin penetration enhancement in vivo in humans. *Journal of Controlled Release, 37,* 299–306.

5. Prausnitz, M. R., & Langer, R. (2008). Transdermal drug delivery. *Nature Biotechnology, 26,* 1261–1268.
6. Zhang, Z., Tsai, P. C., Ramezanli, T., & Michniak-Kohn, B. B. (2013). Polymeric nanoparticles-based topical delivery systems for the treatment of dermatological diseases. *Wiley Interdisciplinary Reviews: Nanomedicine and Nanobiotechnology, 5,* 205–218.
7. Yoo, H. S., Kim, T. G., & Park, T. G. (2009). Surface-functionalized electrospun nanofibers for tissue engineering and drug delivery. *Advanced Drug Delivery Reviews, 61,* 1033–1042.
8. Jin, H. Q., & Min, X. (2008). Characterization and application in bio adhesive drug delivery system of chitosan. *Central South Pharmacy, 6,* 324–327.
9. Wang, J. J., Zeng, Z. W., Xiao, R. Z., Xie, T., Zhou, G. L., & Zhan, X. R. (2011). Recent advances of chitosan nanoparticles as drug carriers. *International Journal of Nanomedicine, 6,* 765–774.
10. Bodnar, M., Hartmann, J. F., & Borbely, J. (2005). Preparation and characterization of chitosan-based nanoparticles. *Biomacromolecules, 6,* 2521–2527.
11. Shabouri, M. H. E. (2002). Positively charged nanoparticles for improving the oral bioavailability of cyclosporin-A. *International Journal of Pharmaceutics, 249,* 101–108.
12. de Moura, M. R., Aouada, F. A., & Mattoso, L. H. (2008). Preparation of chitosan nanoparticles using methacrylic acid. *Journal of Colloid and Interface Science, 321,* 477–483.
13. Bravo-Osuna, I., Vauthier, C., Farabollini, A., Palmieri, G. F., & Ponchel, G. (2007). Mucoadhesion mechanism of chitosan and thiolated chitosan-poly (isobutyl cyanoacrylate) core-shell nanoparticles. *Biomaterials, 28,* 2233–2243.
14. Cheng, H. Y., Zhou, H. J., Liu, Z., & Li, J. F. (2010). Rapid preparation and characterization of chitosan nanoparticles for oligonucleotide. *Current Applied Physics, 10,* 797–800.
15. Kim, D. G., Jeong, Y. I., Choi, C., Roh, S. H., Kang, S. K., Jang, M. K., & Nah, J. W. (2006). Retinol-encapsulated low molecular water-soluble chitosan nanoparticles. *International Journal of Pharmaceutics, 319,* 130–138.
16. Prabaharan, M., & Mano, J. F. (2005). Chitosan-based particles as controlled drug delivery systems. *Drug Delivery, 12,* 41–57.
17. Kashyap, N., Viswanad, B., Sharma, G., Bhardwaj, V., Ramarao, P., & Ravi Kumar, M. N. (2007). Design and evaluation of biodegradable, bio sensitive in situ gelling system for pulsatile delivery of insulin. *Biomaterials, 28*(15), 2051–2060.
18. Ibrahim, M. M., Abd-Elgawad, A. H., Soliman, O. A., & Jablonski, M. M. (2015). Natural Bioadhesive Biodegradable Nanoparticle-Based Topical Ophthalmic Formulations for Management of Glaucoma. *Translational Vision Science & Technology, 4*(2), 12.
19. Friedman, A. J., Phan, J., Schairer, D. O., Champer, J., Qin, M., Pirouz, A., Blecher-Paz, K., Oren, A., Liu, P. T., Modlin, R. L., & Kim, J. (2013). Antimicrobial and anti-inflammatory activity of chitosan-alginate nanoparticles: A targeted therapy for cutaneous pathogens. *The Journal of Investigative Dermatology, 133*(5), 1231–1239.
20. Das, S., Bellare, J. R., & Banerjee, R. (2012). Protein based nanoparticles as platforms for aspirin delivery for ophthalmologic applications. *Colloids and Surfaces. B, Biointerfaces, 93,* 161–168.
21. Panyam, J., & Labhasetwar, V. (2003). Biodegradable nanoparticles for drug and gene delivery to cells and tissue. *Advanced Drug Delivery Reviews, 55*(3), 329–347.

22. Nardin, C., Bolikal, D., & Kohn, J. (2004). Nontoxic block copolymer nanospheres: Design and characterization. *Langmuir, 20*(26), 11721–11725.
23. Sheihet, L., Dubin, R. A., Devore, D., & Kohn, J. (2005). Hydrophobic drug delivery by self-assembling triblock copolymer-derived nanospheres. *Biomacromolecules, 6*(5), 2726–2731.
24. Sheihet, L., Chandra, P., Batheja, P., Devore, D., Kohn, J., & Michniak, B. (2008). Tyrosine-derived nanospheres for enhanced topical skin penetration. *International Journal of Pharmaceutics, 350*(1–2), 312–319.
25. Sheihet, L., Garbuzenko, O. B., Bushman, J., Gounder, M. K., Minko, T., & Kohn, J. (2012). Paclitaxel in tyrosine-derived nanospheres as a potential anti-cancer agent: In vivo evaluation of toxicity and efficacy in comparison with paclitaxel in Cremophor. *European Journal of Pharmaceutical Sciences: Official Journal of the European Federation for Pharmaceutical Sciences, 45*(3), 320–329.
26. Batheja, P., Sheihet, L., Kohn, J., Singer, A. J., & Michniak-Kohn, B. (2011). Topical drug delivery by a polymeric nanosphere gel: Formulation optimization and in vitro and in vivo skin distribution studies. *Journal of Controlled Release: Official Journal of the Controlled Release Society, 149*(2), 159–167.
27. Patel, A. S. N., Dinizo, M., Goyal, R., Hannan, J., Lee, G. T., Kohn, J., Bivalacqua, T. J., & Kim, I. Y. (2014). Development of a nanosphere based delivery of Sildenafil to sites of nerve injury for prevention of post-prostatectomy erectile dysfunction using a rat model of cavernous nerve injury. *The Journal of Urology, 4,* e481.
28. Patel, A. S. N., Dinizo, M., Faiena, I., Goyal, R., Lee, G. T., Hannan, J. L., Bivalacqua, T. J., Kohn, J., & Kim, I. Y. (2015). Efficacy of Rolipram loaded nanospheres in localized delivery to sites of nerve injury for prevention of post-prostatectomy erectile dysfunction using a rat model of cavernous nerve injury. *The Journal of Urology, 4,* e766–e767.
29. Bushman, J., Vaughan, A., Sheihet, L., Zhang, Z., Costache, M., & Kohn, J. (2013). Functionalized nanospheres for targeted delivery of paclitaxel. *Journal of Controlled Release: Official Journal of the Controlled Release Society, 171*(3), 315–321.
30. Sheihet, L., Piotrowska, K., Dubin, R. A., Kohn, J., & Devore, D. (2007). Effect of tyrosine-derived triblock copolymer compositions on nanosphere self-assembly and drug delivery. *Biomacromolecules, 8*(3), 998–1003.
31. Kapoor, D. N., Bhatia, A., Kaur, R., Sharma, R., Kaur, G., & Dhawan, S. (2015). PLGA: A unique polymer for drug delivery. *Therapeutic Delivery, 6*(1), 41–58.
32. Shahin, M., & Lavasanifar, A. (2010). Novel self-associating poly (ethylene oxide)-b-poly(epsilon-caprolactone) based drug conjugates and nano-containers for paclitaxel delivery. *International Journal of Pharmaceutics, 389*(1-2), 213–222.
33. Muller, R. H., Mader, K., & Gohla, S. (2000). Solid lipid nanoparticles (SLN) for controlled drug delivery – A review. *European Journal of Pharmaceutics and Biopharmaceutics: Official Journal of Arbeitsgemeinschaft fur Pharmazeutische Verfahrenstechnik e.V, 50*(1), 161–177.
34. Mehnert, W., & Mader, K. (2001). Solid lipid nanoparticles: Production, characterization and applications. *Advanced Drug Delivery Reviews, 47*, 165–196.
35. Jeon, H. S., Seo, J. E., Kim, M. S., Kang, M. H., Oh, D. H., Jeon, S. O., Seong Hoon, J., Choi, Y. W., Lee, S. (2013). A retinyl palmitate-loaded solid lipid nanoparticle system: Effect of surface modification with dicetyl phosphate on skin permeation in vitro and anti-wrinkle effect in vivo. *International Journal of Pharmaceutics, 452*, 311–320.

36. Kraft, J. C., Freeling, J. P., Wang, Z., & Ho, R. J. (2014). Emerging research and clinical development trends of liposome and lipid nanoparticle drug delivery systems. *Journal of Pharmaceutical Sciences, 103*, 29–52.
37. Scalia, S., Franceschinis, E., Bertelli, D., & Iannuccelli, V. (2013). Comparative evaluation of the effect of permeation enhancers, lipid nanoparticles and colloidal silica on in vivo human skin penetration of quercetin. *Skin Pharmacology and Physiology, 26*, 57–67.
38. Bangham, A. D., Standish, M. M., & Watkins, J. C. (1965). Diffusion of univalent ions across the lamellae of swollen phospholipids. *Journal of Molecular Biology, 13*, 238–252.
39. Szoka, F., Jr., & Papahadjopoulos, D. (1978). Procedure for preparation of liposomes with large internal aqueous space and high capture by reverse-phase evaporation. *Proceedings of the National Academy of Sciences of the United States of America, 75*, 4194–4198.
40. Gentine, P., Bourel-Bonnet, L., & Frisch, B. (2013). Modified and derived ethanol injection toward liposomes: Development of the process. *Journal of Liposome Research, 23*, 11–19.
41. Lasch, W. V., & Brandl, M. (2003). *Preparation of Liposomes*. Oxford University Press. https://global.oup.com/academic/product/liposomes-a-practical-approach-9780199636549?cc=ir&lang=en&.
42. Frederiksen, L., Anton, K., van Hoogevest, P., Keller, H. R., & Leuenberger, H. (1997). Preparation of liposomes encapsulating water-soluble compounds using supercritical carbon dioxide. *Journal of Pharmaceutical Sciences, 86*, 921–928.
43. Barnadas-Rodriguez, R., & Sabes, M. (2001). Factors involved in the production of liposomes with a high-pressure homogenizer. *International Journal of Pharmaceutics, 213*, 175–186.
44. Rameez, S., Bamba, I., & Palmer, A. F. (2010). Large scale production of vesicles by hollow fiber extrusion: A novel method for generating polymersome encapsulated hemoglobin dispersions. *Langmuir, 26*, 5279–5285.
45. Hina Shrestha, R. B., & Arora, S. (2014). Lipid-Based Drug Delivery Systems. *Journal of Pharmaceutics, 2014*, 10.
46. Marin, S., Vlasceanu, G. M., Tiplea, R. E., Bucur, I. R., Lemnaru, M., Marin, M. M., & Grumezescu, A. M. (2015). Applications and toxicity of silver nanoparticles: A recent review. *Current Topics in Medicinal Chemistry, 15*, 1596–1604.
47. Huang, Y., Yu, F., Park, Y. S., Wang, J., Shin, M. C., Chung, H. S., & Yang, V. C. (2010). Co-administration of protein drugs with gold nanoparticles to enable percutaneous delivery. *Biomaterials, 31*, 9086–9091.
48. Robichaud, C. O., Uyar, A. E., Darby, M. R., Zucker, L. G., & Wiesner, M. R. (2009). Estimates of upper bounds and trends in nano-TiO2 production as a basis for exposure assessment. *Environmental Science & Technology, 43*, 4227–4233.
49. Chigurupati, S., Mughal, M. R., Okun, E., Das, S., Kumar, A., McCaffery, M., Seal, S., & Mattson, M. P. (2013). Effects of cerium oxide nanoparticles on the growth of keratinocytes, fibroblasts and vascular endothelial cells in cutaneous wound healing. *Biomaterials, 34*, 2194–2201.
50. Anselmann, R. (2001). Nanoparticles and nanolayers in commercial applications. *Journal of Nanoparticle Research, 3*, 329–336.

51. Mignani, S., El Kazzouli, S., Bousmina, M., & Majoral, J. P. (2013). Expand classical drug administration ways by emerging routes using dendrimer drug delivery systems: A concise overview. *Advanced Drug Delivery Reviews, 65*, 1316–1330.
52. Shi, X., Thomas, T. P., Myc, L. A., Kotlyar, A., & Baker, J. R., Jr. (2007). Synthesis, characterization, and intracellular uptake of carboxyl-terminated poly(amidoamine) dendrimer-stabilized iron oxide nanoparticles. *Physical Chemistry Chemical Physics: PCCP, 9*, 5712–5720.
53. Parat, A., Bordeianu, C., Dib, H., Garofalo, A., Walter, A., Begin-Colin, S., & Felder-Flesch, D. (2015). Dendrimer-nanoparticle conjugates in nanomedicine. *Nanomedicine, 10*, 977–992.
54. Vyas, A., Kumar Sonker, A., & Gidwani, B. (2014). Carrier-based drug delivery system for treatment of acne. *TheScientificWorldJournal, 2014*, 276260.
55. Castro, G. A., & Ferreira, L. A. (2008). Novel vesicular and particulate drug delivery systems for topical treatment of acne. *Expert Opinion on Drug Delivery, 5*, 665–679.
56. Reis, C. P., Martinho, N., Rosado, C., Fernandes, A. S., & Roberto, A. (2014). Design of polymeric nanoparticles and its applications as drug delivery systems for acne treatment. *Drug Development and Industrial Pharmacy, 40*, 409–417.
57. Webster, G. F., Leyden, J. J., & Gross, J. A. (2014). Results of a Phase III, double-blind, randomized, parallel-group, non-inferiority study evaluating the safety and efficacy of isotretinoin-Lidose in patients with severe recalcitrant nodular acne. *Journal of Drugs in Dermatology: JDD, 13*, 665–670.
58. Muller, S., Ebnother, M., & Itin, P. (2014). Green Nail Syndrome (Pseudomonas aeruginosa Nail Infection): Two Cases Successfully Treated with Topical Nadifloxacin, an Acne Medication. *Case Reports in Dermatology, 6*, 180–184.
59. Basnet, P., & Skalko-Basnet, N. (2013). Nanodelivery systems for improved topical antimicrobial therapy. *Current Pharmaceutical Design, 19*, 7237–7243.
60. Franci, G., Falanga, A., Galdiero, S., Palomba, L., Rai, M., Morelli, G., & Galdiero, M. (2015). Silver nanoparticles as potential antibacterial agents. *Molecules, 20*, 8856–8874.
61. Kalishwaralal, K., BarathManiKanth, S., Pandian, S. R., Deepak, V., & Gurunathan, S. (2010). Silver nanoparticles impede the biofilm formation by Pseudomonas aeruginosa and Staphylococcus epidermidis. *Colloids and Surfaces. B, Biointerfaces, 79*, 340–344.
62. Jain, J., Arora, S., Rajwade, J. M., Omray, P., Khandelwal, S., & Paknikar, K. M. (2009). Silver nanoparticles in therapeutics: Development of an antimicrobial gel formulation for topical use. *Molecular Pharmaceutics, 6*, 1388–1401.
63. Zheng, Y., Xiao, M., Jiang, S., Ding, F., & Wang, J. (2013). Coating fabrics with gold nanorods for colouring, UV-protection, and antibacterial functions. *Nanoscale, 5*, 788–795.
64. Addae, E., Dong, X., McCoy, E., Yang, C., Chen, W., & Yang, L. (2014). Investigation of antimicrobial activity of photothermal therapeutic gold/copper sulfide core/shell nanoparticles to bacterial spores and cells. *Journal of Biological Engineering, 8*, 11.
65. Dizaj, S. M., Lotfipour, F., Barzegar-Jalali, M., Zarrintan, M. H., & Adibkia, K. (2014). Antimicrobial activity of the metals and metal oxide nanoparticles. *Materials Science & Engineering. C, Materials for Biological Applications, 44*, 278–284.
66. Cui, Y., Zhao, Y., Tian, Y., Zhang, W., Lu, X., & Jiang, X. (2012). The molecular mechanism of action of bactericidal gold nanoparticles on *Escherichia coli*. *Biomaterials, 33*, 2327–2333.

67. Leung, Y. H., Ng, A. M., Xu, X., Shen, Z., Gethings, L. A., Wong, M. T., Chan, C. M., Guo, M. Y., Ng, Y. H., Djurisic, A. B., Lee, P. K., Chan, W. K., Yu, L. H., Phillips, D. L., Ma, A. P., & Leung, F. C. (2014). Mechanisms of antibacterial activity of MgO: Non-ROS mediated toxicity of MgO nanoparticles towards *Escherichia coli*. *Small, 10*, 1171–1183.
68. Ahamed, H. A. A. M., Khan, M. M., Karuppiah, P., & Aldhabi, N. A. (2014). Synthesis, characterization and antimicrobial activity of copper oxide nanoparticles. *J. Nanomater., 2014*, 4.
69. Usman, M. S., El Zowalaty, M. E., Shameli, K., Zainuddin, N., Salama, M., & Ibrahim, N. A. (2013). Synthesis, characterization, and antimicrobial properties of copper nanoparticles. *International Journal of Nanomedicine, 8*, 4467–4479.
70. Carre, G., Hamon, E., Ennahar, S., Estner, M., Lett, M. C., Horvatovich, P., Gies, J. P., Keller, V., Keller, N., & Andre, P. (2014). TiO2 photocatalysis damages lipids and proteins in *Escherichia coli*. *Applied and Environmental Microbiology, 80*, 2573–2581.
71. Liu, Y., He, L., Mustapha, A., Li, H., Hu, Z. Q., & Lin, M. (2009). Antibacterial activities of zinc oxide nanoparticles against *Escherichia coli* O157. *Journal of Applied Microbiology, 107*, 1193–1201.
72. Cooley, J. F. (1900). Improved methods of and apparatus for electrically separating the relatively volatile liquid component from the component of relatively fixed substances of composite fluids. *United States Patent Office.*
73. Bhardwaj, N., & Kundu, S. C. (2010). Electrospinning: A fascinating fiber fabrication technique. *Biotechnology Advances, 28*(3), 325–347.
74. Verreck, G., Chun, I., Peeters, J., Rosenblatt, J., & Brewster, M. E. (2003). Preparation and characterization of nanofibers containing amorphous drug dispersions generated by electrostatic spinning. *Pharmaceutical Research, 20*(6), 810–817.
75. Macri, L. K., Sheihet, L., Singer, A. J., Kohn, J., & Clark, R. A. (2012). Ultrafast and fast bioerodible electrospun fiber mats for topical delivery of a hydrophilic peptide. *Journal of Controlled Release: Official Journal of the Controlled Release Society, 161*(3), 813–820.
76. Jiang, H., Fang, D., Hsiao, B., Chu, B., & Chen, W. (2004). Preparation and characterization of ibuprofen-loaded poly(lactide-co-glycolide)/poly (ethylene glycol)-g-chitosan electrospun membranes. *Journal of Biomaterials Science. Polymer Edition, 15*(3), 279–296.
77. Nair, L. S., Bhattacharyya, S., & Laurencin, C. T. (2004). Development of novel tissue engineering scaffolds via electrospinning. *Expert Opinion on Biological Therapy, 4*(6), 659–668.
78. Kong, H., He, J. (2013). A modified bubble electrospinning for fabrication of nanofibers. *Journal of Nano Research, 23*, 125–128.
79. Varabhas, J. S., Chase, C. G., & Reneker, D. H. (2008). Electrospun nanofibers from a porous hollow tube. *Polymer, 49*(19), 4226–4229.
80. Katti, D. S., Robinson, K. W., Ko, F. K., & Laurencin, C. T. (2004). Bioresorbable nanofiber-based systems for wound healing and drug delivery: Optimization of fabrication parameters. *Journal of Biomedical Materials Research. Part B, Applied Biomaterials, 70*(2), 286–296.
81. Verreck, G., Chun, I., Rosenblatt, J., Peeters, J., Dijck, A. V., Mensch, J., Noppe, M., & Brewster, M. E. (2003). Incorporation of drugs in an amorphous state into electrospun

nanofibers composed of a water-insoluble, nonbiodegradable polymer. *Journal of Controlled Release: Official Journal of the Controlled Release Society, 92*(3), 349–360.
82. Huang, Z. M., He, C. L., Yang, A., Zhang, Y., Han, X. J., Yin, J., & Wu, Q. (2006). Encapsulating drugs in biodegradable ultrafine fibers through co-axial electrospinning. *Journal of Biomedical Materials Research. Part A, 77*(1), 169–179.
83. Thakur, R. A., Florek, C. A., Kohn, J., & Michniak, B. B. (2008). Electrospun nanofibrous polymeric scaffold with targeted drug release profiles for potential application as wound dressing. *International Journal of Pharmaceutics, 364*(1–2), 87–93.
84. Macri, L. K., & Clark, R. A. (2009). Tissue engineering for cutaneous wounds: Selecting the proper time and space for growth factors, cells and the extracellular matrix. *Skin Pharmacology and Physiology, 22*(2), 83–93.
85. Weldon, C. B., Tsui, J. H., Shankarappa, S. A., Nguyen, V. T., Ma, M., Anderson, D. G., & Kohane, D. S. (2012). Electrospun drug-eluting sutures for local anesthesia. *Journal of Controlled Release, 161*(3), 903–909.
86. Chen, D. W., Liao, J. Y., Liu, S. J., & Chan, E. C. (2012). Novel biodegradable sandwich-structured nanofibrous drug-eluting membranes for repair of infected wounds: An in vitro and in vivo study. *International Journal of Nanomedicine, 7*, 763–771.
87. Goyal, R. M., & Kohn, J. (2015). Formulation strategy for the delivery of Cyclosporine A: Comparison of two polymeric nanospheres. *Scientific Reports,*
88. Chang, R. K., Raw, A., Lionberger, R., & Yu, L. (2013). Generic development of topical dermatologic products: Formulation development, process development, and testing of topical dermatologic products. *The AAPS Journal, 15*(1), 41–52.
89. Zheng Zhang, Ophir Ortiz, Ritu Goyal, & Joachim Kohn (2014). *Handbook of Polymer Applications in Medicine and Medical Devices Plastics Design Library*, pp. 303–335. https://doi.org/10.1016/B978–0-323–22805–3.00013-X; https://www.sciencedirect.com/science/article/abs/pii/B978032322805300013X (accessed on 15 July 2024).
90. Kim, H. Y., & Gladyshev, V. N. (2004). Methionine sulfoxide reduction in mammals: characterization of methionine-R-sulfoxide reductases. *Mol Biol Cell 15,* 1055–1064.

CHAPTER 11

Electrospun Polymer Nanofibers: A Promising Technique for Drug Delivery Systems

M. R. BADWAR,[1] S. R. CHAUDHARI,[1] V. T. PATHAN,[2] SWATI T. TALELE,[2] C. K. SHINDE,[2] ASHWINI SHELKE,[3] and LAXMIKANT B. BORSE[2]

[1]*Department of Quality Assurance, Sandip Institute of Pharmaceutical Sciences, Mahiravani, Nashik, Maharashtra, India*

[2]*Department of Pharmaceutics, Sandip Institute of Pharmaceutical Sciences, Mahiravani, Nashik, Maharashtra, India*

[3]*Sir Dr. M.S. Gosavi College of Pharmaceutical Education and Research, Nashik, Maharashtra, India*

ABSTRACT

This review on floating drug delivery systems (FDDS) was written with the intention of gathering the most recent literature that focuses on the main mechanism of flotation to achieve gastric retention. Drug delivery systems are those that float right away when they come into contact with gastric fluids, offering promising methods for increasing the bioavailability of medication. They are unstable in the intestinal or colonic environment and have low solubility at high pH levels. They also have absorption windows in the stomach or upper small intestine. It is a brand-new method of drug delivery that maximizes efficiency and adherence. The adoption of floating dose forms, which allow for both local and systemic effects, helped to solve physiological issues such as short stomach duration and unpredictable gastric emptying time. The bioavailability of medication with a narrow absorption window is improved by a floating drug delivery

Sustainability in Polymer Technology and Plastics Engineering: Concepts, Strategies, and Opportunities. Tamara Tatrishvili, Neha Kanwar Rawat, Swati Gokul Talele, and A. K. Haghi (Eds.)

device, which allows prolonged and continuous drug input to the upper part of the gastric retention tract. This review aims to provide extensive information on the pharmaceutical foundation of their design, classification, benefits, *in vitro* and *in vivo* evaluation parameters, and use of floating systems, as well as applications of such systems. These systems can help with a variety of issues that arise with drug dosage forms and the potential for FDDS in the future. The goal of this review article is to inform readers of the progress being made in the field of floating drug delivery.

11.1 INTRODUCTION

Due to its noninvasiveness, usability, increased patient compliance, and cost-effectiveness, oral medication delivery is the preferred method of drug administration [1]. Chronic diseases with repeated dosage intake respond well to oral medication administration with controlled-release kinetics. Multiple administrations of instant-release formulations can be used in controlled drug delivery systems to reduce the drug concentration in plasma below the toxic level [2]. The gastro-retentive delivery system (GRDS) is a significant controlled medication delivery mechanism in oral formulations [3]. The use of GRDS can demonstrate the prolonged release of medications in the gastrointestinal environment while resisting stomach contractions and peristaltic waves [4]. Therefore, lengthening some medications' time in the stomach or upper GI system may boost their bioavailability since these medications are stable at an acidic pH. Some medications, including baclofen and metformin HCl, are absorbed primarily in the stomach or the top section of the small intestine, and they are severely destroyed in the colon. As a result, the stomach retention characteristic is quite useful for delivering such pharmaceuticals for a long time [5].

The floating drug delivery system (FDDS) or the non-floating system are two methods for developing GRDS (mucoadhesive, swelling, or high-density systems).

Either an effervescent system or a non-effervescent system can produce FDDS (hydrodynamical system, alginate beads, or matrix layered tablets).

The effervescent FDDS is designed in such a way that when it comes into contact with the acidic gastric contents (pH of 1.5–3), CO_2 gas is produced and trapped in swollen hydrocolloids, giving the dosage form buoyancy. In contrast, the non-effervescent FDDS uses a swelling method [6].

It has been reported that fibers generated by the electrospinning technique offer good controlled release of medicines and other useful compounds [7]. Nanofibers that provide controlled release through a variety of processes have been created as a result of the increased interest in the development of new electrospinning techniques (such as coaxial, triaxial, and side-by-side electrospinning) [8]. Electrospun nanofiber membranes have a microscopic structure that is sufficiently small for it to catch produced gas when it swiftly expands. In addition to their many other benefits for drug delivery systems, electrospun nanofibers have particular advantages [3] for creating an effervescent drug delivery system because (a) gastric fluid can pass through the pores into the embedded sodium carbonate-PEO film; and (b) generated CO_2 gas is unable to diffuse out of the nanofiber system, causing drug-incorporated nanofiber pouches to float [10]. Electrospun nanofiber membranes offer prolonged, dual, or site-specific drug release patterns, making them potential platforms for oral medication delivery [11].

To retain nanofibers in the stomach for sustained medication release, several nanofiber-based drug delivery methods have been created [12]. For instance, Malik et al. created a diacerein-loaded gastro-retentive nanofiber system using poly L-(lactic acid) and tripolyphosphate (TPP)-crosslinked chitosan/poly(ethylene oxide)-ranitidine hydrochloride as an FDDS. The prepared nanofibers remained floating for more than 24 hours and released 70% of the drug at an acidic pH [13]. They also created mucoadhesive nanofibers for gastric retentive drug delivery in acidic settings using chitosan and reported that 61.3% of the medication was released in 30 hours at an acidic pH [14].

In the disciplines of medicine and biotechnology, a great deal of work is currently being done in a wide range of areas, showing an exceptionally tremendous potential for miraculous miracles. The main goal of developing new drug delivery systems is to enhance a drug's therapeutic effects while minimizing its harmful effects by increasing the drug's concentration at the target sites and decreasing the exposure of non-target sites to the drug. To obtain and maintain the optimal medication concentration in the body, a unique drug delivery system is designed to carry the drug directly to the site of action. To enhance treatment efficacy and avoid the medicine's related negative effects, numerous drug delivery systems have been created [15].

Because of its many benefits, including its low cost of therapy, convenience of administration, regulated drug distribution, flexibility in formulation, and increased patient compliance, the oral route is the most popular, promising, and successful way to provide therapeutic agents. The creation of innovative gastro-retentive drug delivery systems (GRDDS) has attracted more attention in recent years [16, 17].

GRDDS are particularly effective for medications with a limited window of absorption, quick absorption from the gastrointestinal system, short half-lives, and intended for local application [18]. In these situations, administration of a typical dosage form might require repeated doses to produce the desired therapeutic effect. Oral sustained controlled-release formulations are created in an effort to release the medication gradually into the gastrointestinal tract and maintain an effective drug concentration for an extended length of time. Regarding their bioavailability, therapeutic effectiveness, potential dose reduction, and improved drug solubility that is less soluble in a high pH alkaline environment, GRDDS is advantageous [19].

GRDDS are superior to traditional drug delivery methods in a number of ways. As a result, it is essential to create a system that enables extended stomach residence, which will increase the window of opportunity for medication absorption. To extend the drug's stay in the stomach, various strategies have been developed in the past, such as changed density formulations, bio-adhesive formulations, swelling systems, ion exchange resins, porous hydrogels, osmoregulated formulations, and effervescent. Nanofibers appear to have greater gastro-retentive qualities compared to other approaches developed [20]. Nanofiber systems offer a variety of functional capabilities to govern the floating behavior of the carrier system because of their special properties. Surface-to-volume ratio, specific density, adsorption characteristics, fiber diameter, porosity, surface hydrophilicity, etc., are all characteristics of nanofibers that are very helpful in predicting the floating behavior [21, 22]. Application areas for electrospun nanofibers as drug delivery systems include tissue scaffolds, antibacterial sheets, and anticancer therapy [23]. Even delicate medications like proteins, peptides, antibodies, and other tiny molecules have been made with nanofibers [24]. For the transmucosal administration of insulin, our team has created a biodegradable poly (vinyl alcohol) and sodium alginate (Na-Alg) electrospun composite nanofiber. The *in-vivo* tests confirmed the sustained and controlled drug release pattern that was demonstrated by the *in-vitro* activity from the nanofiber patch. Insulin was administered in its active state, according to the *in-vivo* activity, and significant results were attained in comparison to the commercial formulation [25]. The tertiary structure of proteins and peptides can be restored thanks to the natural 3D scaffold architecture of nanofibers, preserving their biological function in the process. Additionally, the straightforward fabrication method makes the process affordable and simple to scale up. Nanofibers made from various polymers can be successfully created using the electrospinning technology. Proteins and peptides can be delivered by polysaccharides alone or in conjunction with hydrocolloids, in particular.

Protein medicines have improved storage stability thanks to the polyhydroxyl groups of polysaccharides, which enable extensive hydrogen bonding to interact with proteins and peptides, improving their thermostability [26]. In a different work, we created a new diclofenac sodium transmucosal patch utilizing electrospun polycaprolactone (PCL) nanofibers. Release tests showed that PCL nanofibers have prolonged release behavior, which can be used to administer NSAIDs transmucosally with increased therapeutic efficacy [27]. Due to their distinct physicochemical characteristics, nanofibers present a variety of opportunities to enhance the solubility of medications that are not very soluble in water. Using the XRD technique, it is possible to assess the drug molecules' molecular configuration in the nanofiber system and investigate higher drug solubility in nanofibers. Additionally, nanofibers have a significant surface area for the processes of solvation and dissolution, which are thought to be rate-limiting processes in determining solubility. Additionally, a wide variety of polymers are available that can be employed in the creation of nanofibers. Biodegradable, non-biodegradable, and hybrid materials can all be used to create nanofibers. High drug loadings and site-specific drug transfer are made possible by the high surface areas [28]. The surface-to-volume ratio increases up to 1,000 times greater than microfiber as the fiber diameter gets smaller and smaller until it reaches the nanoscale. Surprisingly, nanofibers allow for the full exploration of the aforementioned features.

Additionally, there are countless opportunities to manipulate the physical properties of the nanofibers to obtain the desired floating behavior due to the formulation and process parameters used in the fabrication of nanofibers. Due to these special qualities, nanofibers can be used to create fantastic gastro-retentive formulations that appear to be an improvement over currently available formulations [29].

The route of administration known as oral administration is extensively used due to its high rate of patient compliance, low price, simplicity of administration, and adaptability in dosage form. The primary issue with oral delivery is the brief drug residence time at the site of absorption [30]. Drugs that have a short half-life and are rapidly absorbed in the gastrointestinal tract (GIT) are quickly removed from the systemic circulation, resulting in inadequate drug absorption. Therefore, scientists continue to look for ways to create oral sustained-controlled release formulations that release the medication gradually in the gastrointestinal tract while maintaining its efficacy for a long time in the systemic circulation. The main obstacle to the development of oral sustained-release medicines is altering GI transit time. Drug absorption from the digestive system is a complex process that varies depending on a number of factors. The duration of contact with the small intestinal mucosa affects how much of a medicine is absorbed from the digestive system. Thus, small intestine transit time is a crucial parameter for drugs that don't get absorbed completely. Therefore, to achieve more therapeutic benefits of the drug, increasing the gastric retention time of such dosage forms is advisable [31]. Davis originally made the FDDS available in 1968.

These systems float for a long time in the stomach because they are less dense than gastric fluid. FDDSs are appropriate delivery systems for medications with a limited window of absorption in the upper small intestine or stomach. They also acquire adequate stomach retention and medication absorption [31]. The set of oral controlled drug delivery systems that can be retained in the stomach without passing through the gastric tract includes gastroretentive drug delivery systems, sometimes referred to as floating drug delivery systems. The main difficulty in developing FDDS is not just maintaining drug release, but also extending the duration of the drug's residence in the stomach and upper GIT until all of the medication is finally released. This prolonged period of gastro retention increases bioavailability, lowers wasteful drug use, and improves solubility for medications that are less soluble at high pH levels [32]. The drug is released from the system slowly and at the prescribed pace when the drug delivery system is floating on the contents of the stomach. Increased stomach retention time and improved management of medication concentration changes are the results of this [33].

11.2 ACCORDING TO THE BUOYANCY MECHANISM, FDDS CAN BE CATEGORIZED AS FOLLOWS:

1. Single Unit FDDS
   i. Effervescent devices
   ii. Systems without effervescence
2. FDDS for multiple units
   i. Systems without effervescence

ii. Systems that bubble up
iii. Microspheres with holes

Stomach anatomy [34].

The fundus, body, and antrum are the three compartments that make up the stomach (pylorus).

The fundus and body, which make up the posterior portion, serve as a storage area for uneaten food.

The pyloric end, in contrast, serves as the primary location for mixing motion and functions as a pump for stomach emptying by moving motions.

Both when onc is fasting and eating, the stomach empties. Two states have the same motility pattern. Interdigestive myoelectric cycle the migratory myoelectric cycle is the term used to describe this occurrence (MMC).

It has four additional phases. Wilson and Washington characterize these periods [35] (Figure 11.1).

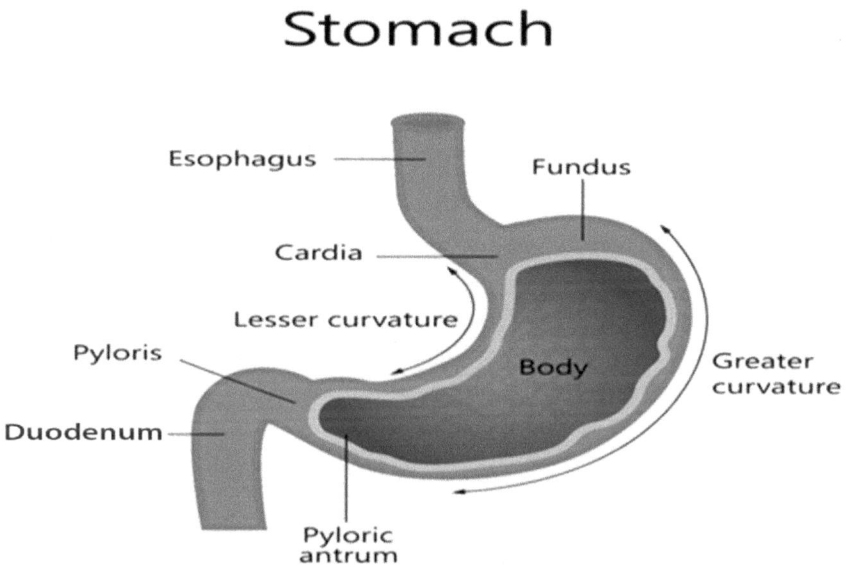

FIGURE 11.1 Gastrointestinal tract physiology.

Both when one is fasting and when one is fed, the process of stomach emptying takes place, but the patterns of motility in the two states are very different.

An inter-digestive series of electrical events that cycle through the stomach and small intestine every two to three hours characterizes the fasting state. The interdigestive myoelectric cycle, also known as the migrating myoelectric complex (MMC), is the term used to describe this action [9].

1. 40 to 60 minutes are spent in the basal phase (phase I), which hardly ever has contractions.
2. Phase II of the preburst lasts for 40 to 60 minutes. This phase is characterized by intermittent contractions and action potential. As the phase develops, the intensity and frequency likewise steadily rise.
3. Phase III's burst phase lasts for 4 to 6 minutes. Short-lived, regular, and intense contractions take place. All of the undigested material is pushed out of the stomach and into the small intestine as a result of this wave. The housekeeping wave is another name for this.
4. The fourth phase lasts 0 to 5 minutes.

Between phases III and I of two successive cycles, it occurs [35] (Figure 11.2).

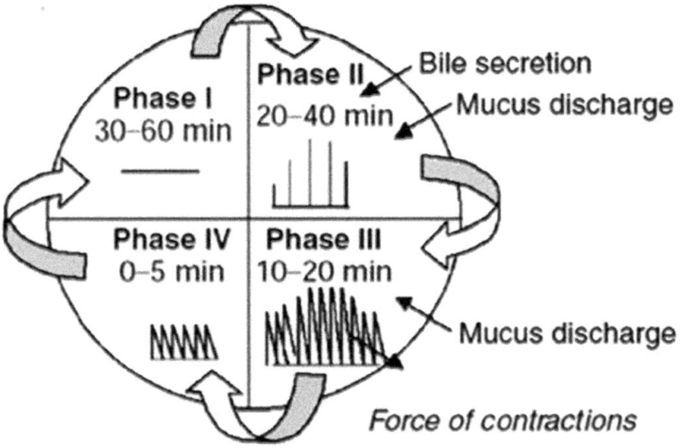

FIGURE 11.2 The interdigestive myoelectric cycle.

11.2.1 FDDS-AFFECTING VARIABLES (FLOATING DRUG DELIVERY SYSTEMS)

- **Density:** Always keep in mind that gastric contents (1.004 g/ml) should be less dense.
- **Size and Shape:** According to reports, unit dosage forms with a diameter of more than 7.5 mm have a higher GRT than those with a diameter

of 8.9 mm. Tetrahedron-shaped dosage forms and rings with flexural moduli of 48 and 22.5 kilo pound per square inch (KSI) are said to have greater GIT retention for 90 to 100% at 24 hours than other shapes.

- GRT can differ between a patient's supine and upright ambulatory phases.
- **Caloric Content:** A high-protein, high-fat meal can extend GRT by 4 to 10 hours. Due to the low frequency of MMC, the GRT can grow for more than 400 minutes when multiple meals are given instead of just one.
- **Age:** GRT is noticeably longer in adults over the age of 70.
- **Biological Factors:** Crohn's disease and diabetes
- **Nature of the Meal:** Consuming indigestible polymers of fatty acid salts might cause the stomach to enter a fed state, slowing down gastric emptying and delaying the release of drugs.
- **Gender:** Regardless of weight, height, or body surface, mean ambulatory GRT in males is lower than in age- and race-matched female counterparts.
- Anticholinergic drugs like atropine and propantheline opiates like codeine, and prokinetic drugs like metoclopramide and cisapride are all administered together.
- **Fed or Unfed State:** GI motility is characterized by bursts of vigorous motor activity or the migrating myoelectric complexes that occur every 1.5 to 2 hours while one is fasting.

The MMC removes undigested matter from the stomach; therefore, if the formulation is administered at the same time as the MMC, the unit's GRT should be very brief. However, in the fed condition, GRT is significantly longer and MMC is delayed [32].

11.2.2 FLOATING MEDICATION DELIVERY SYSTEM MECHANISM

Systems for delivering floating medications: These uncommon devices have enough buoyancy to float above the contents of the stomach and remain there for an extended amount of time. The drug is released slowly at the prescribed rate while the system floats over the gastric contents, lengthening the period the drug is retained in the stomach and lowering fluctuation.

A minimal degree of floating force (F) is additionally needed to keep the dosage form consistently buoyant on the surface of the meal, in addition to the minimal stomach content necessary to allow the appropriate attainment of the buoyancy retention principle.

A novel apparatus for calculating the resultant weight has been described in the literature to live the floating force kinetics. The device works by continually measuring the force F (expressed as a function of time) needed to retain the submerged object. If F is on the upper positive side, as seen in the figure, the object floats better. This device aids in FDDS optimization in terms of the stability and longevity of floating forces generated in order to prevent the negative effects of unforeseen intragastric buoyancy capability variations [9, 35] (Figure 11.3).

$$F = F\ buoyancy - F\ gravity = (DF - Ds)\ GV \quad (1)$$

Where; F is the total vertical force; DF is the fluid density; Ds is the object density; v is the volume; and g is the acceleration thanks to gravity (4).

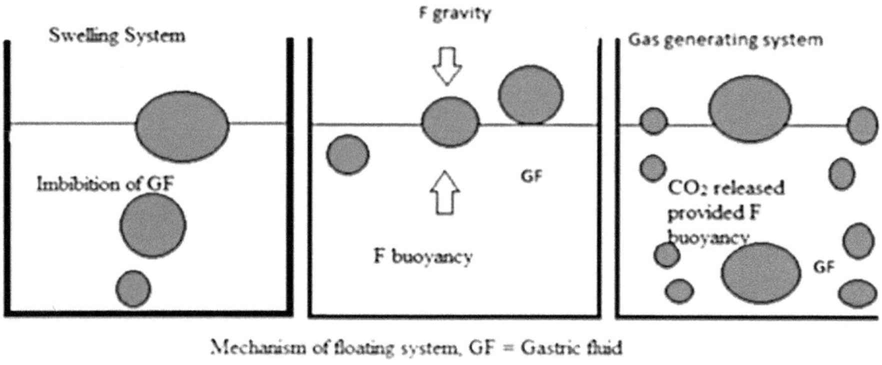

FIGURE 11.3 Mechanism of FS (floating system).

11.2.3 FLOATING MEDICATION DELIVERY SYSTEM BENEFITS

Floating dose systems have a number of benefits for drug administration, such as:

- Improved therapeutic efficacy, dose economy, and bioavailability of medicines.
- Increased stomach residence time will boost drug absorption from floating dosage forms like tablets or capsules, and the drug will stay

in the solution for a long period at its absorption site, even at the alkaline pH of the intestine.

- Decreased dose frequency and convenience of administration improve patient compliance.
- For medications that are absorbed through the stomach, such as antacids and ferrous salts, FDDS is helpful.
- Controlled drug delivery by releasing the medication slowly and at a controlled rate reduces stomach or mucosal irritation.
- Aspirin and other similar medications can be administered more effectively when using FDDS formulation because acidic compounds like aspirin irritate the stomach wall when they come into contact with it.
- Gastroesophageal reflux disease, among other gastrointestinal conditions, can be treated using FDDS.
- Provide a drug delivery system that is site-specific.
- By keeping the medication in a floating state in the stomach to get a substantially better reaction, floating dose forms are beneficial in cases of frequent diarrhea and bowel movements.
- Simple and conventional equipment is required for manufacture.
- Slow release of the drug from the FDDS dosage form into the body minimizes the counteractivity leading to higher drug efficiency.
- FDDS enhances the pharmacological effects and improves clinical outcomes by reducing the drug concentration fluctuations over a critical concentration.
- Through FDDS, better therapeutic effects of short half-life drugs can be achieved [31].

11.2.4 FLOATING MEDICATION DELIVERY SYSTEM DRAWBACKS

- The main drawback of the floating system is that for the medication delivery to float and function effectively, the stomach must have a lot of fluid.
- These systems are not practical for medicines with GIT solubility or stability issues.
- Drugs that are well absorbed throughout the GIT and go through first-pass metabolism, such as nifedipine, propranolol, etc., are not a good selection.

- Drugs that can irritate and damage the gastric mucosa, such as NSAIDs, should not be designed as floating drug delivery devices.
- The FDDS systems should not contain pharmacological candidates that are unstable in the acidic environment of the stomach.
- The dosage form's level of hydration affects its capacity to float.
- The mucus that coats the stomach walls is constantly renewing, which causes unexpected adhesion.
- The system must fully swell in dosage forms before the gastric emptying period in order to obtain faster-swelling qualities [31].

11.2.5 MEDICATIONS THAT WORK WELL WITH FLOATING DRUG DELIVERY SYSTEMS

- Drugs only have local effects in the stomach, e.g., Medicines and antacids for *H. pylori*, Misoprostol for Pylori.
- Medications that are mostly absorbed in the stomach, e.g., Amoxicillin.
- Drugs that don't dissolve well at an alkaline pH, such as Verapamil, Diazepam, Furosemide, etc.
- Medications having a small window for absorption, e.g., Levodopa, methotrexate, cyclosporine, etc.
- Drugs taken orally and are quickly absorbed. Tetracycline with metronidazole.
- Medications that break down in the gut, e.g., Metformin HCl, ranitidine.
- Drugs that disrupt the natural microbes in the colon. For instance, antibiotics for *Helicobacter pylori* [32].

11.2.6 MEDICINES THAT SHOULDN'T BE ADMINISTERED WITH GASTRORETENTIVE DRUG DELIVERY SYSTEMS

- Drugs like phenytoin and others that only dissolve very little in acid.
- Medications that are unstable in the stomach environment, such as erythromycin, etc.
- Drugs designed for the colon, such as corticosteroids and 5-amino-2-hydroxybenzoic acid, etc.

The stomach's residual system is emptied after the medication has been released. As a result, the GRT is elevated, and the oscillations in plasma drug concentration are better managed. A minimum level of floating force (F) is necessary to keep the dosage form consistently buoyant on the surface of the meal in addition to the minimal stomach content necessary to allow the proper attainment of the buoyancy retention principle. Granules, powders, capsules, tablets, laminated films, and hollow microspheres have all been used to support a variety of buoyant systems [32].

11.2.6 APPLICATIONS FDDS

Due to the limited absorption window in the upper gastrointestinal tract, medicines with poor bioavailability have a number of possibilities for floating drug delivery. It increases bioavailability by keeping the dose form at the site of absorption. They are given below:

1. **Long-Term Drug Delivery:** HBS systems can stay in the stomach for extended periods of time, allowing the medicine to be released gradually. With these approaches, the issue of a brief gastric residence period that arises with an oral CR formulation can be resolved. These systems can float on the gastric contents because of their bulk density of 1. These systems are not allowed to pass through the pyloric aperture because of their size, which is relatively considerable.
2. **Drug Delivery at a Specific Location:** These systems are especially useful for medications like riboflavin and furosemide that are primarily absorbed from the stomach or the proximal small intestine. The stomach absorbs furosemide the most, followed by the duodenum. According to reports, a monolithic floating dosage form with an extended stomach residence duration was created, increasing the bioavailability. The floating tablets' AUC was around 1.8 times greater than that of regular furosemide tablets.
3. **Amplification of Absorption:** Potential possibilities for formulation as floating drug delivery systems include medications with low bioavailability due to site-specific absorption from the upper gastrointestinal tract, which would maximize their absorption. For instance, a considerable improvement in the bioavailability of floating dosage forms (42.9%) compared to enteric-coated LASIX-long products (29.5%) and commercially available LASIX tablets (33.4%) could be made.

KEYWORDS

- **floating drug delivery systems**
- **gastrointestinal tract**
- **gastro-retentive delivery system**
- **polycaprolactone**
- **tripolyphosphate**

REFERENCES

1. Viswanathan, P., Muralidaran, Y., Ragavan, G. (2017). Challenges in oral drug delivery: A nano-based strategy to overcome. In *Nanostructures for Oral Medicine*, 173–201. Elsevier. https://doi.org/10.1016/B978-0-323-47720-8.00008-0.
2. Moodley, K., Pillay, V., Choonara, Y. E., du Toit, L. C., Ndesendo, V. M. K., Kumar, P., Cooppan, S., & Bawa, P. (2011). Oral drug delivery systems comprising altered geometric configurations for controlled drug delivery. *International Journal of Molecular Sciences, 13*, 18–43. https://doi.org/10.3390/ijms13010018.
3. Malik, R., Garg, T., Goyal, A. K., & Rath, G. (2015a). Polymeric nanofibers: Targeted gastro-retentive drug delivery systems. *Journal of Drug Targeting, 23*, 109–124. https://doi.org/10.3109/1061186X.2014.965715.
4. Awasthi, R., & Kulkarni, G. T. (2016). Decades of research in drug targeting to the upper gastrointestinal tract using gastroretention technologies: Where do we stand? *Drug Delivery, 23*, 378–394. https://doi.org/10.3109/10717544.2014.936535.
5. Mandal, U. K., Chatterjee, B., & Senjoti, F. G. (2016). Gastro-retentive drug delivery systems and their in vivo success: A recent update. *Asian Journal of Pharmaceutical Sciences, 11*, 575–584. https://doi.org/10.1016/j.ajps.2016.04.007.
6. Singh, B. (2000). Floating drug delivery systems: An approach to oral controlled drug delivery via gastric retention. *Journal of Controlled Release, 63*, 235–259. https://doi.org/10.1016/S0168-3659(99)00204-7.
7. Han, D., Yu, X., Chai, Q., Ayres, N., & Steckl, A. J. (2017). Stimuli-responsive self-immolative polymer nanofiber membranes formed by coaxial electrospinning. *ACS Applied Materials & Interfaces, 9*, 11858–11865. https://doi.org/10.1021/acsami.6b16501.
8. Jiang, H., Wang, L., & Zhu, K. (2014). Coaxial electrospinning for encapsulation and controlled release of fragile water-soluble bioactive agents. *Journal of Controlled Release, 193*, 296–303. https://doi.org/10.1016/j.jconrel.2014.04.025.
9. Chowdary, K. P. R., & Chaitanya, CH. K. L. (2014). Recent research on floating drug delivery systems – A review. ISSN: 2230–7346. *Journal of Global Trends in Pharmaceutical Sciences, 5*(1), 1361–1373.

10. Adibkia, K., Hamedeyazdan, S., & Javadzadeh, Y. (2011). Drug release kinetics and physicochemical characteristics of floating drug delivery systems. *Expert Opinion on Drug Delivery, 8*, 891–903. https://doi.org/10.1517/17425247.2011.574124.
11. Hamori, M., Yoshimatsu, S., Hukuchi, Y., Shimizu, Y., Fukushima, K., Sugioka, N., Nishimura, A., & Shibata, N. (2014). Preparation and pharmaceutical evaluation of nano-fiber matrix supported drug delivery system using the solvent-based electrospinning method. *International Journal of Pharmaceutics, 464*, 243–251. https://doi.org/10.1016/j.ijpharm.2013.12.036.
12. Darbasizadeh, B., Motasadizadeh, H., Foroughi-Nia, B., & Farhadnejad, H. (2018). Tripolyphosphate-crosslinked chitosan/poly (ethylene oxide) electrospun nanofibrous mats as a floating gastro-retentive delivery system for ranitidine hydrochloride. *Journal of Pharmaceutical and Biomedical Analysis, 153*, 63 75. https://doi.org/10.1016/j.jpba.2018.02.023.
13. Malik, R., Garg, T., Goyal, A. K., & Rath, G. (2015). Diacerein-loaded novel gastroretentive nanofiber system using PLLA: Development and in vitro characterization. *Artificial Cells, Nanomedicine, and Biotechnology, 44*, 928–936. https://doi.org/10.3109/21691401.2014.1000492.
14. Suvannasara, P., Praphairaksit, N., & Muangsin, N. (2014). Self-assembly of mucoadhesive nanofibers. *RSC Advances, 4*, 58664–58673. https://doi.org/10.1039/c4ra09329a.
15. Garg, T., Singh, O., Arora, S., & Murthy, R. S. R. (2012). Scaffold: A novel carrier for cell and drug delivery. *Critical Reviews in Therapeutic Drug Carrier Systems, 29*, 1–63.
16. Dixit, N. (2011). Floating drug delivery system. *Journal of Current Pharmaceutical Research, 7*, 6–20.
17. Waterman, K. C. (2007). A critical review of gastric retentive controlled drug delivery. *Pharmaceutical Development and Technology, 12*, 1–10.
18. Garg, T., & Goyal, A. K. (2014). Liposomes: Targeted and controlled delivery system. *Drug Delivery Letters, 4*, 62–71.
19. Garg, R., & Gupta, G. (2008). Progress in controlled gastroretentive delivery systems. *Tropical Journal of Pharmaceutical Research, 7*, 1055–1066.
20. Garg, T., Singh, S., & Goyal, A. K. (2013). Stimuli-sensitive hydrogels: An excellent carrier for drug and cell delivery. *Critical Reviews in Therapeutic Drug Carrier Systems, 30*, 369–409.
21. Parnami, N., Garg, T., Rath, G., & Goyal, A. K. (2013). Development and characterization of nanocarriers for topical treatment of psoriasis by using combination therapy. *Artificial Cells, Nanomedicine, and Biotechnology, 41*(4), 276–281.
22. Kaur, M., Malik, B., Garg, T., et al. (2014). Development and characterization of guar gum nanoparticles for oral immunization against tuberculosis. *Drug Delivery*. Advance online publication. https://doi.org/10.3109/10717544.2014.947859.
23. Lee, J. A., Krogman, K. C., Ma, M., et al. (2009). Highly reactive multilayer-assembled TiO2 coating on electrospun polymer nanofibers. *Advanced Materials, 21*(12), 1252–1256. https://doi.org/10.1002/adma.200801646.
24. Son, Y. J., Kim, W. J., & Yoo, H. S. (2014). Therapeutic applications of electrospun nanofibers for drug delivery systems. *Archives of Pharmacal Research, 37*(1), 69–78. https://doi.org/10.1007/s12272-013-0171-y.

25. Sharma, A., Gupta, A., Rath, G., et al. (2013). Electrospun composite nanofiber-based transmucosal patch for anti-diabetic drug delivery. *Journal of Materials Chemistry B, 1*(25), 3410–3418. https://doi.org/10.1039/c3tb20198a.
26. Kaur, M., Garg, T., Rath, G., & Goyal, A. K. (2014). Current nanotechnological strategies for effective delivery of bioactive drug molecules in the treatment of tuberculosis. *Critical Reviews in Therapeutic Drug Carrier Systems, 31*(1), 49–88. https://doi.org/10.1615/CritRevTherDrugCarrierSyst.2014006716.
27. Grewal, H., Dhakate, S. R., Goyal, A. K., et al. (2012). Development of transmucosal patch using nanofibers. *Artificial Cells, Blood Substitutes, and Biotechnology, 40*(3), 146–150. https://doi.org/10.3109/10731199.2011.620193.
28. Ramakrishna, S., Fujihara, K., Teo, W. E., et al. (2006). Electrospun nanofibers: Solving global issues. *Materials Today, 9*(9), 40–50. https://doi.org/10.1016/S1369-7021(06)71572-8.
29. Garg, T., Goyal, A. K., Arora, S., & Murthy, R. S. R. (2012). Development, optimization & evaluation of porous chitosan scaffold formulation of gliclazide for the treatment of type-2 diabetes mellitus. *Drug Delivery Letters, 2*(4), 251–261.
30. Shariat, R., Kouchak, M., Feizoleslam, F., & Veysi, M. (2021). An Overview on Floating Drug Delivery Systems (FDDS); Conventional and New Approaches for Preparation and In Vitro – In Vivo Evaluation. *FABAD J. Pharm. Sci., 46*(3), 345–362.
31. Patial, K., Dua, J. S., Menra, M., & Prasad, D. N. (n.d.). A review: Floating drug delivery system (FDDS). eISSN 2319–1082.
32. Mali, A., & Bathe, R. (n.d.). A review on gastroretentive floating drug delivery system. ISSN: 2249–3387.
33. Reddy, B. V., Navaneetha, K., & Sandeep, P. A. D. (n.d.). Gastro retentive drug delivery system – A review. ISSN: 2230–7346.
34. Tabassum, N., Naqash, A., & Masoodi, M. H. (n.d.). Floating drug delivery system: A novel acceptable approach.
35. Gupta, P., Gnanarajan, P., & Kothiyal, P. (n.d.). Floating drug delivery system: A review. ISSN: 2278–6074.

CHAPTER 12

Seaweed Extracted Biopolymer-Based Sustainable Films for Packaging Applications

RINKY GHOSH,[1] KIRAN RANA,[2] and NEHA KANWAR RAWAT[3,4]

[1]*Department of Materials Science and Engineering, Indian Institute of Technology Kanpur, Kanpur, Uttar Pradesh, India*

[2]*Department of Textile Fiber and Engineering, Indian Institute of Technology Delhi, New Delhi, India*

[3]*Materials Research Laboratory, Department of Chemistry, Jamia Millia Islamia, New Delhi, India*

[4]*Department of Materials Engineering, Indian Institute of Science, Bangalore, Karnataka, India*

ABSTRACT

The demand for sustainable packaging materials has been increasing due to the growing awareness of environmental issues and the need to reduce plastic waste. Seaweed, a renewable and abundant resource, has emerged as a promising alternative for the development of biodegradable and biocompatible materials. This study focuses on the utilization of seaweed extracts as the main ingredient to produce biopolymer-based films for packaging applications. Moreover, seaweed extracts contain bioactive compounds that possess antimicrobial properties, making the films suitable for food packaging applications by extending shelf life and preventing microbial growth. The performance of seaweed-based films can be further enhanced by incorporating additives such as plasticizers, antioxidants, and reinforcement

Sustainability in Polymer Technology and Plastics Engineering: Concepts, Strategies, and Opportunities.
Tamara Tatrishvili, Neha Kanwar Rawat, Swati Gokul Talele, and A. K. Haghi (Eds.)

materials. These additives improve the flexibility, stability, and overall functionality of the films, ensuring their suitability for different packaging requirements. Additionally, the films can be modified through surface treatments to enhance their water resistance and oxygen barrier properties. Environmental impact assessments reveal that seaweed-based films have a significantly lower carbon footprint compared to conventional petroleum-based films. Furthermore, the biodegradability of seaweed-based films ensures their compatibility with composting and recycling systems, further reducing waste accumulation. In this context, this chapter showcases the highlighted properties of seaweed-based biopolymers, classification based on their origin, and their various applications in the field of food packaging.

12.1 INTRODUCTION

The current challenges and global concerns related to plastic waste are outrageous to human health as well as to society. Plastic packaging dominates, accounting for around 70% of our plastic waste [1]. As a result, millions of tons of plastic-based packaging materials are buried in landfills every year. This generates a massive amount of plastic waste due to its notorious properties, which require proper management and has led to an urgent development of biodegradable packaging films. Recycling more plastics and preventing them from leaking into the environment requires sorting, recycling, and an urgent need for waste treatment plants. The annual consumption rate of packaging materials made from petroleum-based plastics is increasing day by day, with an annual growth rate of 17.3%, according to the survey forecasts in 2012, named the compound annual growth rate (CAGR) [2–5]. They have a significant impact on packaging materials as they possess outstanding mechanical stability and exceptional flexibility. With rising petro-plastics costs and degradation point of view, there is concern in finding a cost-effective solution with faster degradation kinetics as a substitute for petroleum-based products for packaging applications. Therefore, biodegradable polymers have emerged as the most desirable materials as an alternative to non-biodegradable polymers as they do not leach toxins into the environment.

The growing interest in seaweed-based polysaccharides in the field of packaging applications has recently escalated (Figure 12.1). They show promising versatility due to their biocompatibility, abundance, and excellent gelling capability, and therefore have proven to be effective alternatives to petroleum-based products. Synthetic thermoplastics such as polyethylene, polypropylene, and polyethylene terephthalate are widely used as packaging materials due to their low price and easy processing, but they pose a huge threat to the environment and have been banned worldwide by environmental legislation [6–11]. The use of versatile biopolymers has gained scientific attention and is being widely explored, not only reducing waste disposal but also being beneficial in terms of sustainability and socioeconomic benefits [12–17]. Therefore, biodegradable polymers have emerged as the most desirable materials as an alternative to non-biodegradable polymers, as they do not leach toxins into the environment and address the shortcomings discussed above [18–21]. The term biopolymer is derived from polymeric biomolecules, consisting of repeating units of monomeric substituents that undergo degradation or easily break down into resulting byproducts such as CO_2 and H_2O under the action of aerobic microorganisms. Biopolymers are widely utilized as packaging materials due to their eco-friendly nature and irrefutability. However, due to their hydrophilic nature and poor barrier properties (water vapor permeability), which are the two major obstacles of biopolymers, they are less likely to be used as packaging material. Moreover, their strength at high humidity reduces drastically. Barrier properties of polymers are very important for product-package shelf-life. The product characteristics are related to the specific barrier needed. Biopolymers are generally permeable to gases and water vapor. Water vapor and oxygen are the two main important permeants studied in packaging applications. The permeations are generally possible through microcavities between polymer chains. Therefore, the drawbacks can be solved by fabricating composite films. This is one of the most frequently employed methods to overcome the shortcomings of bio-based packaging films [22]. The water resistance properties as well as the elongation at break can be tailored with the gradual incorporation of various water-soluble plasticizers, namely glycerol, sorbitol, mannitol, or PEG, and cross-linkers (mild organic acids) to significantly enhance the desirable mechanical strength in extreme weather conditions (humidity, pH, temperature, etc.). To exploit their beneficial advantages, modification of these biopolymers has been done with remarkable success.

Chen et al. [23] have reported the inevitable toxicological effect of poligeenan (degraded carrageenan), which became evident through colonic inflammation, overall suppressing its use in biomedical applications. In view of this, researchers around the globe have reported successive improvements in the mechanical and barrier properties of packaging films by chemically and physically modifying bioplastics using cross-linking agents – low-cost and non-toxic chemicals that substantially increase their stability against harsh processing environments. Additionally, surface modifications provide a new platform for biodegradable polymers to significantly enhance their physicochemical properties and effectively play an important role in terms of sustainability and economics.

Cross-linking of bioplastics can be performed in two different ways: through chemical and physical methods. Physical cross-linking involves the utilization of UV rays [24], Gamma rays [12], and electron beam irradiation [26–28] to generate a free radical site to initiate cross-linking reactions, leading to improvements in the functional properties of bioplastics. On the other hand, chemical cross-linking has been employed to improve surface hydrophobic properties, and stress transferability, and significantly reduce water uptake capacity. Chemical cross-linking agents or any moieties consisting of two or more reactive functional groups are preferably formed into an interconnected structure with the bulk matrix of the polymeric materials [29, 30].

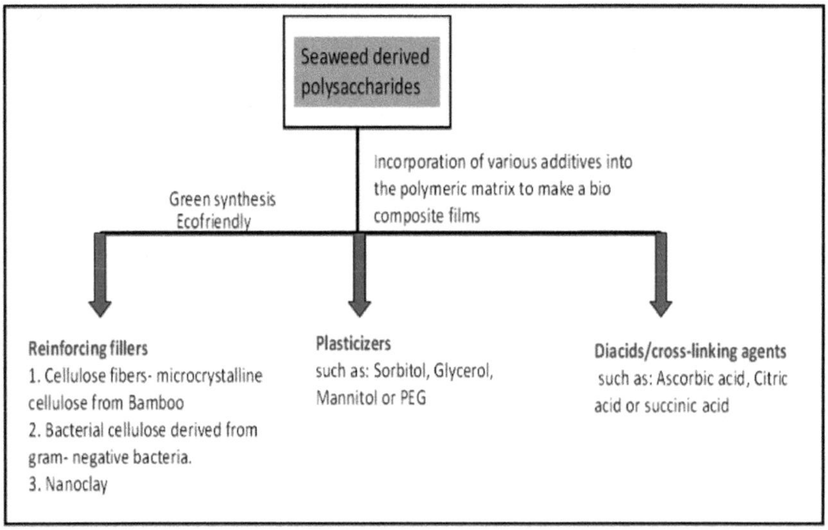

FIGURE 12.1 Seaweed-derived polysaccharides and their composite.

Miguel et al. have successfully demonstrated the cross-linking effect of epichlorohydrin on enhancing the durability and thermal stability of bioplastics [31]. Various cross-linkers including aldehyde, thiol, amine, glutaraldehyde, genipin, etc., are employed to enhance the surface rigidity of polysaccharides [32–38]. In this envision, this chapter discusses the highlighted properties of seaweed-based biopolymers, classification based on their origin, and their various applications in the field of food packaging, pharmaceutical, and biomedical arenas.

12.2 SEAWEED AND ITS CLASSIFICATION

In recent years, the rapid advancements of marine resources are considered as biomass feedstock due to the production of renowned biopolymers with numerous miscellaneous applications. Seaweed is a prominent source of polysaccharides, is biologically degradable, and could be metabolized by the action of microorganisms into CO_2 and H_2O, and hence is non-toxic to the environment and wildlife. It has recently gained renewed attention due to its high abundance and is considered a major component of aquatic biomass. Returning to nature in a short period of time for production further indicates great potential for the use of algae as a renewable source of energy and other high-value products [39].

Seaweeds are macroalgae, either unicellular or multicellular heterogeneous plants, mainly harvested from oceans, and found to grow at a depth of up to 180 meters on rocky solid substrates. Seaweeds, as captivating eukaryotic organisms, have gained attention due to their ability to synthesize a wide array of intricate polysaccharides, antioxidants, and other biologically active compounds. These compounds possess remarkable structural and functional properties, making seaweeds ecologically significant and instrumental in enhancing biodiversity within their native habitats [40].

Seaweeds exhibit remarkable diversity and are classified into three main groups: green algae (*Chlorophyta*), brown algae (Phaeophyta), and red algae (*Rhodophyta*) (Figure 12.2). These classifications are based on various factors including pigmentation, morphological characteristics, reproductive structures, and molecular analyses [41] (Table 12.1).

TABLE 12.1 Showing Various Seaweeds and its Respective Polysaccharides

Seaweed	Algae Species	Polysaccharides	References
Brown seaweed (Phaeophyta)	*Fucus, Laminaria pallida,* and *Zonaria species*	Laminarin, Alginate, Fucoidan	[61]
Red seaweed (*Rhodophyta*)	*Aeodes orbitosa, Notogenia stiriata,* and *Poryphyra capensis*	Agar, Carrageenan	[62]
Green Seaweed (*Chlorophyta*)	*Ulva, Cladophora* and *Monostroma species*	Ulvan	[61]

Green algae, belonging to the Chlorophyta group, are the most common type of seaweed. They owe their characteristic green color to the presence of chlorophyll. Green algae are often found in intertidal zones and shallow waters, and they range in size from microscopic forms to larger, leaf-like structures. They play a significant ecological role by producing a substantial portion of the world's oxygen through photosynthesis. Brown algae, or Phaeophyta, are predominantly found in colder regions and are recognized by their olive to dark brown color. They are the largest and most structurally complex seaweeds, with species such as kelps and rockweeds reaching impressive sizes. Brown algae possess unique adaptations to survive in harsh marine environments, including gas-filled bladders that help them float near the water's surface. Red algae, or *Rhodophyta*, are characterized by their reddish hue, which results from the presence of phycobilins, pigments that allow them to absorb light efficiently in deeper waters. They range in size from microscopic to large, branching forms and are often found in deeper marine habitats. Red algae have significant economic importance as they are a valuable source of carrageenan and agar, which are widely used in the food industry as thickeners and stabilizers.

The classification of seaweeds is essential for understanding their ecological roles, studying their biodiversity, and exploring their potential applications in various industries. Researchers utilize morphological characteristics, such as cell structure and arrangement, as well as molecular analyses to further differentiate and classify seaweeds accurately. With ongoing advancements in genetic and molecular techniques, scientists can delve deeper into the evolutionary relationships and genetic diversity among

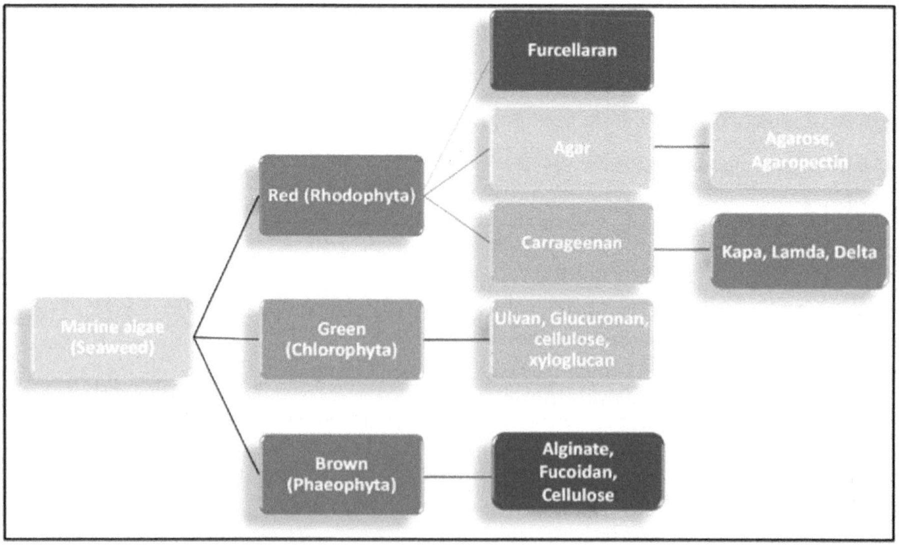

FIGURE 12.2 Classification of seaweeds used in packaging films.

Source: Reproduced with permission from Ebrahim Zadeh et al. [63]. Copyright 2023, Elsevier.

different seaweed groups. Furthermore, understanding seaweed classifications aids in the identification of specific species, allowing researchers to explore their unique biochemical compositions and potential applications in fields such as pharmaceuticals, cosmetics, and biofuels [40].

12.2.1 ALGAE

Algae, comprising both unicellular and multicellular forms, exhibit distinctive characteristics that set them apart from plants, such as the absence of a vascular system. As eukaryotic organisms, they play a critical role in maintaining the delicate balance of carbon dioxide and oxygen in our atmosphere [42]. By acting as a carbon sink, algae contribute to the regulation of carbon levels, while simultaneously releasing oxygen through the process of photosynthesis. This paper explores the vital ecological function of algae in the global carbon cycle and the pivotal role they play in oxygen production.

The etymology of Algae represents a diverse group of photosynthetic organisms and mainly grows in the benthic and littoral zones of the seashore [43, 44]. Seaweed algae have gained tremendous recognition in terms of availability, sustainability, and renewability, compared to other biomass resources. Their growing interest is attributed to exhibiting promising features such as moderate strength, biocompatibility, barrier properties, availability, simple processing, and allowing for surface modification and physicochemical changes. Therefore, these algae-based polysaccharides are of paramount interest in the field of packaging and biomedical applications.

12.2.2 CARRAGEENAN

Carrageenan has been considered one of the versatile polymers with good gelling properties, biological activities, and thermoreversible gelling characteristics. Hence, it can attain various structural confirmations according to the equilibrium of the system. The change in conformation from random coiled to a double helical structure determines the gelling ability [12] and imparts renewability and good mechanical strength, which are highly in demand for packaging film forming. Additionally, they are widely used as gelling and thickening agents in food and other dairy products. Carrageenan is a water-soluble hydrocolloid consisting of alternating units of D-galactose and 3,6-anhydro-galactose connected by sulfate groups. The degree of sulfation primarily determines the different forms of carrageenan (kappa, iota, and lambda) [45, 46]. It is derived from the cell walls of various seaweeds such as the Rhodophyceae family, *Kappaphycus alvarezii*, and *Eucheuma denticulatum*, mostly cultivated on the seashores of Malaysia and Southeast Asia, and exhibits high export performance either through ships or airways [47]. Thus, carrageenan is highly abundant and promisingly attractive for large-scale production.

12.2.3 ULVAN

Ulvan, a captivating macromolecule discovered in the cell walls of green algae, is a desirable complex sulfated polysaccharide belonging to the *Ulva lactuca* class. It serves as an edible polysaccharide and is widely used worldwide for bioplastic production. Comprising α- and β-(1,4)-linked monosaccharides like xylose, rhamnose, sulfate, glucuronic acid, and iduronic acid, as reported by Lahaye et al., Ulvan is categorized as an anionic sulfated polysaccharide [48]. Its distinctive chemical structure and various physicochemical properties make it an indispensable component of seaweed with numerous potential applications. Remarkably, Ulvan exhibits remarkable gelling and film-forming abilities, making it highly suitable for packaging purposes. Recent research has indicated that Ulvan possesses potential antioxidant properties, as well as favorable barrier and optical characteristics [49]. It is important to note that the properties of Ulvan depend primarily on the specific species it is sourced from, along with the eco-physiology of the algae and the methods employed for extraction.

12.2.4 ALGINATE

Alginate is a naturally occurring polysaccharide derived from brown seaweeds, particularly from the cell walls of species belonging to the Phaeophyta group. It is a linear copolymer composed of two monomers: D-mannuronic acid (M) and L-guluronic acid (G). The relative proportion of M and G monomers in the alginate molecule determines its physical and chemical properties [50].

Alginate-based packaging offers several advantages over traditional packaging materials. First and foremost, alginate is biodegradable, which means it can break down naturally over time, reducing environmental pollution. This characteristic aligns with the growing demand for eco-friendly packaging solutions. Alginate films and coatings provide an alternative to petroleum-based plastics, contributing to the reduction of plastic waste. Alginate films have excellent barrier properties, effectively protecting food products from moisture, oxygen, and light [51]. These films can be used as wraps or liners for fresh produce, extending their shelf life by preventing dehydration and spoilage. Alginate coatings can also be applied to fruits and vegetables, creating a protective layer that preserves their freshness and appearance. Furthermore, alginate-based packaging materials are edible and tasteless, making them safe for direct contact with food. This quality eliminates the need for additional packaging removal steps during food preparation, reducing waste and simplifying the consumer experience (Figure 12.3).

12.2.5 AGAR

Agar, a naturally abundant polymer derived from seaweed belonging to the *Rhodophyta* phylum, has gained significant attention due to its unique properties and potential applications. Extracted from agarophyte algae, agar is a polysaccharide that offers numerous advantages as a sustainable alternative to materials like cellulose and starch. It is non-toxic, biodegradable, and water-soluble, making it environmentally friendly. The structure of agar consists of 1,3-linked D-galactose and 1,4-linked 3,6-anhydro-L-galactose units. It is a mixture of agarose, which forms gels, and agaropectin, a slightly

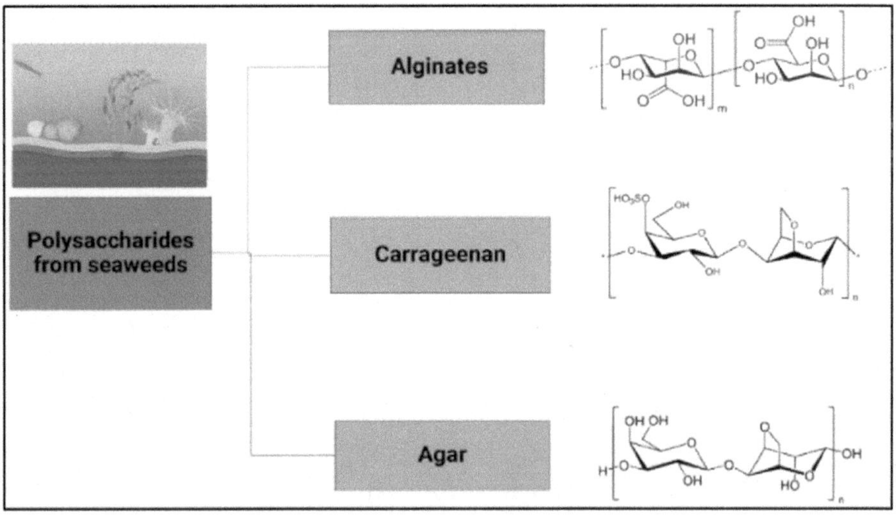

FIGURE 12.3 Shows the structure of seaweed macromolecules.

Source: Reproduced with permission from Ebrahim Zadeh et al. [63]. Copyright 2023, Elsevier.

branched and sulfated fraction. This composition gives agar good film-forming properties [52, 53].

One of the potential applications of agar is in packaging materials. It has been studied by various research groups for its ability to create biodegradable and sustainable packaging options. However, the high water absorption (swelling) and moderate mechanical strength of agar present challenges for its use as a packaging material. Despite these limitations, ongoing research aims to overcome these obstacles and explore ways to enhance its properties.

12.3 SEAWEEDS IN PACKAGING APPLICATIONS

The migration of chemicals from packaging materials to food products can pose potential health risks, highlighting the need for natural alternatives. As a result, research efforts are now shifting towards the use of natural compounds as active agents in packaging. Seaweeds, with their abundance of polysaccharides, have emerged as a promising option. Seaweeds offer a rich source of polysaccharides that can be utilized as active agents or raw materials in packaging applications. By combining seaweeds with biodegradable polymers, they become an excellent alternative to traditional materials, enhancing sustainability, functionality, and sensory properties. Seaweeds can be employed in various forms of packaging, including sustainable packaging, biodegradable plastics, active packaging, edible packaging, and sachets. By harnessing the properties of seaweeds, these packaging solutions become more environmentally friendly and reduce the risk of chemical migration into food. The incorporation of seaweeds into packaging materials improves sustainability by utilizing renewable resources that do not require arable land or excessive resources. Additionally, seaweed-based packaging can biodegrade naturally, minimizing environmental impact.

Moreover, seaweeds provide functional benefits such as moisture resistance and can be modified to possess antimicrobial or antioxidant properties, thereby extending the shelf life of packaged products. Their use also enhances the sensory experience by imparting natural flavors or aromas to the packaged items.

Seaweed-based packaging materials have several advantages. Firstly, they are biodegradable, meaning they can naturally break down without leaving harmful residues behind. This is in stark contrast to plastics, which can persist in the environment for hundreds of years. Seaweed packaging also offers good moisture resistance, ensuring the preservation of packaged goods. Furthermore, seaweed packaging can be tailored to meet specific requirements. It can be processed to enhance its strength, flexibility, and barrier properties, making it suitable for a wide range of products. Seaweed-based packaging has been used for applications such as food and beverage packaging, cosmetics, and even 3D printing.

12.3.1 ACTIVE PACKAGING

Active packaging refers to a dynamic system that promotes a positive interaction between the product, packaging, and the surrounding environment. Its primary purpose is to extend the shelf life, enhance safety, maintain product quality, and improve sensory properties. This is achieved by incorporating components that facilitate the exchange of substances between the product, packaging, and environment. Active packaging can be categorized into monolayer and multi-layer systems. In the monolayer system, the active compound is integrated with the packaging polymer, while in the multi-layer system, the active compound is trapped between different layers of the packaging material to regulate its release [54].

There are several types of active food packaging systems that have been developed. These include oxygen scavengers, carbon dioxide generators, antimicrobial packaging, moisture control packaging, ethylene scavengers, and flavor and odor-absorbent packaging. Oxygen scavengers are widely used to remove oxygen from the packaging environment, preventing the growth of aerobic microbes, and color changes, and extending the product's shelf life. Commonly employed oxygen scavengers involve chemical reactions with compounds such as iron powder, catechol, ascorbic acid, oxidative enzymes, or unsaturated hydrocarbons. Compared to packaging techniques like vacuum packaging and modified atmosphere packaging, which do not guarantee complete oxygen removal, active packaging systems offer more controlled and reliable oxygen reduction. Iron powder-based systems, contained in permeable sachets, are commonly used to react with and remove oxygen from the surrounding atmosphere [55].

Active packaging is a system in which product, packaging, and environment interact in a positive way to extend the shelf life, improve safety, as well as sensory properties, and maintain the quality of the product. It incorporates components that enable the exchange of substances from the product into the environment, the environment to the product, or the packaging to the product. The active packaging could be monolayered or multi-layered [56]. In the monolayer system, the active compound is incorporated with the polymer, whereas in the multilayer system, the active compound is entrapped between the layers of the polymer to control its release. This process relies on strategies like temperature control and the addition of chemicals. Currently, there are numerous active food packaging systems developed such as oxygen scavengers, carbon dioxide-generating systems, antimicrobial active packaging, moisture control packaging, ethylene scavengers, and flavor and odor absorbent packaging [57]. Also, the chemicals used as active agents in these systems are shown in (Figure 12.4).

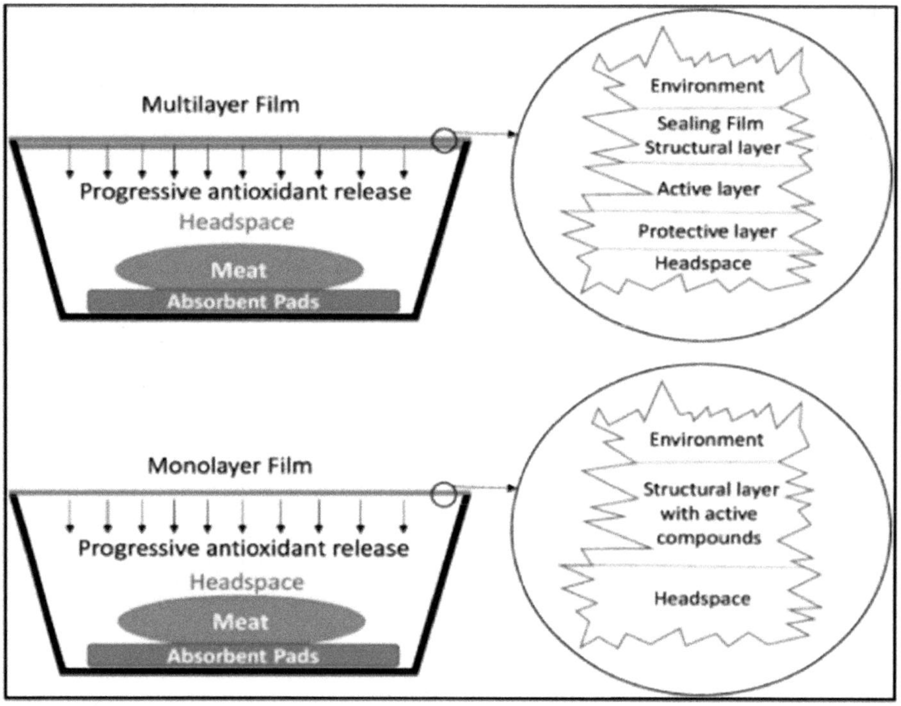

FIGURE 12.4 Active packaging in graphical form (both multilayer and monolayer)
Source: Adapted from Ref.[25]

Oxygen scavengers (OS) are the most widely used forms of active packaging with the purpose of absorbing oxygen within the environment, thereby preventing oxygen from reacting with the product. This would otherwise cause the growth of aerobic microbes, color change, the reduction of shelf life, and the survival of insects. In general, OS is defined as "materials incorporated into the package structure which form a chemical bond with, and thus effectively remove, oxygen from the inner package environment." Packaging techniques such as vacuum packaging and modified atmosphere packaging are based on the same goal; however, with these methods, complete oxygen removal is not assured. The operating principle behind O_2 scavengers varies depending on the active compounds in use, such as iron, catechol, ascorbic acid, oxidative enzymes, unsaturated hydrocarbons polyamides, etc. A common and commercially used system is iron powder, which is stored in a permeable sachet and enters an oxidation-reduction reaction with oxygen, thereby removing it from the atmosphere.

12.3.2 ANTIMICROBIAL PACKAGING

Antimicrobial packaging is designed to hinder the growth of microorganisms by incorporating or coating antimicrobial agents onto or into the food packaging materials. It offers a means to extend the lag phase and inhibit the proliferation of microorganisms, thus preserving the product. Antimicrobial packaging can be categorized into four main types: antimicrobial sachets or pads, direct integration into polymers, antimicrobial coatings, and inherently antimicrobial polymers. In the case of antimicrobial sachets or pads, an antimicrobial substance is contained within a sachet or pad that is added to the packaging. Direct integration involves incorporating the antimicrobial agent directly into the polymer, enabling a controlled release of the agent into the headspace or onto the food surface. Coating the packaging with a carrier matrix allows the antimicrobial substance to evaporate into the headspace or diffuse into the food. Lastly, inherently antimicrobial polymers possess inherent antimicrobial properties due to the composition of the polymer itself [58].

Several antimicrobial agents are utilized in antimicrobial packaging, such as ethanol, which is effective against various microorganisms and can be emitted as vapor from sachets. Chlorine dioxide is another agent used to combat viruses, fungi, and bacteria, where it is released by the reaction of sodium chlorite and acid precursors upon contact with moisture. Examples of antimicrobial polymers include Nisin-loaded chitosan/poly L-lactic acid films, which exhibit antibacterial properties against *Staphylococcus aureus*. Other antimicrobial agents employed in packaging include enzymes (e.g., lysozyme), essential oils (e.g., cinnamon essential oil), organic acids (e.g., citric acid), and nanoparticles (e.g., silver) [59, 60]. Antimicrobial packaging provides a proactive approach to maintaining product safety and extending shelf life by inhibiting microbial growth. It offers versatility in terms of application methods and antimicrobial agents, contributing to the overall quality and preservation of packaged food products.

12.4 CONCLUSION AND FUTURE PERSPECTIVE

The incorporation of seaweed into natural polymers holds great potential for sustainable food packaging, aligning with the increasing consumer demand for eco-friendly products. By utilizing seaweed, the reliance on synthetic packaging materials can be reduced, contributing to a decrease in plastic pollution. The antioxidant and antimicrobial properties of seaweed offer benefits in terms of extending the shelf life of food and reducing food waste caused by spoilage. However, further research is needed to determine the specific impact of seaweed on the shelf life of different food products. To ensure the safe and risk-free use of seaweed in food packaging, it is essential to address concerns regarding legislation on toxic metals and potentially harmful compounds that may accumulate. Adjustments in the mechanical properties of seaweed-based films can be achieved using polymer blends or natural plasticizers, tailored to the specific application requirements.

Seaweed-based biodegradable packaging offers a promising alternative to conventional methods, and efforts should be made to optimize the manufacturing process to make it a viable competitor in the market. Detailed research should focus on identifying specific areas where seaweed-based packaging can provide significant advantages over traditional materials, considering its mechanical strength, antibacterial properties, antioxidant activity, and controlled release capabilities. Enhancements in existing polysaccharide membranes can be achieved through the incorporation of additives like lipids, diverse polymer blends, and a wider range of possibly modified polysaccharides, all contributing to a more sustainable approach to food packaging.

KEYWORDS

- **carrageenan**
- **compound annual growth rate**
- **food packaging applications**
- **oxygen scavengers**
- **seaweed polysaccharides**
- **sustainable films**

REFERENCES

1. Wang, Z., Ganewatta, M. S., & Tang, C. (2020). Sustainable polymers from biomass: Bridging chemistry with materials and processing. *Progress in Polymer Science*, *101*, 101197.
2. Johann, T. (2020). *Polymer Chemistry*, *101*, 101–197.
3. Valdes, A., Mellinas, A. C., Ramos, M., Garrigos, M. C., & Jimenez, A. (2014). Natural additives and agricultural wastes in biopolymer formulations for food packaging. *Frontiers in Chemistry*, *2*, https://people.phys.ethz.ch/~andreoli/%C3%BCK_2/Zusatz/Kunststoffe/Kunststoff_Facts_2015.pdf.
4. Mahalik, N. P. (2014). Advances in packaging methods: processes and systems. *Challenges*, *5*(2), 374–389.
5. PlasticsEurope. (2015). *Plastics—the Facts 2015: An analysis of European plastics production, demand and waste data.*
6. Plastics – the Facts 2019. (2019).
7. Tavassoli-Kafrani, E., Shekarchizadeh, H., & Masoudpour-Behabadi, M. (2016). Development of edible films and coatings from alginates and carrageenans. *Carbohydrate Polymers*, *137*, 360–374.
8. Mu, C., Guo, J., Li, X., Lin, W., & Li, D. (2012). Preparation and properties of dialdehyde carboxymethyl cellulose crosslinked gelatin edible films. *Food Hydrocolloids*, *27*, 22–29.
9. Hasan, M. R., & Chakrabarti, R. (2009). Use of algae and aquatic macrophytes as feed in small-scale aquaculture. *FAO Fisheries and Aquaculture Technical Paper*, *531*, 123.
10. El Gamal, A. A. (2012). Biological importance of marine algae. In S. K. Kim (Ed.), *Handbook of Marine Macroalgae: Biotechnology and Applied Phycology* (p. 567). John Wiley & Sons.
11. Noreen, A., Zia, K. M., Zuber, M., Ali, M., & Mujahid, M. (2016). *International Journal of Biological Macromolecules*, *86*, 937–949.
12. Rhein-Knudsen, N., Ale, M. T., & Meyer, A. S. (2015). Seaweed hydrocolloid production: An update on enzyme-assisted extraction and modification technologies. *Marine Drugs*, *13*, 3340–3359.
13. Liu, J., Zhan, X., Wan, J., Wang, Y., & Wang, C. (2015). Review for carrageenan-based pharmaceutical biomaterials: Favourable physical features versus adverse biological effects. *Carbohydrate Polymers*, *121*, 27–36.
14. Kirk, R. E., Othmer, F. D., Kroschwitz, J. I., & Howe-Grant, M. (1992). *Encyclopedia of Chemical Technology* (Vol. *4*). John Wiley & Sons.
15. Chen, Y., Liao, M. L., & Dustan, D. E. (2002). The rheology of K+-κ-carrageenan as a weak gel. *Carbohydrate Polymers*, *50*, 109–116.
16. Wolfrom, M. L., Tipson, R. S., & Horton, D. (Eds.). (1969). *Advances in Carbohydrate Chemistry and Biochemistry* (Vol. *24*). Academic Press.
17. Rochas, C., Rinaudo, M., & Landry, S. (1989). Relation between the molecular structure and mechanical properties of carrageenan gels. *Carbohydrate Polymers*, *10*, 115–127.
18. Snoeren, T. H. M. (1976). Kappa-carrageenan: A study of its physicochemical properties, sol–gel transition and interactions with milk (PhD thesis). Wageningen, The Netherlands.
19. Stoloff, L. (1959). Carrageenan. In R. I. Whistler (Ed.), *Industrial Gums* (Vol. *84*, pp. 85). Academic Press.

20. Guiry, M. D., & Guiry, G. M. (2012). *AlgaeBase*. World-wide electronic publication, National University of Ireland, Galway.
21. Velde, V. D. F., Knutsen, S. H., Usov, A. L., Rollema, H. S., & Cerezo, A. S. (2002). 1H and 13C high resolution NMR spectroscopy of carrageenans: Application in research and industry. *Trends in Food Science & Technology, 13*, 73–92.
22. Hurtado, A. Q., Gerung, G. S., Yasir, S., & Critchley, A. T. (2013). Cultivation of tropical red seaweeds in the BIMP-EAGA region. *Journal of Applied Phycology, 26*, 707–718.
23. Chen, H. M., Yan, X. J., Wang, F., Xu, W. F., & Zhang, L. (2010). Assessment of the oxidative cellular toxicity of a κ-carrageenan oxidative degradation product towards Caco-2 cells. *Food Research International, 43*(10), 2390–2401.
24. Liu, J., Zhan, X., Wan, J., Wang, Y., & Wang, C. (2015). Review for carrageenan-based pharmaceutical biomaterials: Favorable physical features versus adverse biological effects. *Carbohydrate Polymers, 121*, 27–36.
25. Carina, D., Sharma, S., Jaiswal, A. K., & Jaiswal, S. (2021). Seaweeds polysaccharides in active food packaging: A review of recent progress. *Trends in Food Science & Technology, 110*, 559–572. https://doi.org/10.1016/j.tifs.2021.01.014.
26. Zhou, J., Ma, Y., Ren, L., Tong, J., Liu, Z., & Xie, L. (2009). Preparation and characterization of surface crosslinked TPS/PVA blend films. *Carbohydrate Polymers, 76*(4), 632–638.
27. Lee, S., Lee, M., & Song, K. (2005). Effect of gamma-irradiation on the physicochemical properties of gluten films. *Food Chemistry, 92*(4), 621–625.
28. Gohil, J. M., Bhattacharya, A., & Ray, P. (2006). Studies on the crosslinking of poly (vinyl alcohol). *Journal of Polymer Research, 13*(2), 161–169.
29. Yeom, C.-K., & Lee, K.-H. (1996). Pervaporation separation of water-acetic acid mixtures through poly (vinyl alcohol) membranes crosslinked with glutaraldehyde. *Journal of Membrane Science, 109*(2), 257–265.
30. Zhang, Y., Zhu, P. C., & Edgren, D. (2010). Crosslinking reaction of poly (vinyl alcohol) with glyoxal. *Journal of Polymer Research, 17*(5), 725–730.
31. De Miguel, I., Rieumajou, V., & Betbeder, D. (1999). New methods to determine the extent of reaction of epichlorohydrin with maltodextrins. *Carbohydrate Research, 319*, 17–23.
32. Jeong, S. H., Huh, K. M., & Park, K. (2006). Hydrogel drug delivery systems. *Journal of Biomaterials and Nanobiotechnology, 3*(2), 185.
33. Hernandez-Munoz, P., Kanavouras, A., Lagaron, J. M., & Gavara, R. (2005). Development and characterization of films based on chemically cross-linked gliadins. *Journal of Agricultural and Food Chemistry, 53*(21), 8216–8223.
34. Zhang, C., Wang, Z., Li, Y., Yang, Y., Ju, X., & He, R. (2019). The preparation and physiochemical characterization of rapeseed protein hydrolysate-chitosan composite films. *Food Chemistry, 272*, 694–701.
35. Reddy, N., & Yang, Y. (2010). Citric acid cross-linking of starch films. *Food Chemistry, 118*, 702–711.
36. Fakhouri, F. M., Costa, D., Yamashita, F., Martelli, S. M., Jesus, R. C., Alganer, K., Collares-Queiroz, F. P., & Innocentini-Mei, L. H. (2013). Comparative study of processing methods for starch/gelatin films. *Carbohydrate Polymers, 95*, 681–689.
37. Garavand, F., Rouhi, M., Razavi, S. H., Cacciotti, I., & Mohammadi, R. (2017). Improving the integrity of natural biopolymer films used in food packaging by

crosslinking approach: A review. *International Journal of Biological Macromolecules*, *104*, 687–707.
38. Lei, Y. L., Wu, H. J., Jiao, C., Jiang, Y., Liu, R., Xiao, D., et al. (2019). Investigation of the structural and physical properties, antioxidant and antimicrobial activity of pectin-konjac glucomannan composite edible films incorporated with tea polyphenol. *Food Hydrocolloids*, *94*, 128–135.
39. Jayakody, M. M., & Vanniarachchy, M. P. G. (2022). Seaweed derived alginate, agar, and carrageenan based edible coatings and films for the food industry: A review. *Food Measure*, *16*, 1195-1227.
40. Salehi, B., Sharifi-Rad, J., Seca, A. M. L., Pinto, D. C. G. A., Michalak, I., Trincone, A., Mishra, A. P., Nigam, M., Zam, W., & Martins, N. (2019). Current trends on seaweeds: Looking at chemical composition, phytopharmacology, and cosmetic applications. *Molecules*, *24*, 4182.
41. Lim, C., Yusoff, S., Ng, C. G., Lim, P. E., & Ching, Y. C. (2021). Bioplastic made from seaweed polysaccharides with green production methods. *Journal of Environmental Chemical Engineering*, *9*, 105895.
42. Abdo, S. M., & Ali, G. H. (2019). Analysis of polyhydroxybutrate and bioplastic production from microalgae. *Bulletin of the National Research Centre*, *43*, 97.
43. Dang, B. -T., Bui, X. -T., Tran, D. P. H., Hao, N., Nghiem, L. D., Hoang, T. -K. -D., Nguyen, P. -T., Nguyen, H. H., Vo, T. -K. -Q., Lin, C., Yi, A. L. K., & Varjani, S. (2022). Current application of algae derivatives for bioplastic production: A review. *Bioresource Technology*, *347*, 126698.
44. Nanda, N., & Bharadvaja, N. (2022). Algal bioplastics: Current market trends and technical aspects. *Clean Technologies and Environmental Policy*, *24*, 2659–2679.
45. Sedayu, B. B., Cran, M. J., & Bigger, S. W. (2019). A review of property enhancement techniques for carrageenan-based films and coatings. *Carbohydrate Polymers*, *216*, 287–302.
46. Suryanto, H., Rahmawan, A., Solichin, Sahana, R., Muhajir, M., & Yanuhar, U. (2019). Influence of carrageenan on the mechanical strength of starch bioplastic formed by extrusion process. *IOP Conference Series: Materials Science and Engineering*, *494*, 012075.
47. Suryanto, H., Sahana, R. T., Aminnudin, Suyetno, A., Widiyanti, Yanuhar, U. (2019). Effect of carrageenan on the structure of cassava starch bioplastic after extrusion process. Presented at the *International Conference on Biology and Applied Science (ICOBAS)*, 125–130, Malang, Indonesia, p. 080023.
48. Guidara, M., Yaich, H., Benelhadj, S., Adjouman, Y. D., Richel, A., Blecker, C., Sindic, M., Boufi, S., Attia, H., & Garna, H. (2020). Smart Ulvan films responsive to stimuli of plasticizer and extraction condition in physico-chemical, optical, barrier and mechanical properties. *International Journal of Biological Macromolecules*, *150*, 714–726.
49. Yang, Y. -C., Mei, X. -W., Hu, Y. -J., Su, L. -Y., Bian, J., Li, M. -F., Peng, F., & Sun, R. -C. (2019). Fabrication of antimicrobial composite films based on xylan from pulping process for food packaging. *International Journal of Biological Macromolecules*, *134*, 122–130.
50. Łopusiewicz, Ł., Macieja, S., Śliwiński, M., Bartkowiak, A., Roy, S., & Sobolewski, P. (2022). Alginate biofunctional films modified with melanin from watermelon seeds and zinc oxide/silver nanoparticles. *Materials*, *15*, 2381.

51. Jayakumar, A., Radoor, S., Siengchin, S., Shin, G. H., & Kim, J. T. (2023). Recent progress of bioplastics in their properties, standards, certifications and regulations: A review. *Science of The Total Environment, 878*, 163156.
52. Mostafavi, F. S., & Zaeim, D. (2020). Agar-based edible films for food packaging applications – A review. *International Journal of Biological Macromolecules, 159*, 1165–1176.
53. Roy, S., & Rhim, J. -W. (2021). Preparation of pectin/agar-based functional films integrated with zinc sulfide nano petals for active packaging applications. *Colloids and Surfaces B: Biointerfaces, 207*, 111999.
54. Qian, M., Liu, D., Zhang, X., Yin, Z., Ismail, B. B., Ye, X., & Guo, M. (2021). A review of active packaging in bakery products: Applications and future trends. *Trends in Food Science & Technology, 114*, 459–471.
55. Suppakul, P., Miltz, J., Sonneveld, K., & Bigger, S. W. (2003). Active packaging technologies with an emphasis on antimicrobial packaging and its applications. *Journal of Food Science, 68*, 408–420.
56. Vermeiren, L., Devlieghere, F., Van Beest, M., De Kruijf, N., & Debevere, J. (1999). Developments in the active packaging of foods. *Trends in Food Science & Technology, 10*, 77–86.
57. Yildirim, S., Röcker, B., Pettersen, M. K., Nilsen-Nygaard, J., Ayhan, Z., Rutkaite, R., Radusin, T., Suminska, P., Marcos, B., & Coma, V. (2018). Active packaging applications for food: Active packaging applications for food. *Comprehensive Reviews in Food Science and Food Safety, 17*, 165–199.
58. Cha, D. S., & Chinnan, M. S. (2004). Biopolymer-based antimicrobial packaging: A review. *Critical Reviews in Food Science and Nutrition, 44*, 223–237.
59. Sofi, S. A., Singh, J., Rafiq, S., Ashraf, U., Dar, B. N., & Nayik, G. A. (2018). A comprehensive review on antimicrobial packaging and its use in food packaging. *Current Nutrition & Food Science, 14*, 305–312.
60. Zhong, Y., Godwin, P., Jin, Y., & Xiao, H. (2020). Biodegradable polymers and green-based antimicrobial packaging materials: A mini-review. *Advanced Industrial and Engineering Polymer Research, 3*, 27–35.
61. Venkatesan, J., Lowe, B., Anil, S., Manivasagan, P., Kheraif, A. A. A., Kang, K. -H., & Kim, S. -K. (2015). Seaweed polysaccharides and their potential biomedical applications. *Starch – Stärke, 67*(5-6), 381–390. https://doi.org/10.1002/star.201400252.
62. El-hefian, E. A., Nasef, M. M., & Yahaya, A. H. (2012). Preparation and characterization of chitosan/agar blended films: Part 1. Chemical structure and morphology. *E-Journal of Chemistry, 9*(3), 1431–1439. https://doi.org/10.1155/2012/894046.
63. Ebrahimzadeh, S., Biswas, D., Roy,s S., & McClements, D. J. (2023). Incorporation of essential oils in edible seaweed-based films: A comprehensive review. *Trends in Food Science & Technology, 135*, 43–56. https://doi.org/10.1016/j.tifs.2023.03.002.

Index

C

D

E

F

G

N

Q

R

T

U